Studies in Computational Intelligence

Volume 549

Series editor

Janusz Kacprzyk, Polish Academy of Sciences, Warsaw, Poland
e-mail: kacprzyk@ibspan.waw.pl

For further volumes:
http://www.springer.com/series/7092

About this Series

The series "Studies in Computational Intelligence" (SCI) publishes new developments and advances in the various areas of computational intelligence—quickly and with a high quality. The intent is to cover the theory, applications, and design methods of computational intelligence, as embedded in the fields of engineering, computer science, physics and life sciences, as well as the methodologies behind them. The series contains monographs, lecture notes and edited volumes in computational intelligence spanning the areas of neural networks, connectionist systems, genetic algorithms, evolutionary computation, artificial intelligence, cellular automata, self-organizing systems, soft computing, fuzzy systems, and hybrid intelligent systems. Of particular value to both the contributors and the readership are the short publication timeframe and the world-wide distribution, which enable both wide and rapid dissemination of research output.

Pierluigi Contucci · Ronaldo Menezes
Andrea Omicini · Julia Poncela-Casasnovas
Editors

Complex Networks V

Proceedings of the 5th Workshop on Complex Networks CompleNet 2014

Springer

Editors
Pierluigi Contucci
Department of Mathematics
Università di Bologna
Bologna
Italy

Andrea Omicini
Department of Computer Science and Engineering
Università di Bologna
Bologna
Italy

Ronaldo Menezes
Department of Computer Sciences
Florida Institute of Technology
Melbourne, Florida
USA

Julia Poncela-Casasnovas
Northwestern University
Evanston Illinois
USA

ISSN 1860-949X ISSN 1860-9503 (electronic)
ISBN 978-3-319-05400-1 ISBN 978-3-319-05401-8 (eBook)
DOI 10.1007/978-3-319-05401-8
Springer Cham Heidelberg New York Dordrecht London

Library of Congress Control Number: 2014933113

Printed on acid-free paper

Springer is part of Springer Science+Business Media (www.springer.com)

Preface

The International Workshop on Complex Networks – CompleNet (www.complenet.org) was initially proposed in 2008 with the first workshop taking place in 2009. The initiative was the result of efforts from researchers from the *BioComplex Laboratory in the Department of Computer Sciences at Florida Institute of Technology, USA*, and the from the *Dipartimento di Ingegneria Informatica e delle Telecomunicazioni, Università di Catania, Italia.*

CompleNet aims at bringing together researchers and practitioners working on areas related to complex networks. In the past two decades we have been witnessing an exponential increase on the number of publications in this field. From biological systems to computer science, from social systems and language to science of science, complex networks are becoming pervasive in many fields of science. It is this interdisciplinary nature of complex networks that CompleNet aims at addressing. CompleNet 2014 was the fifth event in the series and was hosted by the Università di Bologna, Italy, on March 12–14, 2014.

This book includes the some of the peer-reviewed works presented at CompleNet 2014. We received 86 submissions from 32 countries. Each submission was reviewed by at least 3 members of the Program Committee. Acceptance was judged based on the relevance to the symposium themes, clarity of presentation, originality and accuracy of results and proposed solutions. After the review process, 12 papers and 18 short papers were selected to be included in this book.

The 30 contributions in this book address many topics related to complex networks and have been organized in five major groups: (1) Social Networks, Social Media and the Arts, (2) Diffusion, transportation and search on networks, (3) Network theory, structure, growth and community detection, (4) Biological and health-related networks, and (5) Language networks and science of science

We would like to thank the Program Committee members for their time an expertise spent for the refereeing process. We deeply appreciate the efforts of our Keynote Speakers: Raffaella Burioni (Università di Parma, Italy), Tamás Vicsek (Eötvös University, Hungary), Alessandro Vespignani (Northeastern University, Boston, US); their presentations are among the reasons CompleNet 2014 was such a success. We are grateful to our Invited Speakers who enriched CompleNet 2014 with their presentations and insights in the field of Complex Networks: Juyong Park (KAIST, South Korea), Mirkodegli Esposti (Università di Bologna, Italy), Stephen Uzzo (New York Institute of Technology), Giorgio Fagiolo (Sant'Anna School of Advanced Studies, Pisa, Italy), Adriano Barra (Università La Sapienza di Roma, Italy), Bruno Gonçalves (Aix-Marseille Université, Marseille, France).

Special thanks also go to the Local Organizer and Poster Chair, Enrico Denti (Università di Bologna, Italy), the Steering Committee, Giuseppe Mangioni (Università di Catania, Italy) and José Mendes (University of Aveiro, Portugal).

Bologna, Italy
March 2014

Pierluigi Contucci
Università di Bologna, Italy

Ronaldo Menezes
Florida Institute of Technology, US

Andrea Omicini
Università di Bologna, Italy

Julia Poncela-Casasnovas
Northwestern University, US

Contents

Biological and Health-Related Networks

Language Networks and Science of Science

The Network of Western Classical Music Composers

Doheum Park, Arram Bae, and Juyong Park

Abstract. Network science focuses on the connections between the elements of a complex system in order to uncover the nature and the underlying patterns of interaction relationships inside the system. In this paper we apply network theory to understand associations between the composers of western classical music constructed from a comprehensive data of CD recordings. We study the properties of the network of composer-composer ties including the degree distribution, the component structure, clustering, and several types of centralities of the composers. We also investigate the nature of prominent modules found in the network, and show how the tastes of consumers of western classical music manifest themselves in the network. We believe that our work shows how network science can be a useful tool for studying arts and humanities.

1 Introduction

Recently network science has been instrumental in the modeling and understanding of various complex systems, ranging from technical systems such as the Internet and the Worldwide Web [1, 2] to social networks [3] and biological systems [4]. The success and the wide range of applicability of network science is based on the fact that by focusing on the connection patterns of a system's constituents it provides a unified framework for studying diverse systems, allowing developments in one area to quickly find use in others [5].

One area where network science as a methodology is garnering interest is arts and humanities [6, 7] including archaeology, history, and music. Coupled with an accelerating accumulation of so-called "Big Data," network science is advancing

Doheum Park · Arram Bae · Juyong Park
Graduate School of Culture Technology
Korea Advanced Institute of Science and Technology
Daejeon, Republic of Korea
e-mail: {park154,redsin0617,juyongp}@kaist.ac.kr

P. Contucci et al. (eds.), *Complex Networks V*,
Studies in Computational Intelligence 549,
DOI: 10.1007/978-3-319-05401-8_1,

the field of humanities by helping us make sense of the patterns inherent in the data and their significance in the arts and humanities. Some notable large-scale data and network analyses in the field are as follows. Suarez, Sancho and Rosa [8] analyzed the data set of 11 443 works from Spain and Latin America and the linkage patterns of paintings with respect to genre and theme, finding that religious theme is the most dominant factor linking the paintings. Gleiser and Dan [9] studied the topology and the community structure of the collaboration network of Jazz musicians. Their analysis uncovered the presence of communities based on the recording locations of the bands correlated with the racial segregation between the musicians. Park *et al.* [10] highlighted the discrepancies between the network of collaboration and that of similarity in their study of contemporary popular musicians.

In this paper we analyze the network of composers of western classical music that covers its 700 years of development, constructed from a comprehensive data of CD recordings. We study the patterns of composer-composer associations including the degree distribution, the component structure, clustering, and the centralities of the composers. We also investigate the properties of prominent modules (communities) in the network that shed light on the nature of large-scale associations in western classical music, which we believe could prove useful for advances in traditional musicology that have often focused chiefly on understanding the individual composers.

2 Data

We utilized data sets from two prominent providers of information on classical music, ArkivMusic[1] and All Music Guide[2]. ArkivMusic is an online classical music retailer specializing in the distribution of CDs and DVDs. All Music Guide is an online music guide service website. As of early 2013, ArkivMusic lists a total of 96 911 classical music CDs along with their titles, release dates, labels, and the musicians involved, namely the composers of the pieces and the performers (conductors, soloists, and ensembles). The data can thus be represented as a bipartite network between CDs and the musicians in which an edge connects a CD and a musician if the musician has been featured on the CD (see Fig. 1). While there are several interesting possibilities of exploring the patterns of connections between different classes of musicians, in this paper we focus specifically on the network of composers obtained via the one-mode projection onto the set of composers. Therefore in our network an edge between two composers means that they were featured on at least one common CD.

We also processed the data to eliminate so-called "compilation CDs" that are essentially repackaged collections of previously issued CDs that are the root of undesirable effect of most well-known becoming connected to each other, resulting in an effective complete (full) network. We also trimmed out composers whose attribute data (periods and active years) were not available from All Music Guide,

[1] `http://www.arkivmusic.com`
[2] `http://www.allmusic.com`

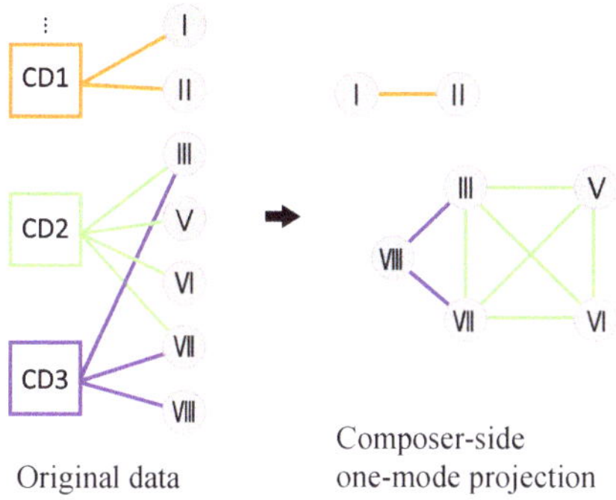

Fig. 1 The network representation of ArkivMusic data. The association between the CDs and performers or composer can be visualized as a bipartite network (left). A one-mode projection of the bipartite network onto the set of composers works by connecting two composers that are associated with a common CD (right).

as we would need them later when we investigate the relationship between node attributes and network properties [11].[3]

3 Network Properties

3.1 Small-World Property and Giant Component

Many networks exhibit the so-called "small-world property," meaning that the distance between two nodes of a network measured by the length of the geodesic (shortest path) connecting them is typically small. Also referred to as showing "six degrees of separation" in common parlance and made famous by Milgram's experiment in 1967 [12], it is now known to be true for many other networks. The average geodesic length between node pairs in our network is 2.6, and the longest geodesic length (also called the diameter of the network) is 7, showing that it also has the small-world property. A component in a network is a set of nodes between which at least one geodesic exists. Many networks possess one giant component that accounts for most of the nodes in the network, and it is true for our network as well: the largest component consists of 99.8% of the nodes.

[3] This process leaves us with 6.5% of all composers in the ArkivMusic data, however, we find that most prominent classical composers – who turn out to be high-degree nodes – that we are most interested in are left intact. The average number of CDs in which the removed composers are featured on is 3.5, significantly lower than that for the composers who remain in our database, which is 64.8.

Table 1 Basic network properties

Number of nodes	878
Number of edges	13,667
Size of the largest component	876
Mean geodesic length	2.62
Diameter	7
Clustering coefficient (random expectation)	0.618 (0.035)

3.2 Clustering Coefficient

Clustering refers to the tendency for triangles to form in a network. It is most commonly quantified by the Clustering Coefficient $C \in [0,1]$ defined as

$$C = \frac{3 \times \text{number of triangles}}{\text{number of connected triples}}, \tag{1}$$

where a connected triple is a set of three nodes $\{u,v,w\}$ such that u and v are connected, and v and w are connected. C is the probability that two nodes connected to a common neighbor are neighbors themselves. C for a network is often compared with the expected value from a random graph of equal n (number of nodes) and m (number of edges), and social networks in particular exhibit a large C [13]. Our composer network exhibits $C = 0.62$, which is also significantly larger than the random expectation 0.035. Thus two composers who have been featured on a CD with a common composer are highly likely to have been featured on a common CD.

3.3 Degree Distribution

The number of a node's neighbors is called its degree, often written as k. It is often the most fundamental quantity underlying many features of a network [5]. The degrees of nodes in a network can sometimes vary widely, and it is represented by the degree distribution $p(k)$ or its cumulative distribution $P(k) = \sum_{k'=k}^{\infty} p(k)$. We show the $P(k)$ for our network in Fig. 2 on a log-log scale. The node degrees vary widely in the network, with Johann Sebastian Bach (1685–1750) having the highest degree with $k = 348$, approximately 11 times that of the average degree $\overline{k} = 31.1$, followed by Wolfgang Amadeus Mozart (1756–1791) with $k = 287$.

3.4 Centralities of Composers

The most outstanding characteristic of a network is the heterogeneity in the structures around each node, and the right-skewed degree distribution in Fig. 2 is one example of it. The differences between each node in a network are exemplified by the nodes' centralities. As its name suggests, centrality is a measure of the importance or influence of a node in a network. The degree is one type of a centrality; in

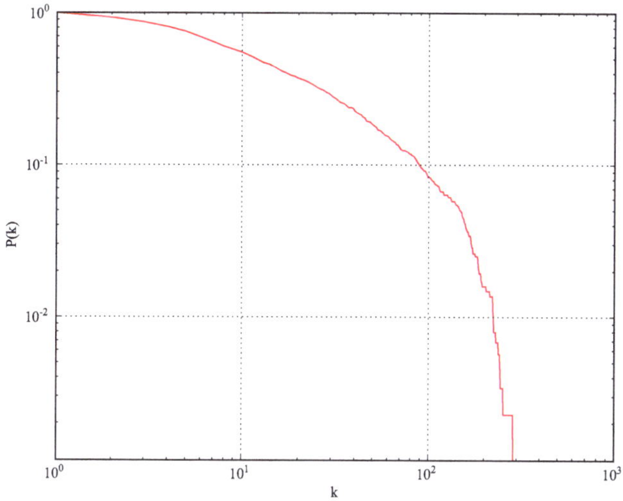

Fig. 2 The cumulative degree distribution $P(k)$ of the composers. The three highest-degree musicians in our network are Johann Sebastian Bach ($k = 348$), Wolfgang Amadeus Mozart ($k = 287$), and George Frideric Handel ($k = 254$).

a social network, for instance, a high-degree person with many friends can be assumed to be more influential than one with few friends. In our network of classical music composers, the degree is the number of composers that one has been featured on a common CD. Since in a projected network the degree is bounded by $\sum_i B_{ij} n_j$ where B_{ij} is the number of connections (either 0 or 1) between a CD i and composer j, and the n_j is the number of composers featured with composer j on the CD, a high degree implies being featured in many CDs or with many composers, i.e. popularity or compatibility with many composers. The top-degree composers are shown in Table 2.

Different centralities capture different types of nodes' importance, and while they are often correlated, noticeable disagreements often point to some interesting aspects of the network. The **Eigenvector centrality** and the **(Freeman) Betweenness centrality** are two popular centrality measures besides the degree [14, 15]. Unlike the degree, the eigenvector centrality takes into consideration the "quality" of a connection. The idea behind it is that not all connections may be equal, and that being neighbors with an important node makes one more important. It is given as $x_i = \kappa^{-1} \sum_{ij} A_{ij} x_j$ which happens to be the definition of the eigenvector of the adjacency matrix $\mathbf{A} = \{A_{ij}\}$, and the eigenvector centrality means the components of the leading eigenvector of $\mathbf{A}$ [5]. This can be thought of as a generalization of the degree, and it often correlates highly with the degree centrality: In our network the Spearman rank correlation between the two is 0.940 ± 0.003, with seventeen common composers in the lists of top 20 composers of each centrality.

Another popular centrality is the (Freeman) betweenness centrality. It measures how often a node sits between two nodes, acting as an intermediary when the two nodes were to, say, exchange messages. It is given by $f_i \equiv \frac{1}{2}\sum_{jk} g_{jik}/g_{jk}$, where g_{jk} is the number of geodesics between j and k, and g_{jik} is the number of geodesics that go through i. The Spearman rank correlation between betweenness and degree centralities is 0.831 ± 0.009, and between betweenness and eigenvector centralities is 0.698 ± 0.012, respectively. While the correlations are positive, an inspection of Table 2 tells us that Modern composers are ranked extraordinarily high in betweenness centrality, whereas they were not so in other centralities. It turns out that modern composers who account for a majority of composers (70.3%) form a close-knit community between themselves, raising the betweenness centrality of the prominent ones such as Leonard Bernstein (1918–1990) and John Cage (1912–1992) despite their low degrees compared with those of composers from other periods. This tells us that investigating how composers in a common period are closely knit amongst themselves could be useful for understanding our network, the results of which we present next.

Table 2 Top 20 composers for the Degree, Eigenvector, and Betweenness centralities. Periods are abbreviated: Baroque (B), Classical (C), Romantic (R), and Modern (M).

Rank	Degree centrality		Eigenvector centrality		Betweenness centrality	
	Name	Period	Name	Period	Name	Period
1	Johann S. Bach	B	Johann S. Bach	B	Johann S. Bach	B
2	Wolfgang A. Mozart	C	Wolfgang A. Mozart	C	George Gershwin	M
3	George F. Handel	B	Claude Debussy	M	Wolfgang A. Mozart	C
4	Felix Mendelssohn	R	Beethoven	R	Leonard Bernstein	M
5	Franz Schubert	R	Franz Schubert	R	John Cage	M
6	Claude Debussy	M	Felix Mendelssohn	R	Ástor Piazzolla	M
7	Johannes Brahms	R	Johannes Brahms	R	Claude Debussy	M
8	Beethoven	R	Tchaikovsky	R	Beethoven	R
9	Tchaikovsky	R	Robert Schumann	R	Aaron Copland	M
10	Maurice Ravel	M	Maurice Ravel	M	Richard Rodgers	M
11	Gabriel Fauré	R	George F. Handel	B	Heitor Villa-Lobos	M
12	George Gershwin	M	Franz Liszt	R	Igor Stravinsky	M
13	Robert Schumann	R	Gabriel Fauré	R	George F. Handel	B
14	Franz Liszt	R	Camille Saint-Saëns	R	Johannes Brahms	R
15	Leonard Bernstein	M	George Gershwin	M	Maurice Ravel	M
16	Camille Saint-Saëns	R	Richard Strauss	R	Franz Schubert	R
17	Franz J. Haydn	C	Antonín Dvořák	R	Felix Mendelssohn	R
18	Igor Stravinsky	M	Franz J. Haydn	C	Alan Hovhaness	M
19	Frédéric Chopin	R	Igor Stravinsky	M	Irving Berlin	M
20	Samuel Barber	M	Sergei Rachmaninoff	M	Tchaikovsky	R

4 Mixing Patterns and Community Structures

Music is one of the oldest art forms created and enjoyed by humans, and accordingly has a rich history of development over time [16, 17, 18, 19]. Historians of music have attempted to break down the evolution of music into stages centered on distinguishable styles [20]. In network science, the study of commonalities between nodes is performed by investigating the mixing patterns between nodes and the community (modular) structures. For our network, in particular, this would allow us to see how well their connection patterns match with the conventional classification scheme.

While any classification scheme of a system can show varying degrees of complexity, a common convention for composers in western classical music is to assign them to certain periods [21]. Here we adopt the period designations used by All Music Guide database that consist of Medieval, Renaissance, Baroque, Classical, Romantic, and Modern, whose musical characteristics are summarized as follows (all years are approximate):

- **Medieval** (500 CE – 1400 CE). It is generally assumed that the primeval shape of musical notation appeared in this period, and several advances over previous practice were shown in regard to tonal material, texture and rhythm. In terms of tonal material, polyphony took a shape, settled down in Renaissance period and has been used in a variety of pieces and even recent ones [16]. Notable composers from this period are Guillaume de Machaut (1300–1377) and Francesco Landini (1325–1397).
- **Renaissance** (1401 – 1600). The main features of music from this period are modes and rich textures in four or more parts that blend strands in the musical texture and harmony with a greater concern with the flow and progression of chords. Polyphony is one of the notable changes that mark the Renaissance from the Middle Ages musically [22]. Notable composers from this period are Thomas Tallis (1505–1585), William Byrd (1540–1623), and John Dowland (1563–1626).
- **Baroque** (1601 – 1750). The creation of tonality distinguishes Baroque music from previous periods. During this period, composers used more elaborate musical ornamentation and made changes in musical notation. Baroque music became more complex in comparison with the songs of earlier periods and expanded the size and range of instrumental performance [18]. Notable composers from this period are Henry Purcell (1659–1695), Antonio Vivaldi (1678–1741), Johann Sebastian Bach (1685–1750), and George Frideric Handel (1685–1759).
- **Classical** (1730 – 1820). Classical music is characterized by a lighter, clearer texture than Baroque music and is less complex. It is mainly homophonic, although counterpoint was used often in later periods. Importance was given to instrumental music. Variety and contrast within a piece became more pronounced than before, and melodies tended to be shorter than those of Baroque music, with clear-cut phrases and clearly marked cadences [23]. Notable composers from this period are Wolfgang Amadeus Mozart (1756–1791) and Franz Joseph Haydn (1732–1809).

- **Romantic** (1815 – 1910). Romanticism, the artistic and literary movement in Europe that occurred in the second half of the 18th century, is a closely-related term with Romantic music [24]. It is characterized by freedom of form, emotions, individuality, dynamic changes and nationalism, a reaction against German influence. It was more personal and emotional than before so there was more freedom in form. Lyrical melodies as well as chromatic harmonies and discords boosted up this situation more along with dramatic contrasts of dynamics and pitches and wide variety of pieces were popular at the same time. Notable composers from this period are Ludwig van Beethoven (1770–1827), Franz Schubert (1797–1828), Frédéric Chopin (1810–1849), Robert Schumann (1810–1856), Franz Liszt (1811–1886) and Pyotr Ilyich Tchaikovsky (1840–1893).
- **Modern** (1900 – current). Modern music is characterized by innovations in the ways of organizing and approaching harmonic, melodic, sonic, and rhythmic aspects of music. Changes in aesthetic views and developments in technology have led to many novel techniques and styles, often called expressionism, abstractionism, neoclassicism, futurism and etc. [25] Besides the aesthetic changes, the rise of American classical music broke the tradition of composers replicating the European classical music. Notable composers from this period are Claude Debussy (1862–1918), Maurice Ravel (1875–1937), Sergei Rachmaninoff (1873–1943), Igor Stravinsky (1882–1971), George Gershwin (1898–1937) and Leonard Bernstein (1918–1990).

4.1 Assortative Mixing

Assortative mixing measures the tendency for similar nodes to be connected, given by the following assortativity measure for discrete node characteristics [26]:

$$r \equiv \frac{\sum_i e_{ii} - \sum_i a_i b_i}{1 - \sum_i a_i b_i} = \frac{\mathrm{Tr}\,\mathbf{e} - \|\mathbf{e}^2\|}{1 - \|\mathbf{e}^2\|}, \tag{2}$$

where $\mathbf{e} = \{e_{ij}\}$ is a matrix whose elements e_{ij} is the fraction of edges in a network that connect a vertex of type i to one of type j, and $\|\mathbf{x}\|$ is the sum of all elements of the matrix $\mathbf{x}$, and a_i and b_i are the fraction of each type of end of an edge that is attached to nodes of type i. For the periods of composers, we have $r = 0.257 \pm 0.005$, meaning that composers belonging to a common period tend to be connected preferentially to one another. The Pearson Correlation Coefficient between connected composers' active years (the middle point between their birth and death years) is even higher, with $\rho = 0.451 \pm 0.009$.

4.2 Communities

A positive assortative mixing that we see above is a symptom of the existence of **communities** or **modules** in a network. A community is commonly defined as a group of nodes of a network where connections between the nodes are denser than to the rest of the network. Algorithms that seek to find communities inside a network are deeply related to the graph partitioning problem, and have seen much development in recent years [27, 28, 29, 30]. Here we used the Louvain algorithm of Blondel *et al.* [31], which returned six communities of which the five largest were studied in more detail that account for 99.4% of the composers in our network. In Fig. 3 we show the period compositions of the member nodes of each module (1A and 1B are submodules of module 1 found by re-running the algorithm on the module, which we discuss later).

We find that the modules represent certain aspects in the history of developments in classical music. First, we see that each module corresponds reasonably well to one single period except for Module 1, which contains composers from four distinct periods – Medieval, Renaissance, Baroque, and Classical. In each of other modules (2 to 5), the majority of nodes belong to a specific period: Module 2 are mainly Romantic, while Modules 3, 4, and 5 are mainly Modern.

To further break down Module 1 we applied the Louvain algorithm one more time, after which we obtained two sizable submodules 1A and 1B. The division along the periods of the nodes is clearer now: Module 1A represents mainly Renaissance and early Baroque composers, while Module 1B represents later composers of Baroque and Classical periods. The Modern composers in 1B, while they appear to be many, are rather insignificant ones with average degree 19.9 in comparison to 77.8, the average degree of later Baroque and Classical composers. They are therefore nicely separated in chronological order. Notable composers in Module 1A

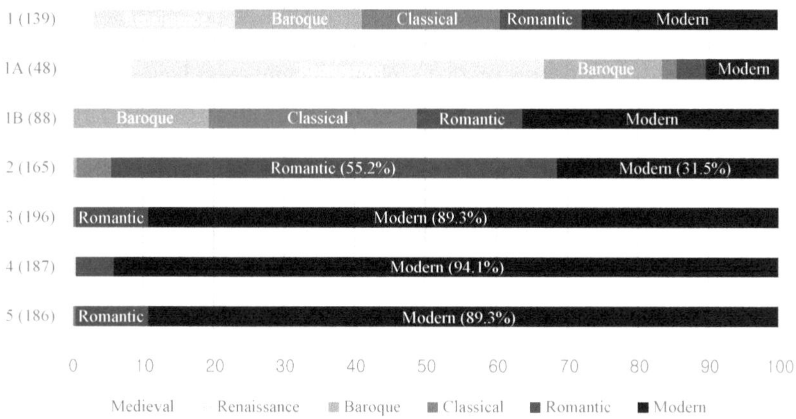

Fig. 3 Period compositions of the network communities. The numbers in parentheses are the modules' sizes. The grayscale color bars show the relative fractions of the periods. Module 1 includes composers from periods between Medieval and Classical. Module 2 represents the Romantic period, while Modules 3, 4, and 5 represent Modern composers. Modules 1A and 1B are submodules of Module 1, and correspond to the earlier and the later periods of Medieval and Classical.

include William Byrd (1540–1623, Renaissance) and Henry Purcell (1659–1695, Baroque). Notable composers in Module 1B include Antonio Vivaldi (1678–1741, Baroque), Johann Sebastian Bach (1685–1750, Baroque), George Frideric Handel (1685–1759, Baroque) from the Baroque period, and Wolfgang Amadeus Mozart (1756–1791, Classical), and Franz Joseph Haydn (1732–1809, Classical) from the Classical period.

Module 2 represents a later time in history, consisting mainly of Romantic (55.2%) and Modern (31.5%) composers. Among these, Romantic composers are generally more prominent (the average degree of the Romantic composers in this module is 104.4, and that of the Modern composers is 38.2), including Robert Schumann (1810–1856, Romantic), Frédéric Chopin (1810–1849, Romantic), Franz Liszt (1811–1886, Romantic), Johannes Brahms (1833–1897, Romantic), and Pyotr Ilyich Tchaikovsky (1840–1893, Romantic). We also note the existence of transitional composers between the Classical and Romantic periods, Ludwig van Beethoven (1770–1827) and Franz Schubert (1797–1828).

Modules 3, 4, and 5 represent the Modern period. In Module 3, the fraction of the Modern composers is 89.3%. The two highest-degree Modern composers are George Gershwin (1898–1937, Modern) of *Rhapsody in Blue* and *Porgy and Bess* and Leonard Bernstein (1918–1990, Modern) of *West Side Story*. Module 3 also includes Jazz composers such as Scott Joplin (1867–1917, Modern) and Billy Strayhorn (1915–1967, Modern), and Broadway composers such as Richard Rodgers (1902–1979, Modern) and Irving Berlin (1888–1989, Modern), reflecting the variety of musical styles of the 20th century.

Module 4 include Charles Ives (1874–1954, Modern) of *The Unanswered Question*, Aaron Copland (1900–1990, Modern) of *Appalachian Spring*, Samuel Barber (1910–1981, Modern) of *Adagio for Strings* and John Cage (1912–1992) of *4'33"*. In fact, composers from the United States account for 86.8% of composers in this module with the average degree of 25.53. Non-US composers have the average degree of 7.33. This module thus represents the growth of American vernacular style of classical music in the 20th century [32]. Ernest Bloch (1880–1959, Modern), Alan Hovhaness (1911–2000, Modern), Ned Rorem (1923–current, Modern), Terry Riley (1935–current, Modern), Steve Reich (1936–current, Modern) and Philip Glass (1937–current, Modern), all from the US, are also in this module.

Module 5 comprises of Modern (89.3%) and Romantic (10.2%) composers. Transitional figures between the periods – e.g. Gabriel Fauré (1845–1924, Romantic), impressionists such as Claude Debussy (1862–1918, Modern), Maurice Ravel (1875–1937, Modern) – are found in this module. In a nice contrast with Module 4, Module 5 appears to represent the non-US branch of modern music, with non-US Modern composers accounting for 79.1% of the composers. The average degree of non-US Modern composers is 41.8, noticeably larger than that of American composers in the module, 8.0. Notable composers include Arnold Schoenberg (1874–1951, Modern) from Austria, Manuel de Falla (1876–1946, Modern) from Spain, Béla Bartók (1881–1945, Modern) from Hungary, Igor Stravinsky (1882–1971, Modern) from Russia, Heitor Villa-Lobos (1887–1959, Modern) from Brazil, Paul Hindemith (1895–1963, Modern) from Germany, Francis Poulenc (1899–1963,

Modern) from France, Ástor Piazzolla (1921–1992, Modern) from Argentina, and Luciano Berio (1925–2003, Modern) from Italy.

In summary, the modules we find algorithmically correspond reasonably well to the developmental history of western classical music. This shows that the associations between composers originally constructed in the academic (musicological) tradition are also reflected deeply in the music recording business, suggesting that a more in-depth exploration of the modular structures could potentially yield new and helpful insights into understanding the landscape of classical music.

5 Conclusion and Discussions

In this paper, we studied the properties of the network of classical music composers. We started by conducting a basic analysis of the structural properties of the network, finding that our network exhibits characteristics common to many real-world networks, including the small-world property, the existence of a giant component, and a high level of clustering. The centrality measurements of the composers showed a reasonable agreement with a common perception of the popularity of the composers. We also explored the global association patterns of composers via assortative mixing and community structure analysis, which showed us the extent to which our network reflected our musicological understanding of the western classical music tradition.

Directions for further research are as follows. First, we note that our work is based on a commercial data archive of classical music, and we believe a similar work based on academic data sources may yield interesting and complementary findings. Since artistic creations serve multiple purposes, as objects of appreciation (consumption) as well as of scholarly study by scholars, both are necessary for a proper understanding for art and culture. Second, we can ask the temporal aspects of the networks in music to understand how a musical style emerges, evolves, and fades in popularity. We believe that our work highlights the potential of network science coupled with advanced data analytics in answering many such pertinent questions in the arts and humanities, playing an instrumental role in the developing field of "digital humanities."

Acknowledgments. We would like to thank Yong-Yeol Ahn, Sebastian Ahnert, and Max Schich for helpful discussions. This work was supported by the National Research Foundation of Korea (NRF-20100004910 and NRF-2010-330-B00028) and IT R&D program of MSIP/KEIT (10045459).

References

1. Albert, R., Jeong, H., Barabási, A.L.: Nature 401(6749), 130 (1999)
2. Faloutsos, M., Faloutsos, P., Faloutsos, C.: ACM SIGCOMM Computer Communication Review, vol. 29, pp. 251–262. ACM (1999)

3. McPherson, M., Smith-Lovin, L., Cook, J.M.: Annual Review of Sociology, 415–444 (2001)
4. Park, J., Lee, D.S., Christakis, N.A., Barabási, A.L.: Molecular Systems Biology 5(1) (2009)
5. Newman, M.: Networks: an introduction. Oxford University Press (2009)
6. Schich, M.: Rezeption und Tradierung als komplexes Netzwerk: der CENSUS und visuelle Dokumente zu den Thermen in Rom. Maximilian Schich (2009)
7. Schich, M., Meirelles, I., Barabási, A.L.: Leonardo 43(3), 212 (2010)
8. Suárez, J.L., Sancho, F., de la Rosa, J.: Leonardo 45(3), 281 (2012)
9. Gleiser, P.M., Danon, L.: Advances in Complex Systems 6(4), 565 (2003)
10. Park, J., Celma, O., Koppenberger, M., Cano, P., Buldú, J.M.: International Journal of Bifurcation and Chaos 17(7), 2281 (2007)
11. Park, J., Barabási, A.L.: Proceedings of the National Academy of Sciences 104(46), 17916 (2007)
12. Milgram, S.: Psychology Today 2(1), 60 (1967)
13. Newman, M.E., Park, J.: Physical Review E 68(3), 036122 (2003)
14. Newman, M.E.: Physical Review E 70(5), 056131 (2004)
15. Freeman, L.C.: Sociometry, pp. 35–41 (1977)
16. Hoppin, R.H.: Medieval music. WW Norton (1978)
17. Reese, G.: Music in the Renaissance. WW Norton New York (1959)
18. Bukofzer, M.F., Bukhofzer, M.F.: Music in the baroque era: from Monteverdi to Bach. WW Norton (1947)
19. Barzun, J.: Classic, romantic, and modern, vol. 255. University of Chicago Press (1961)
20. Taruskin, R.: The Oxford History of Western Music: Music in the Nineteenth Century, vol. 3. OUP USA (2009)
21. Grout, D.J., Palisca, C.V., et al.: A history of Western music, 5th edn. WW Norton & Company, Inc. (1996)
22. Atlas, A.W.: Renaissance music: music in western Europe, pp. 1400–1600. Norton (1998)
23. Rosen, C.: The Classical Style: Haydn, Mozart, Beethoven, vol. 1. WW Norton & Company (1997)
24. Kravitt, E.F.: The Musical Quarterly 76(1), 93 (1992)
25. Albright, D.: Modernism and music: an anthology of sources. University of Chicago Press (2004)
26. Newman, M.E.: Physical Review E 67(2), 026126 (2003)
27. Ahn, Y.Y., Bagrow, J.P., Lehmann, S.: Nature 466(7307), 761 (2010)
28. Mucha, P.J., Richardson, T., Macon, K., Porter, M.A., Onnela, J.P.: Science 328(5980), 876 (2010)
29. Newman, M.E.: Proceedings of the National Academy of Sciences 103(23), 8577 (2006)
30. Sales-Pardo, M., Guimera, R., Moreira, A.A., Amaral, L.A.N.: Proceedings of the National Academy of Sciences 104(39), 15224 (2007)
31. Blondel, V.D., Guillaume, J.L., Lambiotte, R., Lefebvre, E.: Journal of Statistical Mechanics: Theory and Experiment 2008(10), P10008 (2008)
32. Struble, J.W.: The history of American classical music: MacDowell through minimalism. Facts on File (1995)

The Network of Western Classical Musicians

Arram Bae, Doheum Park, and Juyong Park

Abstract. The expanding availability of large-scale data is leading to increased opportunities for applying advanced data analysis and modeling methodology to a wide range of problems and systems, allowing us to deepen our understandings and make novel discoveries. In this paper we use the tools of network science to the network of composers and performers from the western classical music tradition constructed from an extensive data archive of CD recordings. We measure the fundamental characteristics of the network such as the small-world property and the power-law degree distribution. We also investigate the community structures of the musicians, revealing how individual attributes such as musical style, positions, and nationalities factor into the large-scale association patterns of the network. We believe that our work showcases the potential benefits of network science in the study of arts and humanities.

1 Introduction

Advances in information technology have led to numerous fundamental developments in arts and humanities. In the creative process, for instance, artists are making use of such online softwares and infrastructure as Dr Drum[1] and MAGIX[2] to collaborate and produce novel forms of art. Artistic activities taking place online often leave digital "footprints" – data in the form of logs, documentation, or artworks – that may help us observe the creative process in detail that can in turn enhance our understanding of art. Such anticipation is not limited to the study of arts; the

Arram Bae · Doheum Park · Juyong Park
Graduate School of Culture Technology
Korea Advanced Institute of Science and Technology
Daejeon, Republic of Korea, 305-701
e-mail: {redsin0617,park154,juyongp}@kaist.ac.kr

[1] http://beats-maker-software.com
[2] http://www.magix.com

P. Contucci et al. (eds.), *Complex Networks V*,
Studies in Computational Intelligence 549,
DOI: 10.1007/978-3-319-05401-8_2,

accumulation of large-scale detailed data produced from various human activities is accelerating, and in many fields of science and engineering the analysis and modeling of "Big Data" are considered to be providing novel opportunities for developing the fields. Sudden influxes of data have transformed researchers to document and manage their data with advanced data-mining tools, online community collaborations and sophisticated visualization techniques [1].

A framework of data analysis and modeling that has gained attention and seen substantial advances in recent years is network science [2]. Network science focuses on understanding the complex patterns of connections (ties) between the constituent parts of a system. A product of the convergence of the long traditions of graph theory from mathematics, data mining from computer science, and social network analysis (SNA) from sociology, network science has contributed to a deep understanding of a wide range of systems found nature [3,4], engineering [5,6], and society [7].

There have been some notable work on the use of networks in arts, humanities, and culture studies [8]. In music, Gleiser and Danon studied the topology and community structure of the social interaction network of jazz musicians, and showed the presence of communities based on the recording locations of the musicians and racial segregation between musicians [9]. Silva *et al.* studied the Brazilian popular musician network, and showed small-world effect and a power-law degree distribution [10]. Park *et al.* considered two distinct types of ties – one representing musical similarity, and the other representing collaboration – between modern popular musicians and showed how the central musicians determined from the edge types can vary significantly [11]. Besides musician's connection, a large amount of work exploring the structural properties of human networks are studied. Words in human language interact in sentences revealed small world effect [12]. Recently, Ahn *et al.* studied the flavor network, and discussed how different culinary cultures differ in the combination of flavors to produce popular recipes [13].

In this paper, we study the network of Western classical music that covers the period from 5th century to the present, constructed from the comprehensive data of music CD recordings. As a recording is essentially the product of the music industry's response to the demands of consumers of music, we can say that the network constructed from the recordings data reflect what humans perceive as useful, effective combinations of musicians. In order to understand the combinations of the various players, unlike the majority of previous studies where only one type of nodes was considered at once – e.g. artist-to-artist or artwork-to-artwork associations – we preserve the bipartite structure with multiple musician classes. From this we study the distinct positions that each artist occupies in the landscape of classical music, and investigate how they combine to form large discernible communities (modules). We also make use of network-independent musician attribute data to identify possible factors of community formation.

This paper is organized as follows. First, we perform a general network analysis on the network of musicians. Second we explore the finer details of communities in the network by inspecting their musical styles, positions (instruments), and nationalities.

2 The Data and Network Construction

We utilize the database from ArkivMusic[3] (AM), an online vendor of classical music CDs who maintains a comprehensive data on more than 96 000 CDs released in the US as of 2013. For each CD it lists its title, release date, and label, along with the musicians related to the featured music (composers, conductors, solo performers, and ensembles). After cleaning up the data – mainly removing the so-called "compilation CD" that are mostly collections of past recordings – we have at hand 67 305 CDs, 15 214 composers, 6 432 conductors, 47 262 solo performers, and 11 434 ensembles (see Table 1). We can represent this data set as a *bipartite network*, where a tie can exist between the set of CDs and the set of musicians, shown in Fig. 1. While taking a *one-mode projection* of a bipartite network is a common additional step, in our paper we refrain from doing so, and maintain the bipartite nature of the data so that no information is potentially lost due to the projection [14].

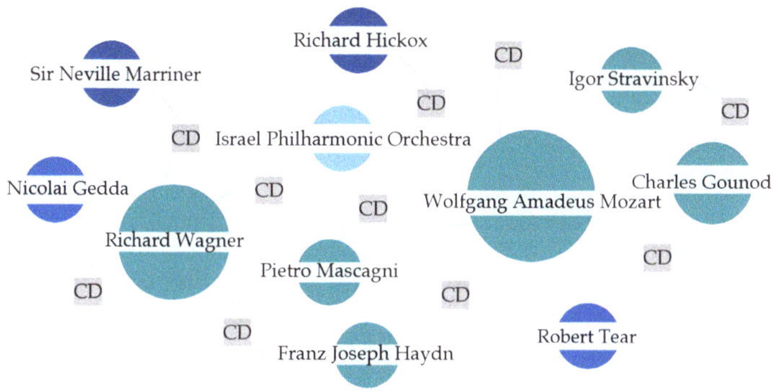

Fig. 1 The construction of the bipartite network from the Arkivmusic dataset. The network shows that Richard Wagner (1813–1883, composer), for instance, was featured on a CD with Sir Neville Marriner (1924–, conductor), the Israel Philharmonic Orchestra (ensemble), and Nicolai Gedda (1925–, tenor).

3 Method and Result

3.1 Basic Network Properties

Here we present some basic, standard network properties of the network, summarized in Table 1.

3.1.1 Mean Geodesic Length, Diameter and Giant Component Size

A geodesic is the shortest path between two nodes in a network. Many networks exhibit the so-called "small-world effect" [3] which means that geodesics of the

[3] http://www.arkivmusic.com

network are typically very small compared with the size of the network, made famous by Milgram's "six degrees of separation" experiment in the 1960's [15]. In our network of over 140 000 nodes the average geodesic length is 5.6. The diameter of the network, the longest geodesic, is 18. The largest component, i.e. the set of nodes that are connected via a path, comprises 98.8% of nodes. The existence of such a giant component is also observed in many networks [16].

Table 1 Basic Network Properties

Total Number of Nodes (including CDs)	142 914
– Number of Composers	15 214
– Number of Conductors	6 432
– Number of Performers	47 262
– Number of Ensembles	11 434
Number of Edges	435 414
Mean Geodesic Length (Diameter)	5.6 (18)
Size of the Largest Component	141 224
Bipartite Clustering Coefficient (Random)	0.0314 (0.00004)

3.1.2 Clustering Coefficient

Clustering, or transitivity, is a measure of how tightly nodes of a network are connected. It is measured in a unipartite network using the clustering coefficient C that is the probability that two neighbors of a node are neighbors themselves, defined as

$$C \equiv \frac{3 \times \text{number of triangles}}{\text{number of connected triples}}, \tag{1}$$

where a connected triple is a set of three nodes $\{u, v, w\}$ such that u and v are connected, and v and w are connected.

This definition cannot be used, however, in a bipartite network since a triangle does not exist in it. An appropriate definition of the clustering coefficient is CC_4 in a bipartite network [17].

$$CC_4 \equiv \frac{4 \times \text{number of full foursomes}}{\text{number of connected foursomes}}, \tag{2}$$

where the number of connected foursomes (L_3) is a set of four nodes $\{U, V, w, x\}$ such that at most one edge is missing between the node sets $\{U, V\}$ and $\{w, x\}$. See Fig. 2. We find in our network $CC_4 = 0.031$, over 700 times larger than the random expectation 0.00004.

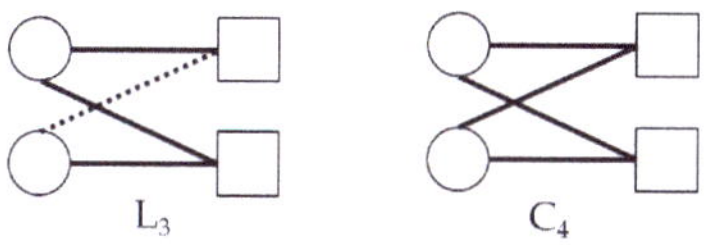

Fig. 2 The clustering coefficient CC_4 for a bipartite network is defined using the concept of a connected foursome (L_3) and a full foursome (C_4)

3.1.3 Degree Distribution

The degree of a node is the number of connections it has. In our network the degree of CD is the number of musicians featured on it, while the the degree of a musician is the number of CDs the are featured on. In Fig. 3 we show the cumulative degree distribution $P(k) = \sum_{k'=k}^{\infty} p(k')$, i.e. the fraction of the nodes that have degree k or larger. We see that, starting from $k \simeq 13$ for more than two decades the $P(k)$ approximates straight line, suggesting a power-law behavior for the degree distribution, i.e. $p(k) \propto k^{-\alpha}$. The maximum likelihood estimate of the power exponent is $\alpha = 2.62 \pm 0.02$ [18]. While Fig. 3 shows the degree of both the CDs and the musicians, we note that the maximum degree of CD is 59, and thus the right-skewed nature of $P(k)$ applies exclusively to the musicians. Also, the degree of a CD does not carry any significant meaning to us, as it is very narrow in range (between 1 and 59) in comparison with the musicians', owing to purely arbitrary technological limitations. For this reason and the fact that it is the musicians that we are primarily interested in, from this point we shall discuss the musicians only.

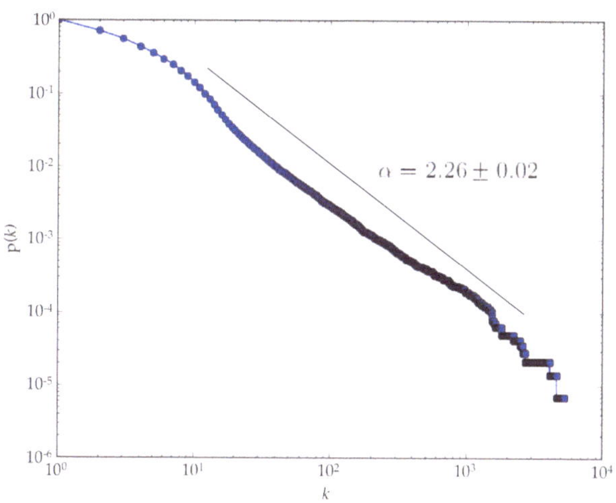

Fig. 3 The cumulative degree distribution in our classical music network. For $k \gtrsim 10$ it appears to follow the power-law. The maximum likelihood estimate of the power exponent in $p(k) \propto k^{-\alpha}$ is $\alpha = 2.26 \pm 0.02$.

The right-skewed degree distribution tells us that a musician's popularity and activity show significant variations, and a few top-degree musicians collectively outweigh the rest. In Table 2, we show the top twenty nodes from the composer and the performer classes. We also indicate an attribute for each class: for the composers, their periods designations (based on All Music Guide[4] and Classical Archives[5]); and for the performers, their positions. For instance, the most-recorded composer is Wolfgang Amadeus Mozart (1756–1791) of the Classical period with 5 288 recordings. Placido Domingo (1941–present), a renowned tenor, is the most-recorded performer with 360 recordings.

The list of musicians on Table 2 may not comes as a surprise to aficionados of western classical music, as they are indeed the most familiar names. The list, however, deserves further investigation since a simple comparison between the musicians based their degrees can be nonsensical across node attributes: For instance, it would be foolish to state that singers are more important to music than all organists are because they exhibit higher degrees.

Table 2 The list of Top 20 Composers & Performers for Degree Centrality

Composer	Composer Period	Performer	Performer Position
Wolfgang Amadeus Mozart	Classical	Placido Domingo	Tenor
Johann Sebastian Bach	Baroque	Andre Previn	Piano
Ludwig Van Beethoven	Romantic	Daniel Barenboim	Piano
Johannes Brahms	Romantic	Dietrich Fischer-Dieskau	Bass
Franz Schubert	Romantic	Maria Callas	Soprano
Giuseppe Verdi	Romantic	Vladimir Ashkenazy	Piano
Peter Ilyich Tchaikovsky	Romantic	Luciano Pavarotti	Tenor
George Frideric Handel	Baroque	Peter Schreier	Tenor
Robert Schumann	Romantic	Fritz Kreisler	Violin
Frederic Chopin	Romantic	Sviatoslav Richter	Piano
Felix Mendelssohn	Romantic	Jeno Jando	Piano
Franz Joseph Haydn	Classical	Nicolai Gedda	Tenor
Richard Wagner	Romantic	Elisabeth Schwarzkopf	Soprano
Claude Debussy	Modern	Bruno Walter	Piano
Franz Liszt	Romantic	Yehudi Menuhin	Violin
Giacomo Puccini	Romantic	Mstislav Rostropovich	Cello
Antonio Vivaldi	Baroque	Dame Joan Sutherland	Soprano
Maurice Ravel	Modern	Jose Carreras	Tenor
Antonin Dvorak	Romantic	Mirella Freni	Soprano
Gioachino Rossini	Romantic	Christa Ludwig	Mezzo-Soprano

Inspecting the degrees of musicians in each attribute class separately and the top rankings covering more than 20 musicians yields more interesting patterns (see Figures 4 and 5). First, we see that within each attribute class (periods for composers and positions for performers) the degrees of musicians are widely distributed (left panels). The right panels in the figures show the fraction of each attribute class as we consider and increasing number of most recorded musicians. For instance, in Fig. 4 when we consider the top 10 composers (x-axis), 70% of them are from the Romantic period, however, when we consider the top 320 composers the fraction

[4] `http://allmusic.com`

[5] `http://classicalarchives.com`

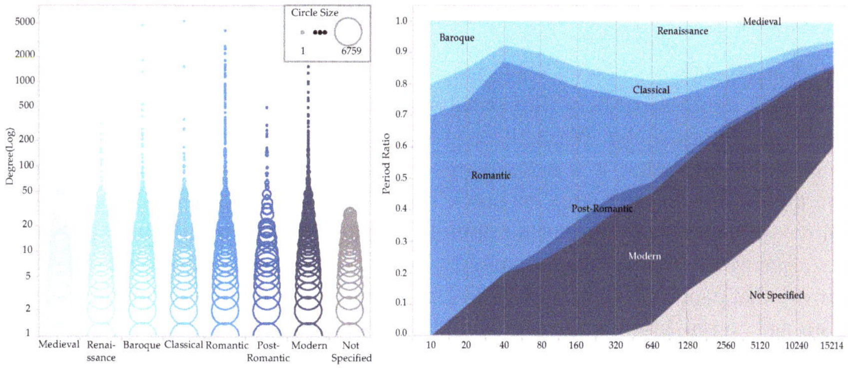

Fig. 4 The degree distributions of composers in each attribute (period) class. The period designations we use consist of Medieval (500–1400), Renaissance (1400–1600), Baroque (1600–1750), Classical (1750–1830), Romantic (1825–1875), Post-Romantic (1875–1900), and Modern (1890–current). The left panel shows that within each period the degree distribution has a heavy tail. The radii of the circles scale logarithmically against the number of musicians in each bin. The right panel shows the fraction of composers of each period belonging to the list of top-degree nodes, as the length of the list is varied from top ten (leftmost) to the entire list.

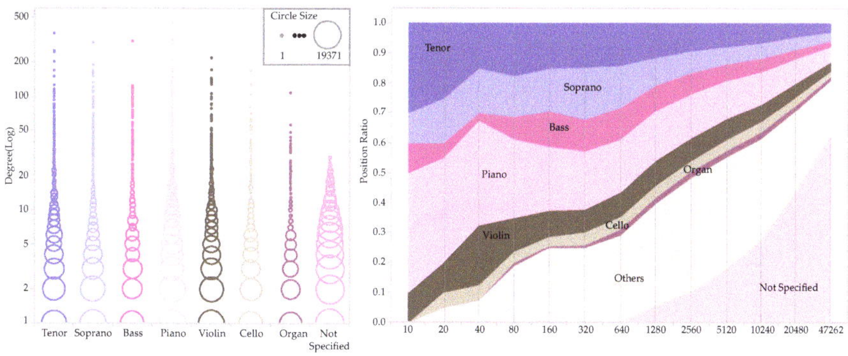

Fig. 5 The degree distributions of performers in each attribute (position) class. The left panel again shows the heavy tails within each position. The right panel shows that tenors and pianists feature highly in the list of top-degree nodes.

drops to below 40%, the value decreasing as we consider more composers, showing that the Romantic composers are indeed disproportionately more recorded in contrast to, say, Medieval composers who do not appear with significant frequency at least until we inspect the top 80 composers. Modern composers are also disproportionately rare near the top of the list. As for the performers (Fig. 5) it is indeed the singers (tenor and soprano in particular) who feature disproportionately high on the list.

3.2 Community Structure of Classical Music Network

Figs. 4 and 5 showcase two ways in which the musicians with certain attributes can be studied, either separately (left panels) or as being in a type of competition for popularity (right panels). While these viewpoints can helpful for understanding the nature of the attributes, it is actually how musicians with different attributes combine that is true to the nature of music on a fundamental level (all compositions and recordings must feature at minimum two musicians with differing attributes). Therefore we now turn to the community, or module, structure of the classical music network that captures quantitatively the large-scale, higher-level patterns of associations and combination of musicians.

Computational methods for identifying communities or modules from a network – a common definition of a community being a set of nodes that have a higher density of connections between themselves to the rest of the network – have seen much advance in recent years [19]. Here we use the popular Louvain algorithm of Blondel *et al.* [20].

The algorithm yields 669 modules, the largest community including 17 871 nodes (7 066 CDs, 2 440 composers, 508 conductors, 1 585 ensembles and 6 272 performers). We show the community size distribution $P(s)$ in Fig. 6. It is right-skewed nonetheless with a steep decline up to the community size $s \simeq 10$, and a slower decline afterwards.

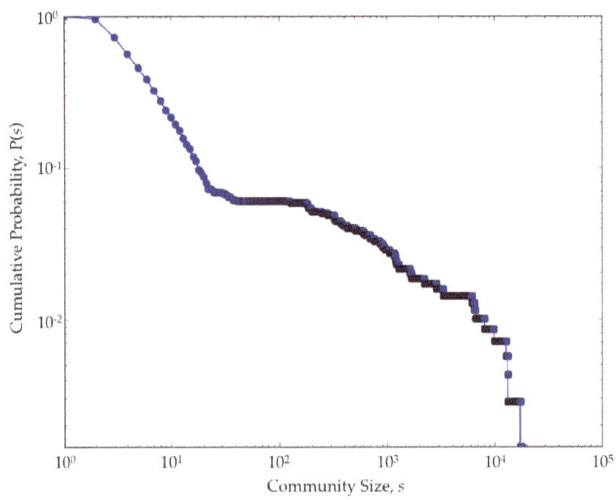

Fig. 6 The cumulative community size distribution $P(s)$. The distribution is right-skewed.

Here we focus on the ten largest communities. We study the composition of the communities via the attribution metadata comprising the period, position (functional roles, such as the major instruments), and nationality of the nodes included in them. The metadata were collected from two online sources, All Music Guide[6] and

[6] `http://allmusic.com`

Classical Archives[7]. The metadata coverage using these data archives for the 75 607 artists nodes (excluding CDs) is 41%, although the missing ones are for the typically low-degree nodes; all nodes with degree 30 or larger have attached attribution metadata.

In Table 3 we show the properties of the ten largest communities (labeled A to J) along with the most frequent attributes of their musician nodes. For instance, the largest one (community *A*) has 10 805 musicians. The most frequent period of the composers is Modern (86.7%), the most common positions are instruments (77.0%), and the most common nationality is USA (75.5%). The most prominent (high-degree) musicians are Leonard Bernstein (1918–1990), George Gershwin (1898–1937), Aaron Copland (1900–1990), Philadelphia Philharmonic, and New York Philharmonic. The second largest one, community B, is of Baroque and Classical composers, including the likes of Wolfgang Amadeus Mozart, Johann Sebastian Bach (1685–1750), and George Friedrich Handel (1685–1759). Famous opera composers Giuseppe Verdi (1813–1901) and Giacomo Puccini (1858–1924) are featured in community C along with popular opera singers Luciano Pavarotti (1935–2007) and Nicolai Gedda (1925–present). Herbert von Karajan (1908–1989), the prolific conductor, is included in community G along with Berlin Philharmonic Orchestra.

Table 3 Basic properties of ten largest communities

Label	Number of Musicians	Major Period	Major Position Group	Major Nationality
A	10 805	Modern (85.7%)	Instrumental (77.0%)	USA (75.0%)
B	6 811	Modern (39.2%)	Instrumental (79.8%)	UK (38.8%)
C	6 726	Romantic (60.2%)	Vocal (90.9%)	Italy (29.7%)
D	5 954	Renaissance (44.0%)	Instrumental (56.7%)	UK (22.1%)
E	5 257	Modern (72.9%)	Instrumental (91.0%)	France (16.1%)
F	5 084	Romantic (46.3%)	Vocal (71.5%)	Germany (43.9%)
G	4 395	Modern (78.2%)	Instrumental (66.8%)	Finland (18.8%)
H	3 802	Modern (75.2%)	Instrumental (87.3%)	USA (18.2%)
I	3 354	Modern (84.3%)	Instrumental (67.5%)	USA (19.2%)
J	3 193	Baroque (27.8%)	Vocal (70.8%)	Germany (27.5%)

While these observations do suggest interesting patterns of community-level associations, simply counting the most frequent attributes may be misleading to characterize each community correctly, since the number of musicians for each attribute may vary widely. To overcome this problem we characterize each community by which attributes are overrepresented. We quantify this using the Z-score for each attribute given as

$$Z(n,p;x) \equiv \frac{x-np}{\sqrt{np(1-p)}}, \tag{3}$$

where x is the observed occurrence of musicians with a specific attribute in a community, n is the size of the community, and p is the overall (global) probability of occurrence of the attribute.

[7] `http://classicalarchives.com`

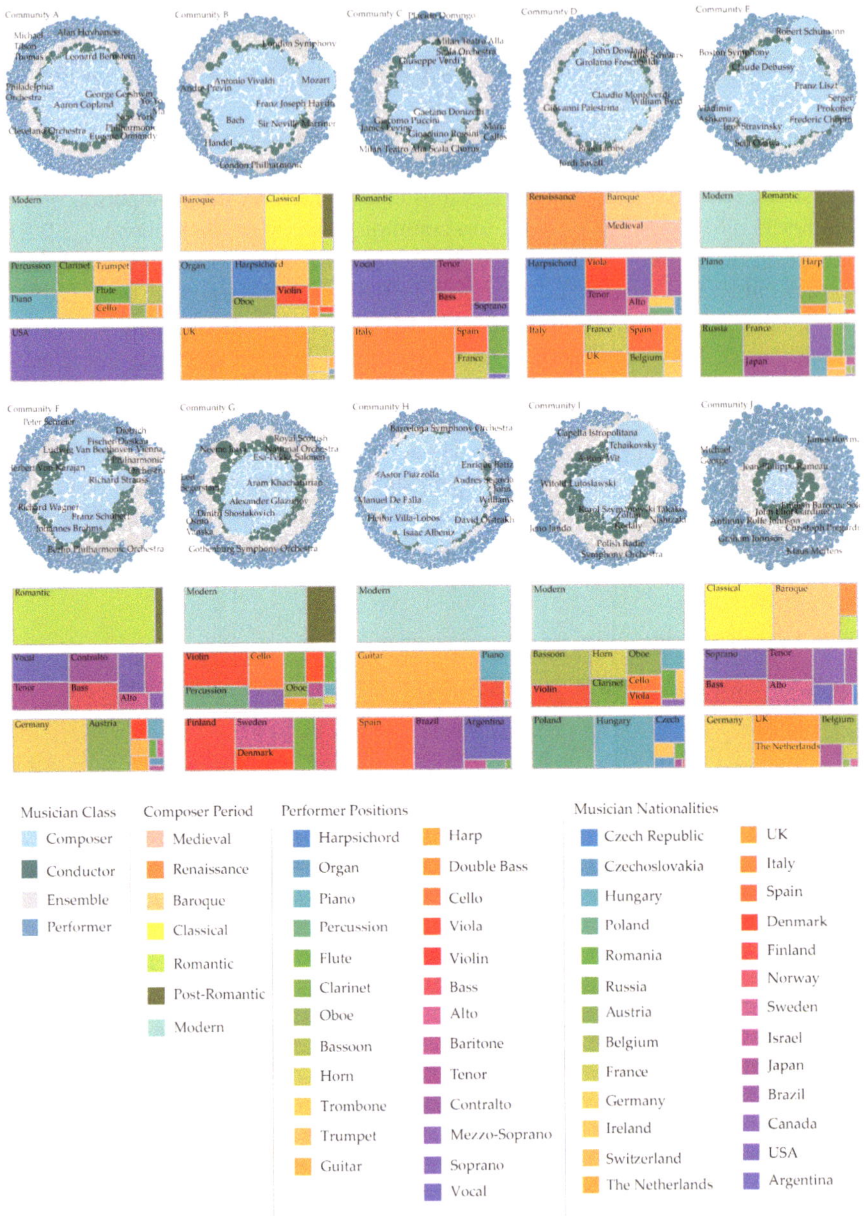

Fig. 7 Major communities found in our classical music network. The boxes show the over-represented node attributes in each community; Community A, for example, is predominantly of Modern-day composers and artists hailing from the USA, including Aaron Copland (1900–1990) and George Gershwin (1898–1937).

Characterizations of the communities using the Z-score for the three attribute class are summarized graphically in Figs. 7. For each community we show four types of graph in the following order, from top to bottom:

1. Visualization of the community. Names of some notable musicians are indicated.
2. Composers' periods. The area of a period represent the relative size of its Z-scores. Under-represented periods ($Z < 0$) are not indicated.
3. Performers' positions.
4. Musicians' nationalities.

According to these four types of information, we can make the following summary observations of the communities. First, as was first observed, community A represents the American classical music scene led by Modern musicians such as Aaron Copland, Leonard Bernstein, New York Philharmonic, etc. Other notable Modern communities are communities E, G, H, and I. Community E is special in that it actually represents the transition between Modern and the (late) Romantic that precedes it; Claude Debussy (1862–1918) and Igor Stravinsky (1882–1971) are well-known figures in the birth of Modern music. Communities G (Northern Europe), H (Latin and Brazilian), and I (Eastern Europe) represent traditions of Modern music developed in different regions. We see two prominently Romantic communities, C and F: Community C is the operatic community led by Italian composers, whereas community F is the Austro-German community led by the likes of Ludwig van Beethoven (1770–1827), Johannes Brahms (1933–1897), and Richard Wagner (1813–1883). Communities B (UK), D (broader Western Europe), and J (Germany-heavy Europe) represent the early Periods, again divided along regions.

Lastly, we can see in the figures that the performers' positions are correlated with the communities' Periods as well. Take communities B and J, for instance: Besides having different nationalities overrepresented, community B is heavily instrumental (representing 79.8% of performers) while community J is vocal (via operas, cantatas, and oratorios). In community H guitarists are highly overrepresented, containing prominent ones including John Williams (1941–present), Andres Segovia (1893–1987), and Julian Bream (1933-present).

4 Conclusion

In this paper, we analyzed the network of western classical music constructed from the comprehensive CD recordings data. We found that our network shows common characteristics of many real-world networks, such as the small-world property, the existence of a giant components, and a high level of clustering. The community structure analysis of the network allowed us to see how the various attribute data (period, position, and nationality) correlate to form groups of musicians that agree well with the development of western classical music.

Our work presented here has naturally answered only a very small fraction of many interesting and crucial questions one can ask in the field of western classical music, which may be answered as more data and methods become available and well understood. We believe that our work showcases how network science and large-scale data analysis can be used for understanding a topic of interest to many in the arts and humanities.

Acknowledgement. We would like to thank Yong-Yeol Ahn, Sebastian E. Ahnert, and Maximilian Schich for data and helpful discussions. This work was supported by the National Research Foundation of Korea (NRF-20100004910 and NRF-2010-330-B00028), and IT R&D program of MSIP/KEIT (10045459).

References

1. Nature 455(1) (2008)
2. Han, J., Kamber, M., Pei, J.: Data mining: concepts and techniques. Morgan kaufmann (2006)
3. Watts, D.J., Strogatz, S.H.: Nature 393(6684), 440 (1998)
4. Jeong, H., Mason, S.P., Barabási, A.L., Oltvai, Z.N.: Nature 411(6833), 41 (2001)
5. Faloutsos, M., Faloutsos, P., Faloutsos, C.: ACM SIGCOMM Computer Communication Review, vol. 29, pp. 251–262. ACM (1999)
6. Barabási, A.L., Albert, R.: Science 286(5439), 509 (1999)
7. Wasserman, S.: Social network analysis: Methods and applications, vol. 8. Cambridge University Press (1994)
8. Schich, M., Meirelles, I.: Leonardo 45(1), 77 (2012)
9. Gleiser, P.M., Danon, L.: Advances in Complex Systems 6(4), 565 (2003)
10. De LimaeSilva, D., Medeiros Soares, M., Henriques, M., Schivani Alves, M., de Aguiar, S., de Carvalho, T., Corso, G., Lucena, L.: Physica A: Statistical Mechanics and its Applications 332, 559 (2004)
11. Park, J., Celma, O., Koppenberger, M., Cano, P., Buldú, J.M.: International Journal of Bifurcation and Chaos 17(7), 2281 (2007)
12. Cancho, R.F.I., Solé, R.V.: Proceedings of the Royal Society of London. Series B: Biological Sciences 268(1482), 2261 (2001)
13. Ahn, Y.Y., Ahnert, S.E., Bagrow, J.P., Barabási, A.L.: Scientific Reports 1 (2011)
14. Newman, M.E., Park, J.: Physical Review E 68(3), 036122 (2003)
15. Milgram, S.: Psychology Today 2(1), 60 (1967)
16. Newman, M.E.: Physical Review Letters 89(20), 208701 (2002)
17. Robins, G., Alexander, M.: Computational & Mathematical Organization Theory 10(1), 69 (2004)
18. Clauset, A., Shalizi, C.R., Newman, M.E.: SIAM Review 51(4), 661 (2009)
19. Newman, M.E., Girvan, M.: Physical Review E 69(2), 026113 (2004)
20. Blondel, V.D., Guillaume, J.L., Lambiotte, R., Lefebvre, E.: Journal of Statistical Mechanics: Theory and Experiment 2008(10), P10008 (2008)

Systematic Dynamic and Heterogeneous Analysis of Rich Social Network Data

Lei Meng, Tijana Milenković, and Aaron Striegel

Abstract. Recent technological advances have lead to increasing amounts of social network data that is longitudinal or encompasses multiple link types. We aim to provide a framework for systematic analysis of such data. We validate the framework on a unique and rich social network, by studying the evolution of network structure over an 18-month period as well as the relationships between different communication types (including both digital (e.g., Facebook) and face-to-face interactions).

1 Introduction

Motivation. Networks (or graphs) have been used to model real phenomena in many domains (e.g., [6, 27, 28]). Here, we focus on social networks, which model communication between people, such as real-world friendships, online social contacts, or electronic communication [5, 16, 18]. Traditionally, due to limitations of techniques for data collection, social network research has focused on studying static networks as well as homogenous networks encompassing a single communication (or link) type [21, 23]. However, with recent advances in data collection techniques, dynamic and heterogeneous (or multiplex) social networks have become available [3, 15, 33]. Consequently, new questions have emerged with regards to network evolution and dependencies between different communication types. Answering these questions is important for many network research applications such as link prediction and others [1, 12, 32]. Thus, in this paper, we focus on comprehensive analysis of dynamic and heterogeneous social network data.

Related Work. Many networks have been typically treated as homogenous data in the sense that different link types are either treated equally or studied in isolation without incorporating their interdependencies [9, 14, 15, 23, 24]. In either case, valuable information encoded into the different network types is lost. Such studies

Lei Meng · Tijana Milenković · Aaron Striegel
Department of Computer Science and Engineering, University of Notre Dame, IN, USA
e-mail: {lmeng,tmilenko,striegel}@nd.edu

P. Contucci et al. (eds.), *Complex Networks V*,
Studies in Computational Intelligence 549,
DOI: 10.1007/978-3-319-05401-8_3, © Springer International Publishing Switzerland 2014

have analyzed network structure in terms of e.g., network density [15], average path length (the small world phenomenon) [23], degree distribution [9], or community structure [10, 14, 24]. Recently, there have been advances in heterogeneous network analysis [3, 4, 29, 30, 33]. However, whereas these studies have demonstrated the potential of using diverse information encoded in multiple communication types, the studies have typically been domain-specific or have not reasoned about the mechanisms underlying the observations [3, 30].

Further, many of the current social network studies (including those listed earlier) have analyzed only the static representation of a social network that is in fact dynamic in nature [21, 24], thus losing valuable temporal information. More recent studies have recognized the potential of accounting for the dynamic nature of the network data [2, 11, 15, 14, 25]. The studies have typically done so by taking multiple snapshots of the evolving network at different times and by using those snapshots to make inferences about the evolution of network structure [14, 15]. Various parameters involved in this process, such as the length of the time interval used to define a snapshot or the criteria for connecting two nodes (i.e., people) within the snapshot, can significantly affect the inferred results [8]. The choice of appropriate values of these parameters is often a neglected issue, as typically the parameters are set to more or less arbitrary values [14].

Our Contributions. Thus, we aim to systematically analyze a dynamic social network encompassing multiple link types in order to gain insights into the evolution of the network as well as dependencies of the different communication types. In the process, we thoroughly evaluate choices of parameters relevant for constructing the dynamic network from longitudinal communication events. The summary of our study and its contributions is as follows:

- We analyze a rich social dataset encompassing various types of digital and face-to-face communication between 150 college students via their smartphones over an 18-month period (starting with the Fall of 2011).
- We explore the effects of the network construction parameters on the results to demonstrate how parameter selection can significantly impact the results.
- We present a novel computational framework for studying dynamic networks with multiple link types. We compare both networks of a given link type over time as well as networks at a given time across multiple link types. Note that in terms of heterogenous analysis, we focus on studying relationships between different communication types; their integration and the analysis of the resulting integrated multuplex network is a subject of future research.
- Moreover, our framework uses four network comparison measures and we comprehensively study the relationships between the different measures.

Our framework reveals the following summary results. In terms of network evolution, networks from consecutive time periods are more similar than non-consecutive networks, and holiday and non-holiday networks are different (users reduce all of their communication during holidays, except for Facebook and email). In terms of dependencies between different communication types, in general, the dependencies are strong between text messaging, phone calls, and physical proximity interactions, each of which relates poorly with Facebook and email interactions. This confirms that Facebook and email might not reflect real-world (personal) interactions to the same extent as the other communication types.

2 Methods

2.1 Dataset

Our data is drawn from the University of Notre Dame's NetSense smartphone study [28] which was launched in August of 2011 with the goal of monitoring the smartphone usage of 200 freshmen entering the university in the Fall of 2011. The study provided each of the students with an Android smartphone (Nexus S) along with plans giving unlimited data, unlimited texting, and unlimited mobile-to-mobile minutes in exchange for complete monitoring privileges on the phone. Full details with regards to the study are available in [28].

For this work, we focus on two broad categories of communication types: *digital communications* (text messages, phone calls, emails, and Facebook postings) and *face-to-face* interactions (proximity observed via Bluetooth). For each communication type, each communication event is associated with the two people involved in the communication, a timestamp, and a length. Face-to-face interactions are detected via Bluetooth and contain the aforementioned parties involved, timestamp, and length, as well as identification of the relative distance between the two smartphones gleaned via observed signal strength. Per the guidelines in [17], we further divide proximity into cases of *close proximity* ($\leq 2.5m, \geq -55dbm$) and *near proximity* ($\leq 5.5m, \geq -65dbm$). We filter random encounters: to form a Bluetooth proximity event, a person needs to detect another person continuously for at least six minutes (at least once in the first three minutes and at least once in the last three minutes); in this case, the timestamp associated with the given event is the first timestamp of the encounter.

Thus, we extract six communication types from the dataset: text (*SMS*), *PhoneCall*, *Email*, *Facebook*, close proximity (*CProximity*), and near proximity (*NProximity*). The data in this paper represents an 18-month period, from September 2011 to March 2013. The pool of users was filtered to 150 users who were involved in the project consistently across the entire period.

2.2 Network Construction

To model the dynamic nature of the data with networks, for each communication type, we take data snapshots at different time points, with all snapshots covering time intervals of equal length, Δt. For each snapshot, we form a corresponding network as follows: nodes are the users (i.e., smartphones) and there is a link of type x between two nodes if there are at least w events of type x between the corresponding users within the given time interval. For simplicity, we treat all networks as unweighted and undirected, but our study can easily be extended to directed and weighted networks. Although the networks are unweighted, we do use w to indicate how strong the connections must be at minimum.

Different values of timescale Δt and link strength threshold w might lead to different network structures. If Δt is very large (Δt=18 months being the extreme), we miss dynamic aspects of the data. On the other hand, if Δt is very small, networks might be largely disconnected. Also, for a small Δt, networks of different type might have zero or weak dependencies with each other, forcing one to study each network type in isolation without the benefit of utilizing across-type information [8]. Similarly, for a given Δt, the larger the value of w, the sparser but more disconnected the network, and the smaller the value of w, the denser but more connected the network. Thus, we aim to find some intermediate Δt and w values that would generate meaningful network structures.

For this purpose, we measure the *density* and the *size of the largest connected component (LCC)* of networks resulting from different Δt and w choices. For graph $G = (V,E)$, the density of G is $\frac{2|E|}{|V| \times (|V|-1)}$, and its LCC size is the number of nodes in the LCC divided by $|V|$. We want our networks to be sparse (just as most of real-world networks are [22]), while the nodes are still as interconnected as possible, with ideally all nodes being in the networks' LCCs. Hence, we want to balance between small network density and large LCC size.

We vary Δt from 1 week to 6 months (Table 1). For each Δt, we vary w as follows (Table 1). For Δt =1 weeks, we only study $w = 1$, since the contact frequency for Δt is already low (and thus, many nodes would be disconnected for a larger w). For $\Delta t = x$ month(s), we evaluate $w = 1$, $w = x$, $w = 2x$, $w = 3x$, and $w = 4x$, since values greater than $4x$ would generate largely disconnected networks.

Table 1 Values for timescale Δt and link threshold w that we evaluate in our study. We do not evaluate $\Delta t > 6$ months, since the resulting networks would fail to capture any holiday information (for details, see Table 2 and Section 3.1).

Timescale (Δt)	Link threshold (w)
1 week	1
1 month	1, 2, 3, 4
3 month	1, 3, 6, 9, 12
6 month	1, 6, 12, 18, 24

2.3 Network Similarity Measures

To study network evolution as well as dependencies between network types, we perform *homogenous* as well as *heterogeneous* network comparisons. By homogenous network comparison, we mean comparing networks of the same communication type but from different time slots, in order to answer how networks of a given type evolve with time. By heterogeneous network comparison, we mean comparing networks from the same time slot but of different types, in order to study dependencies between different data types. We perform network comparison only on networks constructed with same Δt and w values.

To compare any two networks, we use four network similarity measures: 1) *common edges* (CE), i.e., the absolute overlap of the networks' edge sets E_1 and E_2: $|E_1 \cap E_2|$; 2) *adjusted common edges* (ACE), i.e., the relative overlap of the networks' edge sets, as measured by Jaccard index: $\frac{|E_1 \cap E_2|}{|E_1 \cup E_2|}$; 3) *Pearson correlation* of the networks' *degree distributions* (PDD) (the degree distribution of a network is the probability that a node has degree k); and 4) *graphlet degree distribution agreement* (GDD), which generalizes the degree distribution into a spectrum of graphlet degree distributions (graphlets are *induced* subgraphs) [26].

Table 2 Time slots and duration for each holiday (of at least 3 days) within the data, from Fall 2011 to Winter 2013. We do not consider holidays shorter than 3 days, as people are unlikely to change their communication behavior significantly during such short holidays. Hence, including such holidays would have little effect on our results.

Holidays	Duration	Time slots (Δt)			
		1 week	1 month	3 months	6 months
Fall break(2011)	1 week	6-7	1	0	0
Thanksgiving(2011)	5 days	11-12	2	0	0
Winter break(2011)	1 week	14-19	3-4	1	0
Spring break(2012)	1 week	27-28	6	2	1
Easter(2012)	4 days	31	7	2	1
Summer break(2012)	3 months	36-50	8-11	2-3	1
Fall break(2012)	1 week	58	13	4	2
Thanksgiving(2012)	5 days	63-64	14	4	2
Winter break(2012)	1 month	67-71	15-16	5	2

Note that CE and ACE measures explicitly take into account correspondence of node labels between networks: similar topological patterns have to exist in two networks *and* they have to exist between the same nodes for the networks to be similar. But PDD and GDD consider only topological information: it is sufficient for similar topological patterns to exist in networks for the networks to be similar, without the explicit requirement for the patterns to exist between the same nodes. Also, note that of PDD and GDD, GDD is a more constraining similarity measure, as it accounts for more of network topology compared to PDD (GDD includes the degree distribution as the first of its many graphlet degree distributions) [20, 26]. GDD has been used extensively as a state-of-the-art measure for comparison of biological networks [26]. We note though that GDD may perform "suboptimally" at extremely low network connectedness [13] such as with Email networks or Facebook networks during summer time (see below).

3 Results and Discussion

3.1 Effects of the Choice of Network Construction Parameters

We vary values of the two parameters, timescale Δt and link threshold w, in order to empirically choose for further analyses only meaningful parameter values that balance between network sparsity and large LCCs (Table 1 and Section 2.2). Since network density is below 0.35 (and thus satisfactory, as no network is too dense) for all combinations of the studied parameters (Table 1), we next only need to focus on choosing Δt and w values that result in as large of LCCs as possible but without losing valuable information contained in the data.

The effect of Δt is as follows. The LCC size increases as Δt increases, until it reaches a saturating state (Figure 1). However, as Δt increases, useful information about the data is lost. For example, while even the shortest holidays such as Thanksgiving (Table 2) can be captured (in the sense that their effects on the network structure are visible in Figure 1) at Δt=1 week, they cannot be captured at Δt=1 month, Δt=3 months, or Δt=6 months. While the winter break (Table 2) can be captured at Δt=1 week or Δt=1 month, it cannot be captured at Δt=3 months or Δt=6 months. While the summer break (Table 2) can be captured at Δt=1 week, Δt=1 month, or Δt=3 months, it cannot be captured at Δt=6 months. Thus, increase in Δt causes loss of data resolution. In order to capture as much of the relevant (including holiday) information from the data as possible, we leave out from further consideration Δt=3 months and Δt=6 months and henceforth we only focus on Δt=1 week and Δt=1 month.

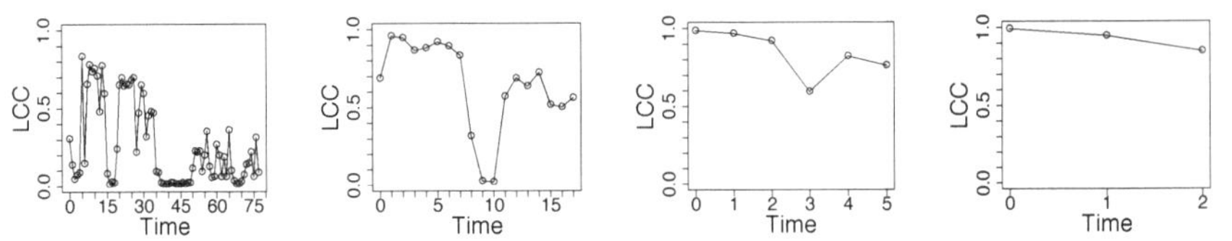

(a) Δt=1 week (b) Δt=1 month (c) Δt=3 months (d) Δt=6 months

Fig. 1 Illustration of the effect of Δt on the LCC size for CProximity and for w=1. The trends are qualitatively similar for other network types and values of w.

The effect of w is as follows. As w increases, more nodes lose links and thus the LCC size decreases. For short time periods such as Δt=1 week, LCCs are not too large even at w=1. Hence, studying larger ws at this Δt yields little insight, as it results in highly disconnected networks. For longer periods such as Δt=1 month, for most network types, the LCC sizes do not decrease significantly when increasing w from 1 to 3, but the LCC sizes do decrease drastically when increasing w from 3 to 4. Therefore, w less than 4 should be used. Of the remaining ws, we henceforth focus only on w=3, because the stronger the value of w, the stronger the communication, and thus, we favor w=3 over w=1 or w=2.

In summary, we continue by studying the following combinations of parameter choices: 1) Δt=1 week and w=1, and 2) Δt=1 month and $w = 3$.

3.2 *Homogenous Network Comparison: Network Evolution with Time*

We compute similarities between networks of a given type from different time slots in order to answer whether networks of consecutive time slots are more similar than networks of non-consecutive time slots, or whether networks from school year periods are different than networks from holiday periods. We use four network similarity measures: CE, ACE, PDD, and GDD (Section 2.3). For measure x and data type y, we form similarity matrix $A_{N\times N}$ where N is the number of time slots and the element $r_{i,j}$ of $A_{N\times N}$ is the similarity value with respect to measure x of two networks of type y from time slots i and j.

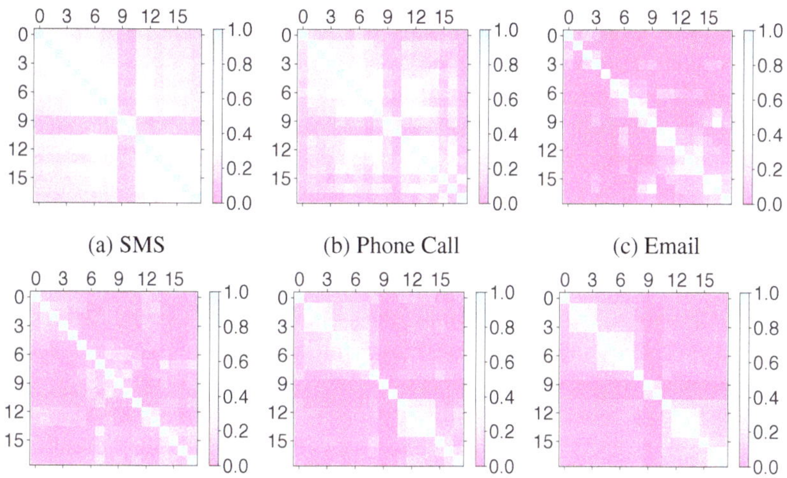

(a) SMS (b) Phone Call (c) Email

(d) Facebook (e) CProximity (f) NProximity

Fig. 2 Illustration of pairwise network similarities for each of the six network types (in the six panels) with respect to ACE for Δt=1 month and w=3. The overall trends are qualitatively similar for other similarity measures and other Δt and w values. Each row/column of a similarity matrix corresponds to a time slot. Values on the diagonals are always one since each network is completely similar to itself. Squares along the diagonals indicate high similarity between consecutive networks covered by the squares.

Indeed, consecutive networks are more similar than non-consecutive networks for SMS, PhoneCall, NProximity, and CProximity (similar has already been observed in proximity networks [7]), whereas this is not as obvious for Email and Facebook (Figure 2). This could be because Email data is extremely sparse in the first place and as such it may be hard to infer meaningful network structure from such data. Note that sparseness of Email is not unique to our dataset as Email is typically not used to maintain personal relationships [19]. Facebook could simply be different from the other network types. Namely, it has already been argued that Facebook does not necessarily capture real-world interactions [31], whereas SMS, PhoneCall, and physical (Bluetooth) proximity likely do.

The difference between Email or Facebook and other network types is further supported with respect to similarities of non-holiday and holiday networks. Networks in summer time (time slots 9-10 in Figure 2) show the lowest similarity with other networks (indicating that non-holiday and holiday networks are very different) for most network types, except for Email and Facebook. Thus, while people reduce SMS, PhoneCall, physical proximity interactions during the break, their Email and Facebook behaviors do not change significantly.

While SMS, PhoneCall, NProximity, and CProximity show similar trends, as discussed above, SMS and PhoneCall are still somewhat different than the other two network types as follows. Non-consecutive network pairs (e.g., networks from time slots 6 and 12 in Figure 2) are overall more similar for SMS and PhoneCall than for NProximity and CProximity. This means that SMS and PhoneCall interactions are more persistent through time than physical proximity.

In Figure 2, we showed trends for ACE. Whereas the trends are *qualitatively* similar for the other network similarity measures, we aim to *quantify* the relationship between the measures. We would expect CE and ACE to be similar, as their only difference is that they capture the absolute versus the relative number of common edges (Section 2.3). Further, we would expect GDD and PDD to be somewhat similar, as GDD incorporates PDD (Section 2.3). Finally, we would expect that GDD, as a more constraining measure of topological similarity than PDD, would be more similar to CE and ACE than PDD, under the assumption that CE and ACE correctly reflect network similarity (which is a reasonable assumption, as the two measures account for node correspondence; Section 2.3).

We illustrate all four network similarity matrices (for the four measures) for one of the network types (Figure 3). But instead of visually inspecting relationships between the different measures, we quantify them by computing for each network type the Pearson correlation between each pair of the four similarity matrices. We

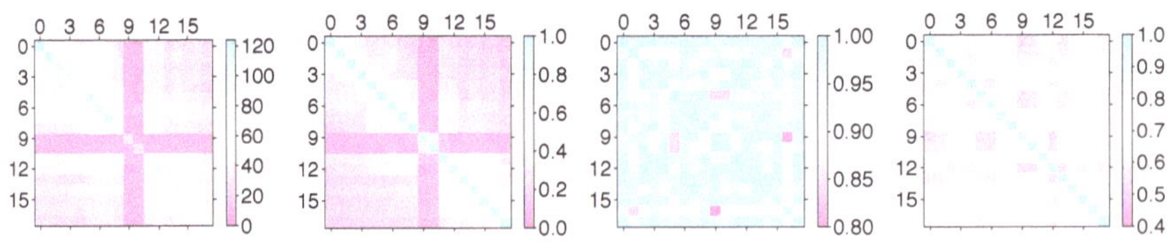

(a) CE (b) ACE (C) PDD (D) GDD

Fig. 3 Illustration of pairwise network similarities with respect to each of the four measures (in the four panels) for SMS, for Δt=1 month and w=3. The trends are qualitatively similar for other network types and other Δt and w values.

Table 3 Pearson correlations between network similarity matrices for each pair of the measures (rows), for each network type (columns), for Δt=1 month and $w = 3$. Similar trends are observed for other Δt and w values. "*", "**", and "***" indicate statistical significance at p-value thresholds of 0.05, 0.01, and 0.001, respectively. For all network types, we ignored holiday time slots 9 and 10, since their networks are extremely small. Email networks are not considered due to their extreme disconnectedness.

Measure I & Measure II	SMS	PhoneCall	Facebook	CProximity	NProximity
CE & ACE	0.884***	0.880***	0.782***	0.849***	0.834***
CE & GDD	0.169	-0.356***	-0.158	0.017	0.450***
CE & PDD	-0.066	0.149	0.156	0.044	0.173
ACE & GDD	0.187*	-0.162	0.328***	0.174	0.413***
ACE & PDD	-0.084	0.068	0.197*	0.213*	0.381***
GDD & PDD	-0.065	0.052	0.194*	0.507***	0.565***

find that indeed CE and ACE are significantly correlated independent of the network type, GDD is significantly correlated to PDD for all but two network types, and GDD is significantly correlated to CE or ACE for all but one network type whereas for PDD this is the case for all but two network types (Table 3). Interestingly, the most significant correlations and between almost all measures are observed for NProximity, which is the densest (i.e., most complete) of all network types. This indicates that the different measures tend to be more robust in denser networks, whereas they tend to give quantitatively less similar (yet qualitatively consistent) results in sparser networks.

3.3 Heterogenous Comparison: Relationships between Network Types

We compute similarities between networks of different types from a given time point to study relationships between the different network types. Since we are dealing with $m = 6$ network types, we perform $M=\binom{m}{2}$ pairwise comparisons for each of N time slots. Thus, for each of the four network similarity measures, we obtain an $A_{M\times N}$ heterogeneous comparison matrix (Figure 4).

Some observations are immediately apparent, such as the holiday effect for each measure (as the trends in time slots 9-10 in Figure 4 are clearly distinguishable from all other trends). While CE is lower during summer time compared to non-holiday time, ACE is higher. This is due to ACE accounting for relative rather than absolute edge intersection between networks having only few edges during summer time. For the same reason, GDD is very high during holidays, as GDD performs "suboptimally" at very low network connectedness (Section 2.3).

We quantify dependencies between different network types by computing, for each matrix (measure) in Figure 4, the average similarity for each row (network type pair). Then, we rank the pairs by the resulting average similarities and

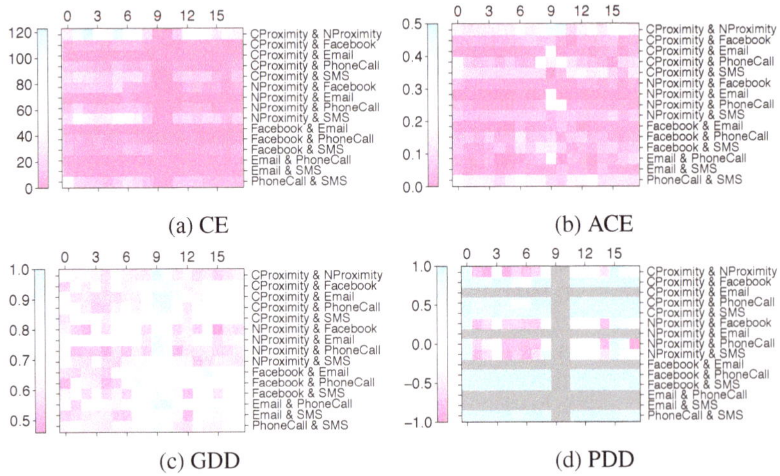

Fig. 4 Pairwise similarities with respect to each of the four measures (in the four panels) between each pair of network types, for Δt=1 month and w=3. The trends are qualitatively similar for other Δt and w values. In a matrix, each row corresponds to a pair of compared network types and each column corresponds to a time slot. A grey color indicates meaningless or undefined Pearson correlation values, resulting from comparing degree distributions of very disconnected networks.

identify top scoring pairs. Interestingly, the most similar pairs for one measure are not necessarily the most similar for another measure (Table 4).

For example, the similarity between CProximity and NProximity, which is ranked the highest with respect to CE and ACE, is not even among the top five with respect to GDD or PDD. This is because CProximity networks are subgraphs of NProximity networks. Thus, all edges within CProximity are also within NProximity, which likely leads in their high CE and ACE scores. However, their GDD and PDD scores are lower, since the overall topologies (when ignoring node label correspondence; Section 2.3) are not as similar.

On the other hand, the top scoring pairs with respect to GDD and PDD include Email and Facebook, none of which are among the top four with respect to CE and ACE (Table 4). This, together with the fact that GDD (and consequently PDD) might not correctly reflect similarities between highly disconnected networks (thus making CE and ACE more trustable measures for such networks; Figure 3.3) supports our finding from Section 3.2 that E-mail and Facebook might not reflect personal interactions to the same extent as the other network types (spatial proximity, SMS, and PhoneCall). Note though that the GDD results should be meaningful for well connected (even if still sparse) networks.

We quantify relationships between the different similarity measures in this context by correlating, for each pair of the measures, their matrices from Figure 4. We observe significant correlations between CE and ACE, as well as between GDD and PDD, but not necessarily between CE or ACE (which account for node correspondence) and GDD or PDD (which do not) (Table 5).

Table 4 Top five network type pairs ranked based on their average similarities by each measure, for Δt=1 month and w=3. The trends are similar for other Δt and w values.

	CE	ACE	GDD	PDD
1	CProximity&NProximity	CProximity&NProximity	Email&Call	Facebook&Call
2	NProximity&SMS	Call&SMS	Email&Facebook	Facebook&SMS
3	CProximity&SMS	CProximity&SMS	Call&Facebook	Call&SMS
4	PhoneCall&SMS	CProximity&Call	CProximity&SMS	CProximity&SMS
5	NProximity&Facebook	NProximity&SMS	CProximity&Facebook	CProximity&Call

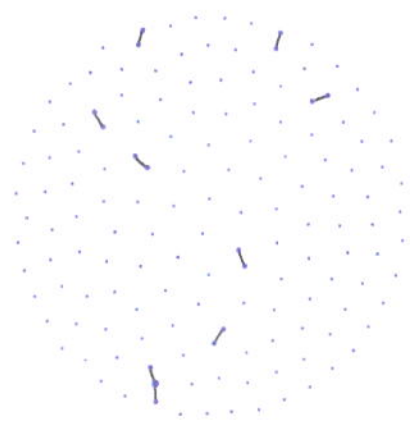
(a) Email

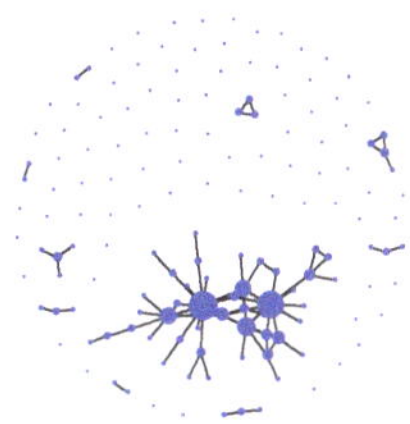
(b) Facebook

Fig. 5 Illustration of highly disconnected Email (left) and Facebook (right) networks from the dataset that despite being structurally different have high GDD score. The size of each node is proportional to its degree.

Table 5 Pearson correlations ("r") between heterogeneous network similarity matrices for each of pair of the measures (columns). "*", "**", and "***" indicate statistical significance at p-value thresholds of 0.05, 0.01, and 0.001, respectively.

Pairs	CE&ACE	CE&GDD	CE&PDD	ACE&GDD	ACE&PDD	GDD&PDD
r	0.530***	-0.132	-0.488***	0.102	0.189*	0.395***

4 Conclusions

We present a framework for the exploration of longitudinal data encompassing multiple link types and validate the framework on a rich dynamic and heterogeneous social network. Our framework reveals that: 1) consecutive networks are more similar than non-consecutive networks; 2) holiday and non-holiday networks are quite different for all network types except Facebook and Email; 3) when studying evolution of networks of a given type, most of the network similarity measures are significantly correlated, especially in more complete networks; 4) when studying dependencies of different network types at a given time, the most similar network types with respect to one measure are not necessarily the most similar with respect to another measure; and 5) Facebook and Email are different than the other network types (SMS, PhoneCall, and physical proximity), suggesting that Facebook and Email might not be capturing personal interactions to the same extent as SMS, PhoneCall, and physical proximity.

Our study has potential to help guide prediction of future links from past/present links or relationships of one type from relationships of another type, thus affecting the link prediction community among many others.

Acknowledgments. This material is based upon work supported by the National Science Foundation under Grant No. IIS-0968529, Grant No. CCF-1243295 and a Google Faculty Research Award.

References

1. Balthrop, J., Forrest, S., Newman, M.E.J., Williamson, M.M.: Technological networks and the spread of computer viruses. Science 304(5670), 527–529 (2004)
2. Berger-wolf, T.Y., Saia, J.: A framework for analysis of dynamic social networks. In: Proceedings of ACM KDD, pp. 523–528 (2006)
3. Davis, D., Lichtenwalter, R., Chawla, N.: Multi-relational link prediction in heterogeneous information networks. In: Proceedings of ASONAM, pp. 281–288 (2011)
4. Dong, Y., Tang, J., Wu, S., Tian, J., Chawla, N.V., Rao, J., Cao, H.: Link prediction and recommendation across heterogeneous social networks. In: Proceedings of ICDM, pp. 181–190 (2012)
5. Eagle, N., Pentland, A.: Reality mining: sensing complex social systems. Personal Ubiquitous Computing 10(4), 255–268 (2006)
6. Eagle, N., Pentland, A., Lazer, D.: Inferring friendship network structure by using mobile phone data. PNAS 106(36), 15,274–15,278 (2009)
7. Eagle, N., Pentland, A., Lazer, D.: Inferring friendship network structure by using mobile phone data. PNAS 106(36), 15,274–15,278 (2009)
8. Emmert-Streib, F., Dehmer, M.: Influence of the time scale on the construction of financial networks. PLoS ONE 5(9), e12,884 (2010), doi:10.1371/journal.pone.0012884
9. Faloutsos, M., Faloutsos, P., Faloutsos, C.: On power-law relationships of the Internet topology. ACM SIGCOMM Computing Communication Review 29(4), 251–262 (1999)
10. Fortunato, S.: Community detection in graphs. Phys. Rev. E 486, 75–174 (2010)
11. Greene, D., Doyle, D., Cunningham, P.: Tracking the evolution of communities in dynamic social networks. In: Proceedings of IEEE/ACM ASONAM, pp. 176–183 (2010)
12. Guy, I., Zwerdling, N., Ronen, I., Carmel, D., Uziel, E.: Social media recommendation based on people and tags. In: Proceedings of ACM SIGIR, pp. 194–201 (2010)
13. Hayes, W., Sun, K., Pržulj, N.: Graphlet-based measures are suitable for biological network comparison. Bioinformatics 29(4), 483–491 (2013)
14. Kumar, R., Novak, J., Tomkins, A.: Structure and evolution of online social networks. In: Link Mining: Models, Algorithms, and Applications, pp. 337–357. Springer (2010)
15. Leskovec, J., Kleinberg, J., Faloutsos, C.: Graphs over time: densification laws, shrinking diameters and possible explanations. In: Proceedings of ACM SIGKDD, pp. 177–187 (2005)
16. Lewis, K., Kaufman, J., Gonzalez, M., Wimmer, A., Christakis, N.: Tastes, ties, and time: A new (cultural, multiplex, and longitudinal) social network dataset using facebook.com. Social Networks (2008)
17. Liu, S., Jiang, Y., Striegel, A.: Face-to-face proximity estimation using bluetooth on smartphones. IEEE Transactions on Mobile Computing, 1 (2013)
18. Masuda, N., Holme, P.: Predicting and controlling infectious disease epidemics using temporal networks. F1000Prime Rep. 5, 6 (2013)

19. Mesch, G., Talmud, I.: The quality of online and offline relationships: The role of multiplexity and duration of social relationships. Information Society 22, 137–148 (2006)
20. Milenković, T., Ng, W.L., Hayes, W., Pržulj, N.: Optimal network alignment with graphlet degree vectors. Cancer Informatics 9, 121 (2010)
21. Mislove, A., Marcon, M., Gummadi, K., Druschel, P., Bhattacharjee, B.: Measurement and analysis of online social networks. In: Proceedings of ACM IMC, pp. 29–42 (2007)
22. Newman, M.: Networks: an introduction. Oxford University Press (2009)
23. Newman, M.E.J.: Scientific collaboration networks. ii. shortest paths, weighted networks, and centrality. Phys. Rev. E 64, 016,132 (2001)
24. Newman, M.E.J.: Modularity and community structure in networks. PNAS 103(23), 8577–8582 (2006)
25. Palla, G., Barabsi, A., Vicsek, T., Hungary, B.: Quantifying social group evolution. Nature 446 (2007)
26. Pržulj, N.: Biological network comparison using graphlet degree distribution. Bioinformatics 23(2), e177–e183 (2007)
27. Solava, R.W., Michaels, R.P., Milenković, T.: Graphlet-based edge clustering reveals pathogen-interacting proteins. Bioinformatics 28(18), i480–i486 (2012)
28. Striegel, A., Liu, S., Meng, L., Poellabauer, C., Hachen, D., Lizardo, O.: Lessons learned from the netsense smartphone study. In: Proceedings of ACM HotPlanet, pp. 51–56 (2013)
29. Sun, Y., Yu, Y., Han, J.: Ranking-based clustering of heterogeneous information networks with star network schema. In: Proceedings of ACM KDD, pp. 797–806 (2009)
30. Wang, X., Sukthankar, G.: Link prediction in multi-relational collaboration networks. In: Proceedings of ACM ASONAM, pp. 1445–1447 (2013)
31. Wilson, C., Boe, B., Sala, A., Puttaswamy, K.P.N., Zhao, B.: User interactions in social networks and their implications. In: Proceedings of ACM EuroSys, pp. 205–218 (2009)
32. Yang, Y., Chawla, N., Sun, Y., Han, J.: Predicting links in multi-relational and heterogeneous networks. In: Proceedings of IEEE ICDM, pp. 755–764 (2012)
33. Ye, J., Chen, K., Wu, T., Li, J., Zhao, Z., Patel, R., Bae, M., Janardan, R., Liu, H., Alexander, G., Reiman, E.: Heterogeneous data fusion for Alzheimers disease study. In: Proceedings of KDD, pp. 1025–1033 (2008)

Moneymakers and Bartering in Online Games

Jane E. Lee, Ah Reum Kang, Huy Kang Kim, and Juyong Park

Abstract. We study the interpersonal trade network from a Massively Multiplayer Online Role-Playing Game (MMORPG), where players actively engage in the exchange and sales of goods and items in a hyperrealistic virtual environment. In this paper we introduce the concept of Standard Price (SP) of items computed from the trade network, which allows us to investigate the relation between the profitability of a trade and the structure of the social networks of the users. We find that the social network is correlated with the outcome of interpersonal trades. For instance, we observe that the margin of profit in a trade correlates with the social distance between trading partners, suggesting that social affinity implies shared information on the value of an item.

1 Introduction

The economic activity is one of the most common and fundamental activities in the human society. The price at which a good gets sold and bought, marked in a common currency of a market, depends upon a wide range of factors including the scarcity of the goods [1], the nature of the distribution channel [2], the relationship between the seller and the buyer [3], to name a few. The complex combination of these variables can cause identical items to be traded at different prices, resulting in the differentiation of profit margins for those involved in the trade. Recently, the characteristics of the complex network of trading partners and their implication on the market structures and dynamics have garnered much interest [4].

In this paper, we study the relationship between the profit generated in a trade and the social network structure of those involved in it. We utilize the economic

Jane E. Lee · Juyong Park
Korea Advanced Institute of Science and Technology, Daejeon, Republic of Korea 305-701
e-mail: {janelee,juyongp}@kaist.ac.kr

Ah Reum Kang · Huy Kang Kim
Korea University, Seoul, Republic of Korea 136-701
e-mail: {armk,cenda}@korea.ac.kr

P. Contucci et al. (eds.), *Complex Networks V*,
Studies in Computational Intelligence 549,
DOI: 10.1007/978-3-319-05401-8_4,

trade and social interaction records from AION, a Massively Multiplayer Online Role-Playing Game (MMORPG) serviced globally. As the goal of MMORPGs is to provide the gamers with a fantastical yet highly lifelike online environment in which they can form social relations and perform realistic actions, they are gaining popularity as "virtual laboratories" for observing human behavior in great detail [5–11].

In collaboration with NC Soft, Inc., the global service provider of AION, we studied the economic trade and the social network data involving more than 50,000 individuals. By introducing the concept of **Standardized Price (SP)** of bartered items, we determined the monetary value of each item, which allowed us to easily quantify the profit or loss that each individual incurred in a barter for universal comparison.

2 Materials and Methodology

We utilized AION data collected during a span of 87 days between April and July of 2010, comprising nearly 1.7 million interactions between 52,757 anonymized players. For our study we considered the following five interaction types, three economic and two social:

1. **Barter** (economic) accounts for 35.6% of all economic transactions. In a barter two players exchange goods for other goods or the in-game currency "Kinah".
2. **Personal Shop** (economic) accounts for 2.7% of all economic transactions. Here a player buys goods from another player who is in a dedicated "merchant mode," functioning only as a shop owner.
3. **Sales Agency** (economic) accounts for 61.7% of all economic transactions. Here a player buys goods from sales agents controlled by the computer acting on behalf of the owners of the goods.
4. **Friendship** (social) indicates that two players have each other on their Friend Lists.
5. **Private Messaging** (social) indicates that two players have communicated with each other.

2.1 *Network Measures*

We measure the following network properties from AION:

1. The **degree** k of a player is the number of his neighbors. In a directed network one has the **in-degree** k^{in} and the **out-degree** k^{out}. A **weighted edge** is an edge with an attached value, e.g. the number of transactions between two players.
2. The **geodesic distance** the length of the shortest path joining two players. The **diameter** of a network is the length of the longest geodesic in the network.
3. The **clustering coefficient** C is the probability that two neighbors of a player are themselves neighbors [12, 13].

2.2 The Standard Price of Goods

Barter in AION comprise both monetary non-monetary transactions between players. Cases where items are traded for items pose a particular challenge, as it is difficult to quantify the value of items involved in the network and therefore determine whether one player has made a "profit." Thus we introduce the concept of the "Standard Price (SP)" of an item (in the unit of Kinah, the in-game currency), to be determined from the transaction network itself. The fundamental idea is that an item's value can be computed from those of other items for which it has been traded. One may set up this problem as solving a system of linear equations but this is unlikely to work, as some equations may simply be contradictory (e.g. when an item has been sold once for 500 Kinahs and another time for 1000 Kinahs). Thus we propose the following method. We start by identifying the items that have been traded for Kinah only at least once. Then we set their prices as the average of the Kinahs they were paid for, which we now use to determine the prices of other items that have been traded for Kinah and the first set of items. We can view the prices as propagating through the network via this iteration. Formally, let us denote by s_x the SP of an item x. We can then represent a barter between two players as

$$\{(n_1,s_1),(n_2,s_2),\ldots\} \longleftrightarrow \{(n_a,s_a),(n_b,s_b),\ldots\}, \tag{1}$$

meaning that one player hands to the other player n_1 of item 1 whose SP is s_1, n_2 of item 2 whose SP is s_2, and so forth, in exchange for n_a of item *a* whose SP is s_a, and so forth. Here we mark a known SP with an asterisk, e.g. s_i^*. The SP of Kinah is its nominal value.

1. Find the transactions that involve only *one* item with the undetermined SP, say item 1 in the barter transaction

$$\{(n_1,s_1),(n_2,s_2^*),\ldots\} \longleftrightarrow \{(n_a,s_a^*),(n_b,s_b^*),\ldots\}, \tag{2}$$

Table 1 Basic network characteristics of AION in comparisons with other online games

Networks	AION			Pardus		
	Trade	PM(9 days)	Friendship	Trade	PM	Friendship
Nodes	21,417	21,771	30,002	18,589	5,877	4,313
Edges	31,539	219,922	100,476	568,923	107,448	21,118
Diameter	23	14	15	NA	NA	NA
Clustering Coefficient / Ratio to Random Network	0.12/872.57	0.04/43.10	0.12/537.50	0.25/109.52	0.28/45.71	0.43/131.95
Average Degree	2.95	20.20	6.70	61.21	36.57	9.79
Average Weighted Degree	6.13	234.00	NA	NA	NA	NA

* measurements of Pardus is from Szell *et al.*'s paper. [8]

from which s_1^* is given as

$$s_1^* = \frac{(n_a s_a^* + n_b s_b^* \cdots) - (n_2 s_2^* + n_3 s_3^* + \cdots)}{n_1}. \quad (3)$$

Perform a similar calculation for all such transactions.

2. Substitute the newly determined SP's for the same items in all the remaining transactions.
3. Repeat.[1]

The SP's of items determined using this method can now be used to quantify the profit incurred by player i in a barter with player j, as

$$R_{ij} = -(\sum_i n_i s_i^*) + (\sum_j n_j s_j^*), \quad (4)$$

where $R_{ij} = -R_{ji}$, and $\sum_i$ is the summation over the items given by i to j, and vice versa. A negative R_{ij} would mean that player i has made a loss (and j a profit).

3 Results

3.1 Basic Network Characteristics

In Table 1 we present the basic properties of networks in AION, along with those of networks from another online game Pardus [8] for comparison. The networks from AION exhibit the so-called "small-world" property (with the diameter being much smaller than the network size), while the Barter and the Friendship network also exhibit high clustering typical of social networks, although the Private Messaging does not, presumably due to the fact that one can send messages freely to anyone on the game. While networks from Pardus generally show higher clustering, the two networks from AION exhibit a larger ratio against the randomized value, possibly resulting from the fact that gameplay in AION are centered around communities called a "Legion." [7]

The cumulative degree distributions are given in Fig. 1. The Barter network is a weighted directed network; the in-degree k^{in} of a node is the number of money-making (profitable) barters, while the out-degree k^{out} is that of lossmaking barters. Friendship and Private Messaging (PM) networks are considered undirected and simple.

[1] For the undetermined s's that remained when the method could no longer be applied, we assumed they were the same price $\tilde{s}$ and solved for it from the linear equation. As they were very few in number and were thus highly likely to be insignificant.

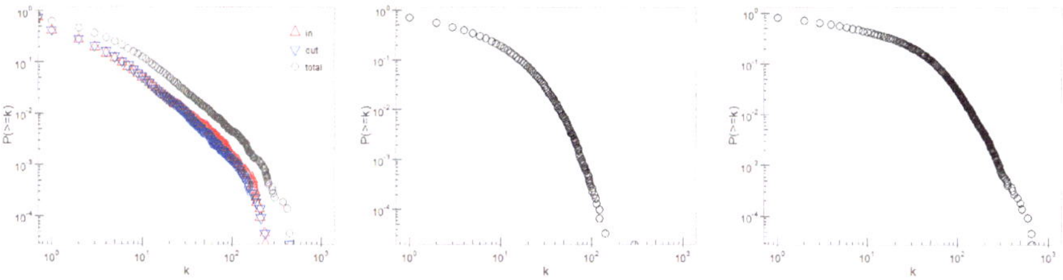

(**a**) Barter (**b**) Friendship (**c**) Private Messaging

Fig. 1 The cumulative degree distributions of networks in AION. The Barter network is weighted and directed, with the in-degree k^{in} defined as the number of profitable transactions, and the out-degree k^{out} defined as that of lossmaking transactions. The Friendship and the Private Messaging (PM) networks are considered undirected and simple.

3.2 *Profits and the Social Network*

In Fig. 2 we show the distribution of net (total) profits of the players based on the SPs calculated using the method above . An intriguing aspect of it is the nearly even split between the "moneymakers" and the "losers", signified by the two peaks on the negative and the positive sides of the x-axis. The near perfect symmetry between the peaks reflects the skewed degree distribution of Fig. 1 (a), where a majority of people conducted one transaction so that for each profit a matching loss exists.

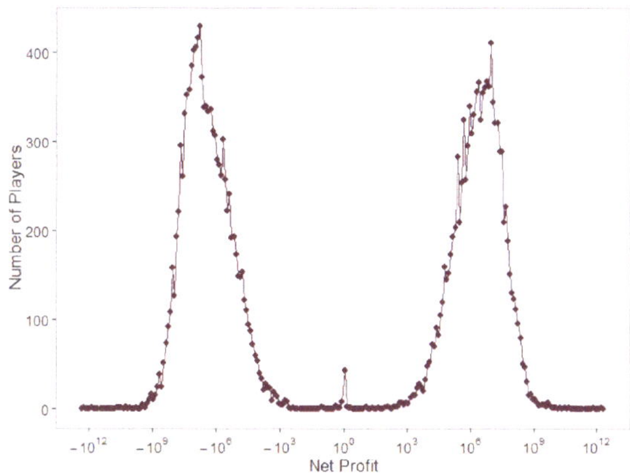

Fig. 2 The distribution of the overall net profits for players who engaged in barter with other players. Interestingly, there exists a split between moneymakers and losers, indicated by the two prominent peaks.

3.2.1 Degree and Profits

Given that the nodes of a network are primarily characterized by their degrees [14], it is interesting to see whether there exists a relationship between the nodes' profits and degrees.

First, presuming that one's skill in deal making increases with one's experience, we can inspect whether one's net profit correlates with the degree itself, i.e. the number of transactions one was involved in. The correlation between degree and net profit turns out to be, however, insignificant: the Pearson correlation coefficient is 0.01 ± 0.02, demonstrating that simply making many trades with others is no indication of good skills[2].

Second, based on the previous realization that a high degree does not necessarily mean a high level of profit, we can ask whether we can distinguish between those who make profits with a high probability and those who do not, given the degree. Indicating the direction of the flow of profit by the direction of the edge in a network (therefore, a player's out-degree is the number of transactions from which the player has suffered a loss), we studied the Pearson correlation coefficient between the in- and out-degrees of the nodes, which turned out to be 0.66 ± 0.09. This means that those with many profitable trades also have many loss-making trades.

These two findings suggest that the degree is generally a poor indicator of profitability in barter, and therefore we may need to investigate other higher-order network properties to understand the nature of the profits better.

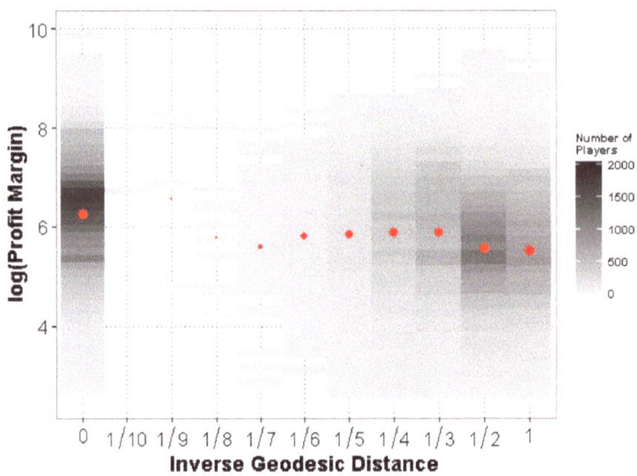

Fig. 3 The profit margin in a trade versus the social distance between the partners in the trade. The red dots represent the average for each inverse distance value. The profit margin generally correlates positively with distance, indicating relative lack of common belief or knowledge about the value of an item between players who are far apart.

[2] To avoid problems from extremely high values, we took the logarithm of the profits in the calculation.

3.2.2 Profitability, Trade Frequency, and Social Network Distance

As mentioned above, an important feature of a network is the concept of the "distance" between nodes. We studied whether there exists a correlation between the profitability (profit margin) of a trade and the social distance between the trading players; from the Friendship network of AION we determined the geodesic distance between the players, and found a Pearson coefficient of -0.25 ± 0.01 between the (logarithmic) profit margin and the inverse geodesic distance[3], implying that the further apart two players are socially, the higher the profit margin and the likelihood of trade to become less "fair." This is a result that strongly suggests the existence of a network effect in assessment of the value of an item in a trade, i.e. an implicitly "shared knowledge" between users that are close in the network.

4 Conclusion

Online gaming environments act as a virtual laboratory in which one can examine detailed socioeconomic behaviors of players, creating interesting opportunities for research. In this paper we analyzed the relationship between the social networks of people and their economic trade behaviors. We found out that a person's profitability is not correlated with their frequency of trade or experience, and that the social distance increased the profit margin, i.e. social affinity lowered the profit margin, indicating some possible role that a social network plays in economic activities.

We believe that our work, along with our method for determining the Standard Price of items from barter data in MMORPGs, lays the foundation for a more in-depth study on the relationship between economic activities and social networks in online environments. We plan to investigate this further, as there are undoubtably many more quantifiable properties of the players' social networks that correlate with their trade behaviors. We also believe that our work has the potential to shed light on the social network-related dynamics of markets in the real world as well.

Acknowledgements. We would like to to thank NCSoft, Inc., for providing us with data. This work was supported by the National Research Foundation of Korea (NRF-20100004910 and NRF-2010-330-B00028), IT R&D program of MSIP/KEIT (10045459, Development of Social Storyboard Technology for Highly Satisfactory Cultural and Tourist Contents based on Unstructured Value Data Spidering), MSIP of Korea under the C-ITRC support program (NIPA-2013-H0301-13-3007) supervised by the NIPA.

References

1. Suri, R., Kohli, C., Monroe, K.B.: Journal of the Academy of Marketing Science 35(1), 89–100 (2007)

[3] Taking the inverse of the geodesic distance makes it possible to deal with node pairs with infinite geodesic distance – when no path exists between the pair – easily.

2. Viswanathan, S., Wang, Q.: European Journal of Operational Research 149(3), 571 (2003)
3. Bolton, R.N., Kannan, P.K., Bramlett, M.D.: Journal of the Academy of Marketing Science 28(1), 95 (2000)
4. Easley, D., Kleinberg, J.: NNetworks, Crowds, and Markets: Reasoning About a Highly Connected World. Cambridge University Press (2010)
5. Boyns, D., Forghani, S., Sosnovskaya, E.: Living virtually: researching new worlds 47, 67 (2009)
6. Lou, J.K., Park, K., Cha, M., Park, J., Lei, C.L., Chen, K.T.: Proceedings of the 22nd International Conference on World Wide Web (International World Wide Web Conferences Steering Committee 2013), pp. 827–836 (2013)
7. Son, S., Kang, A.R., Kim, H.C., Kwon, T., Park, J., Kim, H.K.: PloS One 7(4), e33918 (2012)
8. Szell, M., Lambiotte, R., Thurner, S.: Proceedings of the National Academy of Sciences 107(31), 13636 (2010)
9. Szell, M., Thurner, S.: Social Networks 32(4), 313 (2010)
10. Ducheneaut, N., Yee, N., Nickell, E., Moore, R.J.: Proceedings of the SIGCHI Conference on Human Factors in Computing Systems, pp. 839–848. ACM (2007)
11. Lo, S.K., Wang, C.C., Fang, W.: CyberPsychology & Behavior 8(1), 15 (2005)
12. Ebel, H., Mielsch, L.I., Bornholdt, S.: arXiv preprint cond-mat/0201476 (2002)
13. Newman, M.E.: SIAM Review 45(2), 167 (2003)
14. Newman, M.: Networks: an introduction. Oxford University Press (2009)

Politics Matters: Dynamics of Inter-organizational Networks among Immigrant Associations

Matteo Gagliolo, Tom Lenaerts, and Dirk Jacobs

Abstract. We model the dynamics of the two-mode network among directors and boards of voluntary associations, using a stochastic actor-based model, SIENA [12], including the structural effects proposed in [6], and considering the political orientation of associations as a covariate. Using data from [14], we compare the evolution of interlocks among Turkish associations in two European capitals, and explain the noticeable difference in structure by looking at statistically significant differences among the estimated effects.

1 Introduction

Social capital designs the ensemble of resources which are accessible to a social actor through its relationship with other actors [9]. As such, it is naturally embedded in social networks [1]. In his famous work on the causal relationship between "bridging" social capital (associational life), trust, and civic behavior, Putnam [10] did not investigate the structural aspect of such networks, leaving the question of its relevance unstated. Recently, the relationship between associational life and political participation of ethnic minority groups in Europe has been investigated, obtaining useful insights, yet without reaching uniform conclusions [4, 13, 14]. In this line of work, simple structural properties of the network of interlocking directorates among

Matteo Gagliolo · Dirk Jacobs
GERME, Institute of Sociology, Brussels, Belgium

Matteo Gagliolo · Tom Lenaerts
MLG, Computer Science Department
Université libre de Bruxelles (ULB), Brussels, Belgium

Tom Lenaerts
AI Lab, Computer Science Department, Vrije Universiteit Brussel
Brussels, Belgium

P. Contucci et al. (eds.), *Complex Networks V*,
Studies in Computational Intelligence 549,
DOI: 10.1007/978-3-319-05401-8_5,

ethnic associations have been used as a "proxy" of the social capital of the corresponding minority group. The aim of our research is to pursue this line further, looking at the structure of such networks, but also at the dynamics that produce it. Here, we use a stochastic actor-based model, SIENA [12], which estimates the effect of actor covariates and local structure on network evolution, to analyze data from [14], describing the evolution of the councils of Turkish associations in Amsterdam and Berlin.

2 Background

The concept of social capital designs the ensemble of resources that an actor can access or mobilize via his connections to other actors, regardless of the kind of resource considered (information, control, support, etc.). Most of the early work on social capital focuses on individuals, or small elite groups [9]; its study at the *aggregate* level, for a whole community, has been popularized by Putnam [10], who observed a virtuous circle of causal connections between the amount of associational life, the level of trust, and civic behavior.

More recently, Fennema and Tillie [3] compared the social capital of four migrant communities in Amsterdam, studying structural differences among the networks of *interlocking directorates* of ethnic organizations (i.e., connections among organizations sharing one or more board members), and found a positive correlation among the number of interlocks in the network, and the political participation of the corresponding ethnic minority group. Later studies questioned this hypothesis, finding more subtle relationships among organizational network structures, aggregate indicators of immigrant groups, and civic behavior (see [5], and other papers from the same special issue).

Interlocking directorates [2] are an example of *two-mode* or *bipartite* networks, where two classes of nodes are present (in this case, directors and boards), and links are only possible from one class to the other. A projection onto two distinct *one-mode* networks can be performed, linking two nodes if they are connected to a same node of the other mode. This allows to draw a network among boards that have at least one director in common; or among directors that sit on the same board. As one mode networks have been the subject of a much larger corpus of research, and software development, many scholars prefer to analyze one of the projections, discarding the other mode. While not devoid of interest, this approach has several limitation, in that some of the information in the data is lost during the projection (e.g., in the case of boards, multiple common directors). Moreover, it has been shown that the projection introduces spurious structures in the one-mode network [7], changing the meaning of some of the standard network measures, such as density and clustering [8].

3 Methods

SIENA [12] is a stochastic *actor-oriented* model of network evolution, meaning that the structure of observed network data is assumed to be the result of the actions of a set of agents, each corresponding to a node, and exerting a control over its outgoing ties, by adding or deleting ties to the other nodes. Given a set of potential motives governing social choices, mathematically defined as *effects*, the algorithm estimates a set of parameters, each modulating the impact of the corresponding effect on the probability of forming and deleting ties. Based on a pair of snapshots, or *waves*, of an evolving network, the estimate is performed such that, in a simulation starting from the first snapshot, the final simulated snapshot will be the most similar to the one actually observed. In the following, we provide a simplified description of the model, based on [11] and [6].

Be $\mathbf{x} \in \{0,1\}^N$ the binary matrix representing a network among N nodes, and $\Delta_{ij}\mathbf{x}$ the matrix obtained by switching a single element of $\mathbf{x}$, ($x_{ij} \leftarrow 1 - x_{ij}$). Each node i is an agent, which can perform atomic changes to its outgoing ties x_{ij}, at exponentially distributed points in time, with rate λ. Agents add and remove links according to a "perturbed" utility function: the target index j is drawn with probability $p_i(j|\mathbf{x}) \propto \exp \Delta f_i(j,\mathbf{x})$, where $\Delta f_i(j,\mathbf{x}) = f_i(\Delta_{ij}\mathbf{x}) - f_i(\mathbf{x})$ is the variation in utility that would be obtained by i with a switch of x_{ij}.

The utility function $f_i(j,\mathbf{x})$ is a linear combination of *effects*, which can be arbitrary functions of the current network $\mathbf{x}$, as well as of node covariates:

$$f_i(\mathbf{x}) = \sum_k \theta_k s_{i,k}(\mathbf{x}). \tag{1}$$

Usually, the structural effects can be decomposed according to the outgoing ties of i, as $s_i(\mathbf{x}) = \sum_j s_i(j,\mathbf{x})$. The model is Markovian: at each time step, the probability distribution over the possible next states (all networks at Hamming distance 1) can only depend on the current state. Fixing a set of effects and an initial network $\mathbf{x}$, its further evolution will be a stochastic function of the rate λ and the effect weights $\boldsymbol{\theta} = \{\theta_k\}$. These parameters can be estimated with a Markov chain Monte Carlo approach, selecting those values which produce networks that are the most similar to the ones observed, in terms of the aggregate values of the included effects. The standard deviation of effect estimates can be used to test their significance with a simple t-test. Also differences among an effect estimated on distinct data sets can be tested for significance[1] [11].

[1] Comparing different effects on the same data set is instead not trivial, as the magnitude of an effect is not a relevant index of its actual impact on network evolution, which also depends on the magnitude of the associated effect function. What matters in interpreting the results is the sign of significant effects.

Table 1 Structural effects used in [6] (argument **x** dropped to simplify the notation)

`density`	$s_i(j) = x_{i,j}$
`2-star`	$s_i(j) = x_{i,j} \sum_{h \neq i} x_{h,j}$
`3-path`	$s_i(j) = x_{i,j} \sum_{h \neq i} x_{h,j} \sum_{k \neq j} x_{h,k}$
`4-cycle`	$s_i(j) = x_{i,j} \sum_{h \neq i} x_{h,j} \sum_{k \neq j} x_{h,k} x_{i,k}$

As previously pointed out (Sec. 2), analyzing the one mode projection of two-mode data can be misleading[2]. Luckily for us, the method has been recently extended to two-mode networks by [6], who applied it to model the evolution of interlocking directorates among firms in the Swedish stock market. In the paper, four structural effects were considered as relevant for interlock formation (Table 1): the basic outdegree effect, `density`; simple and double interlocks, labeled `2-star` and `4-cycle`, respectively; and an intermediate structure termed `3-path`. The resulting four effects have distinctly different meanings: while `density` is a "baseline" expressing the tendency of associations of acquiring a new director[3], `2-star` corresponds to the tendency of forming an interlock, which may indicate an underlying social connection among the new director and members of the board, or reflect a strategy of the two boards; and `4-cycle` amounts to adding a second interlock to an existing one, which is interpreted by [6] as a stronger clue of *peer referral*: the fact that at least one of the directors sits already on both boards implies that he knows already all directors involved, and may suggest that one of his colleagues from the first board joins the second one, thus doubling the interlock. The `3-path` is added for completeness, as an alternative explanation to `4-cycle` formation, and may be interpreted as a differential preference for interlocks towards boards with a larger number of directors.

4 Results

The analysis was carried out using `RSiena` [11]: in the following we illustrate the details of the experiments, and the results obtained.

Data. Longitudinal data (1985-2003) of the composition of directorates of Turkish associations in Berlin and Amsterdam was provided by the authors of [14], along with a broad classification of the political orientation of the organizations, on three levels (left, center, right, which we encoded as -1, 0, 1).

[2] It is especially problematic with SIENA, as the basic assumption of atomicity of changes (agents can only change one link at a time) is violated. For example, by joining a new board, a director that sits already on k boards will add k new links at once in the projection, among the new board and the previous ones.

[3] The mode of the organizations is considered the active one, unilaterally deciding when to add/remove directors.

Settings. Following [6], the boards mode was considered the active one, choosing which directors to recruit and release. The default Robbins-Monro approximation was used for estimating the parameters. The estimates were performed on a sequence of data sets, obtained using a shifting window of four consecutive waves (four years).

Effects. The same structural effects of [6] were used (see Table 1). In order to study the impact of political orientation on interlock formation, we implemented a distance 2 similarity effect (`simD2`), to measure the preference for interlocks with similar associations, corresponding to an *homophily* effect in the one-mode projection. `simD2` was defined as[4]:

$$s_i(j) = x_{ij} \frac{\sum_{h \neq i} x_{hj} \mathrm{sim}_{ih}}{\sum_{h \neq i} x_{hj}}. \tag{2}$$

Estimates. In a first set of experiment (Fig. 4), we tested the four structural effect described above. While the numbers involved are much different (interlocks are a much rarer phenomenon in the voluntary sector), the results follow a similar pattern to that observed by [6] on much denser interlock networks in the for profit sector. More precisely, the `density` effect is negative, due to the limited number of directors per board, implying that the vast majority of possible ties are absent. The `2-star` effect is mostly negative, suggesting that boards do not actively search to form interlocks: note that, given that `density` acts as a baseline, a negative `2-star` means that associations prefer to enroll an inactive director (forming a simple tie) rather than someone who is already active in another board (forming an interlock). The `3-path` is often not significant, and when it is, its value is quite small (i.e., when forming an interlock with another board, its size does not matter): in our case, this effect is small but significantly negative in Berlin for some of the waves (1986-1995 and 1997-2003), indicating a moderate preference for smaller boards. This needs further interpretation: it may be related to the existence in Berlin of so-called *umbrella* organizations, which aim at coordinating the activities of several smaller ones, enrolling one director for each member association. `4-cycle` is, instead, significantly positive. In practice this means that, when a simple interlock is already present, an association will prefer to enroll a director from the connected board, rather than a complete outsider. In [6], this is interpreted as a clue of peer referral.

While we cannot compare the magnitudes of different effects on a given data set (see note 1), we can however compare the estimates of the *same* effect on *different* data sets: in particular, we can check for statistically significant differences among the two data-sets. In this case, the first obvious

[4] Here, v_i is the covariate value of i, $\Delta_v = \max_{ih} |v_i - v_h|$ the observed range of the covariate, and $\mathrm{sim}_{ih} = \frac{\Delta_v - |v_i - v_h|}{\Delta_v}$ the *similarity* among two nodes i and h in the same mode. Therefore, `simD2` varies between 0 (for interlocks connecting two boards with maximum covariate difference) and 1 (for identical covariates).

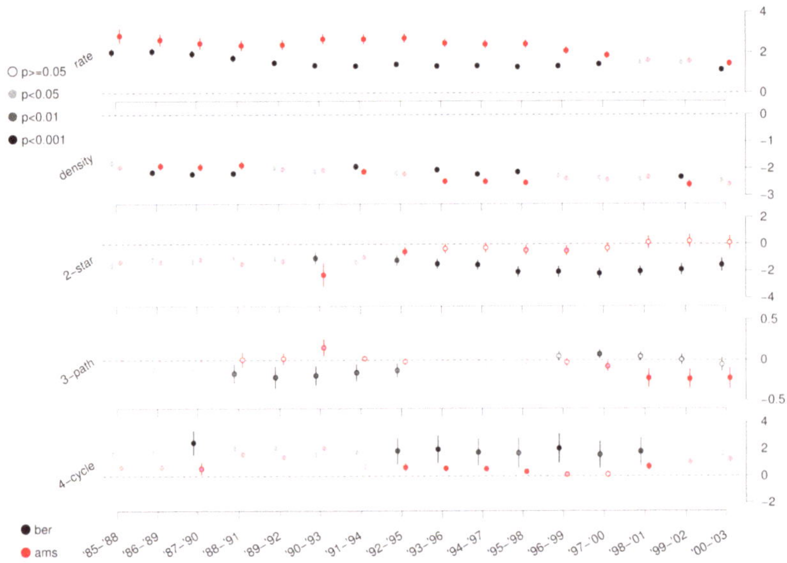

Fig. 1 SIENA estimates, structural effects only. Each horizontal section corresponds to an effect. Ticks on the horizontal axes correspond to four consecutive years of data. For each effect, at each year, two circles are plotted, indicating the estimated values for Berlin (left, black), and Amsterdam (right, red). Full circles indicate significant results (based on *p*-values, see legend at upper left), while empty circles are not significant. Vertical segments represent the standard deviation of each estimate. Significant difference among the two towns is instead highlighted using a transparent tint effect where the difference is not significant.

difference is in the rate parameter, which is almost always higher in Amsterdam, indicating more frequent variations in the composition of the boards. In terms of the structural effects, a comparison among Amsterdam and Berlin reveals that Turkish associations in Berlin have a stronger negative `2-star` and stronger positive `4-cycle`, suggesting that they form interlocks less easy, but are more prone to reinforcing them when present, compared to Turks in Amsterdam: however, the absolute number of four cycles in Berlin is very small (hence the large standard deviations of the estimates).

This difference is reflected in the evolution of these two indicators in the two towns (see Fig. 2 for `2-star`). While similar during the 1980s, both indicators start diverging during the early 1990s: the steeper increase of interlocks in Amsterdam corresponds to a significant difference in the `2-star` effect during this decade. Grouping interlocks according to the similarity of the connected associations, we can remark that political polarization is very strong in both cities, as the vast majority of interlocks connect associations from the same political side.

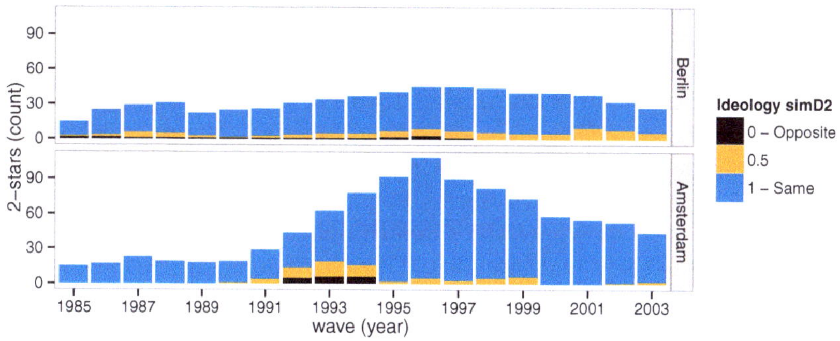

Fig. 2 Simple interlocks (`2-star`), per year, in Amsterdam and Berlin. Colors indicate the ideology similarity among the two interlocked organizations (0: left-right; 0.5: center-left, center-right; 1 left-left, right-right).

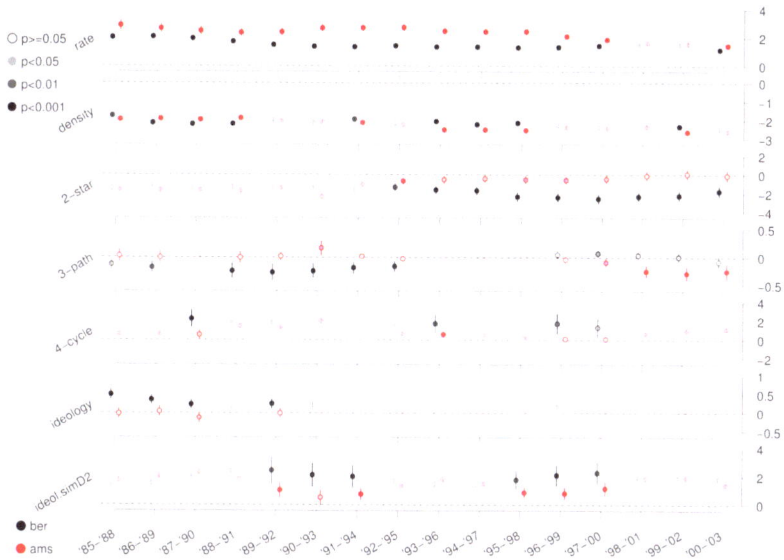

Fig. 3 SIENA estimates, including covariates. See Figure 4 for the legend.

Are these differences sufficient to explain the differences in structure observed in the two cities? Intuitively, politics must play a role, otherwise the interlocks would *not* follow political alignment. After implementing the `simD2` effect, we could test the impact of similarity among associations, including also a net effect of political orientation (Fig. 3). While `density` and `2-star` remain unaffected, the `4-cycle` effect keeps the same pattern, but its difference looses significance in most of the waves, confirming that structural effects alone are not sufficient to explain this data: as the vast majority of four cycles links politically homogeneous organizations, adding this covariate effect renders the structural one superfluous. Ideology by itself is also mostly not significant, except in Berlin during 1985-1992, indicating a greater activity of right-wing associations in recruiting directors. The political affinity among different associations (`ideol.simD2`) is instead relevant. It is mostly significant, and positive, in both towns. Regarding significant differences, we can identify two periods where the similarity effect is significantly lower in Amsterdam, in the early and late 90's. This corresponds to periods characterized by a slight increase of politically heterogeneous interlocks in this town (see Figure 2).

5 Conclusions

While preliminary, our results already highlight the importance of political homophily in interlock formation, and allow to describe the difference among the two communities in quantitative terms. In this sense, the `2-star` effect seems particularly relevant. In the longer term, we intend to relate the dynamics of these networks to the political participation of the corresponding communities, in order to test existing hypotheses on the impact of ethnic social capital on political participation [3, 4].

Acknowledgments. We would like to thank Floris Vermeulen and Jean Tillie (Political Science Dept., University of Amsterdam) for kindly contributing the data on Amsterdam and Berlin; and Johan Koskinen for useful insights on his work. The first author is supported by the *Brussels Institute for Research and Innovation* (Innoviris), under the program "Prospective Research for Brussels".

References

1. Burt, R.: The network structure of social capital. Research in Organizational Behavior 22, 345–423 (2000)
2. Fennema, M., Schijf, H.: Analysing interlocking directorates: Theory and methods. Social Networks 1(4), 297–332 (1978), http://dx.doi.org/10.1016/0378-8733(78)90002-3
3. Fennema, M., Tillie, J.: Civic community, political participation and political trust of ethnic groups. Connections 24(1), 26–41 (2001), http://www.insna.org/PDF/Connections/v24/2001_I-1_26-41.pdf
4. Jacobs, D., Phalet, K., Swyngedouw, M.: Associational membership and political involvement among ethnic minority groups in brussels. Journal of Ethnic and Migration Studies 30(3), 543–559 (2004), http://dx.doi.org/10.1080/13691830410001682089
5. Jacobs, D., Tillie, J.: Introduction: Social capital and political integration of migrants. Journal of Ethnic and Migration Studies 30(3), 419–427 (2004)

6. Koskinen, J., Edling, C.: Modelling the evolution of a bipartite network — peer referral in interlocking directorates. Social Networks 34, 309–322 (2012), http://dx.doi.org/10.1016/j.socnet.2010.03.001
7. Newman, M.E.J., Strogatz, S.H., Watts, D.J.: Random graphs with arbitrary degree distributions and their applications. Physical Review E 64(2), 026118 (2001)
8. Opsahl, T.: Triadic closure in two-mode networks: Redefining the global and local clustering coefficients. Social Networks 35(2), 159–167 (2013), http://dx.doi.org/10.1016/j.socnet.2011.07.001
9. Portes, A.: Social capital: Its origins and applications in modern sociology. Annual Review of Sociology 24(1), 1–24 (1998), http://dx.doi.org/10.1146/annurev.soc.24.1.1
10. Putnam, R.: Making democracy work: civic traditions in modern Italy. Princeton University Press, Princeton (1993)
11. Ripley, R.M., Snijders, T.A., Preciado, P.: Manual for siena version 4.0 (version January 17, 2012). Tech. rep., Oxford: University of Oxford, Department of Statistics; Nuffield College. (January 2012), http://www.stats.ox.ac.uk/siena/
12. Snijders, T.A.B., van de Bunt, G.G., Steglich, C.E.G.: Introduction to stochastic actor-based models for network dynamics. Social Networks 32(1), 44–60 (2010), http://dx.doi.org/10.1016/j.socnet.2009.02.004
13. Tillie, J.: Social capital of organisations and their members: Explaining the political integration of immigrants in amsterdam. Journal of Ethnic and Migration Studies 30(3), 529–542 (2004)
14. Vermeulen, F., Berger, M.: Civic networks and political behavior: Turks in amsterdam and berlin. In: Ramakrishnan, S.K., Bloemraad, I. (eds.) Civic Hopes and Political Realities: Immigrants, Community Organizations and Political Engagement, pp. 160–192. Russell Sage Foundation Press, New York (2008)

A Statistical Mechanics Approach to Immigrant Integration in Emilia Romagna (Italy)

Francesco De Pretis and Cecilia Vernia

Abstract. Integration phenomena are social processes among human beings that take place every day when an autochthone population is experiencing the arrival of new immigrants. Although being a rising phenomenon (involving now over one billion people according to United Nations) which questions societies and policy-makers all over the world, numerical measurements capable to give robust insights over the way immigrant integration occurs are still far from what is usually considered an affordable standard in mathematical and physical sciences. Basing our analysis on previous seminal works, we follow here a statistical physics approach to the analysis of immigrant integration. In specific, we consider a large dataset collected by the Emilia Romagna region office of statistics (Italy), containing information over all marriages occurred amid the regional population during a sixteen years span, from 1995 to 2010. We define as quantifier of integration the percentage of marriages with spouses of mixed origin and we perform several analyses over the dataset, including binning and data fitting. The final outcome consists in an emerging pattern: quantifier's average measurements align around a square root fit when considered with respect to a suitable function of the immigrant density. The theoretical interpretation we offer is that such result agrees with a suitable version of the Curie-Weiss model used in statistical mechanics to describe ferromagnetisms. More explicitly, immigrants living in Emilia Romagna municipalities seem to present mainly imitative behavior's phenomena in making social actions for integration. The result emerged with Emilia Romagna data complies with previous works concerning similar data coming from Spain.

Francesco De Pretis
M2SCS School of Graduate Studies, Università degli Studi di Modena e Reggio Emilia, Italy
e-mail: francesco.depretis@unimore.it

Cecilia Vernia
Dipartimento di Scienze Fisiche, Informatiche e Matematiche, Università degli Studi di Modena e Reggio Emilia, Italy
e-mail: cecilia.vernia@unimore.it

P. Contucci et al. (eds.), *Complex Networks V*,
Studies in Computational Intelligence 549,
DOI: 10.1007/978-3-319-05401-8_6, © Springer International Publishing Switzerland 2014

1 Introduction

Integration of immigrants is a political priority in many countries: there are over one billion migrants all over the world, one quarter of which are international migrants [1]. Even though it is not clear how sensitive integration is to an increase of immigrant density and to what extent social interaction goes into higher integration, it is easy to guess that social interaction between immigrants and autochthonous population is a necessary condition for immigrant integration. Curie-Weiss models have been used in the last years in the quest to model social interactions and processes of decision taken by individual human beings [2,3,4,5,6]. In this paper we follow a statistical physics approach to the study of immigrant integration using methods and models already explored in a previous seminal work concerning a large collection of Spanish data [7]. Given a large dataset described in the subsequent section, we focus on a classical quantifier of integration such as the fraction of marriages with spouses of mixed origin (native – i.e. bearing Italian citizenship – and immigrant)

$$M_m = \frac{number\ of\ mixed\ marriages}{number\ of\ marriages}.$$

Within this framework, our goal is a statistical mechanics theory by which the magnitude of the above-mentioned quantifier can be expressed as a function of the density of immigrants, i.e. the ratio between the number of immigrants N_{imm} and the total population $N = N_{imm} + N_{nat}$ where N_{nat} is the number of natives:

$$\gamma = \frac{N_{imm}}{N_{imm} + N_{nat}} \in [0,1]$$

For a better representation of the integration quantifier – based on combinatorial reasoning [7] – we are interested in studying its dependence on the quantity $\Gamma = \gamma(1-\gamma)$. Afterwards, we seek an empirical function from real data, able to entail the observed collective behavior. As it will be reported in the results section, this work confirms that it is possible to discriminate, using quantitative methods, whether the value of the integration quantifier follows from people acting according to some individual preferences independently of other people (*independent choices*), or whether it follows as a result of social interaction with other ones (*imitative behaviors*). These two opposite cases are described in statistical mechanics theory either as perfect gas of independent particles (in this case, average measurements of the quantifier against Γ follow a linear growth) or interacting theory with possible phase transitions (in this case, average measurements of the quantifier against Γ align around a square root curve).

2 Data Description and Methods

As stated above, the perspective of this work belongs to the statistical mechanics methods used to explain social integration phenomena, starting from the analysis of real data. More precisely, the work has been centered on a large dataset collected by the Emilia Romagna region office of statistics (Italy), containing information recorded in all Emilia Romagna region municipalities (348 cities) regarding marriages occurred amid the population during a sixteen years span, from 1995 to 2010. In particular, for each municipality, the database provides the reference year, the number of marriages between Italians and foreigners, the number of marriages only between foreigners, the number of marriages only between Italians and the total amount of marriages.

Regarding the sources of data, the Emilia Romagna region dataset was somehow freely downloadable from the regional office of statistics website (data were accessible per municipality at given year, so that techniques of automatic web-contents wrapping have been employed to collect the entire dataset). All data were real (i.e. not estimated) and were subsequently matched with the density of immigrants for each municipality at each given year (information freely retrievable from ISTAT – Italian National Institute of Statistics – sources): the density of immigrants was estimated only for two specific years (1999 and 2000) since the recording of immigrant population was suspended during that period. Given the dimension of the considered dataset, the work can be somewhat inscribed in a *big data exploitation*: to give a rough idea of the computational efforts pursued, around 50.000 data have been processed in order to compute the above described quantifier (y-axis) matched with the density of immigrants (x-axis), producing the scatter plot reported below in Figure 1. It is worth to note that according to prescriptions of a time-independent analysis, data have been plotted together independently from the year they were referring to.

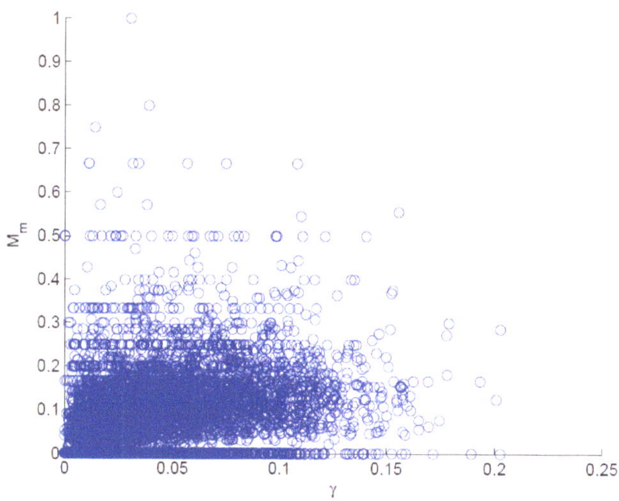

Fig. 1 Raw data versus γ. Blue points represent the fraction of mixed marriages occurred from 1995 to 2010 in all municipalities located in Emilia Romagna region where a percentage γ of migrants is present.

Besides the representation through a scatter plot, we have been interested in the quantifier's average measurements as functions of Γ: since the quantifier M_m is a ratio (i.e. the total number of mixed marriages over the total number of marriages) and we were concerned in looking at patterns in the global scale of the dataset, a natural way to compute averages has been a *mediant* measure. This means that for a given bin of γ, we have computed the ratio between the statistical average of numerators and the statistical average of the denominators. This processing has been performed according to a constant information binning, i.e. each bin contained a fixed number of points. After having compacted the initial information of raw data into bins, a classical procedure of curve fitting has taken place. In particular, following the example of previous work conducted in [7], we have evaluated linear and power functions according to the R^2 coefficient of determination.

3 Results

After having performed the procedures described in the methods section, the following result has been obtained: quantifier's average measurements computed for all data coming from the Emilia Romagna region dataset have fitted around a visible square root pattern with a R^2 coefficient of determination being over 95%, highly reproducing results obtained with similar Spanish data in [7]. Since the analysis of the data density versus Γ shows that only about the 7% of the data are found for Γ greater than 10%, we limit our study below this threshold. It is worth of note that in Emilia Romagna region in 2010 the percentage of immigrants over the total population is about 11%, the highest density of immigrants with respect to any other Italian region.

The result has been verified according to various types of binning (i.e. changing the number of bins) and various families of functions (for instance, linear functions). In the end, a square root fit emerged as the best estimation for the quantifier's average measurements, since with linear and other fittings, the outcomes reported lower R^2 coefficient of determination associated with noisy fits highly depending on the nature of binning.

The mathematical model that supports these results is a generalization of the monomer-dimer model [8] with the addition of an imitative interacting social network component of small world-type [9]. The model, proposed and described in [7], reduces to the classical discrete choice theory [10] (or perfect gas of independent particles) with linear growth of the quantifier as a function of Γ, when imitation is negligible, and to the square root behavior when imitation is dominant. The social network structure explains why the integration starts very close to $\Gamma = 0$ when the choice is dependent on other agent behavior.

Therefore, translated in statistical mechanics terms according to the theoretical interpretation shown in [7], the result of an empirical square root function for the quantifier's average measurements offers an interesting picture of immigrant integration issues in Emilia Romagna region. In specific, even though we do not deal with the possible origins of such cooperative influence, we simply conclude that data suggest that in Emilia Romagna municipalities imitative phenomena mainly take place against the possibility of independent choices carried on by the same immigrants.

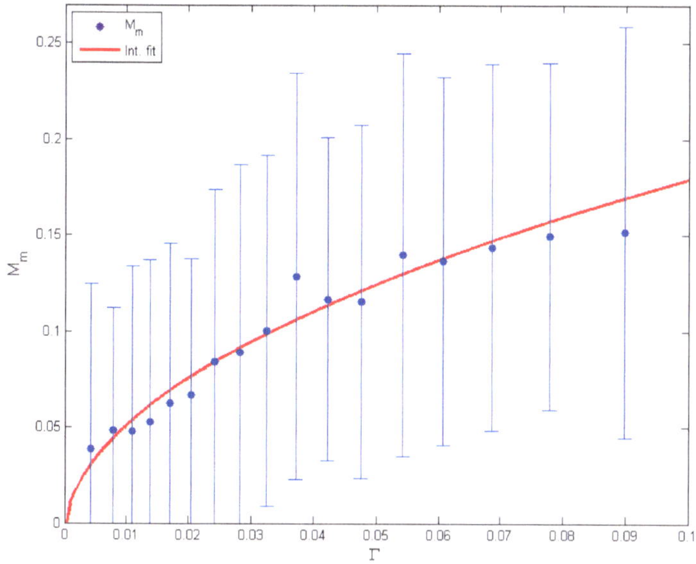

Fig. 2 Emilia Romagna dataset. Dots are average quantities versus Γ, whereas lines denote error bars. Quantifier M_m (blue dots), fraction of mixed marriages occurred from 1995 to 2010 in all the municipalities located in Emilia Romagna region, with the best square root fit (red curve) $a\sqrt{\Gamma} + b$ (a = 0.5943 ± 0.0757, b= -0.008631 ± 0.014019, goodness of fit R^2 = 0.9529 computed for $\Gamma < 0.078$). Parameter b evaluation is compatible with the hypothesis that it can be null, as prescribed by the statistical mechanics model we use for results interpretation.

Acknowledgement. Authors express their gratitude to Pierluigi Contucci and Claudio Giberti for the inspiring insights and comments that helped writing this work.

References

1. The Global Approach to Migration and Mobility. EU Report, Commission 743, sec. 1353 (2011)
2. Durlauf, S.N.: Statistical mechanics approaches to socioeconomic behavior. Technical Working Paper, 203, Natl. Bur. Econ. Res (1996)
3. Barra, A., Contucci, P.: Toward a quantitative approach to migrants social integration. Europhys. Lett. 89, 68001, 68007 (2010)
4. Gallo, I.: An equilibrium approach to modelling social interaction. PhD Thesis, Università di Bologna (2009)
5. Contucci, P., Gallo, I., Menconi, G.: Phase transitions in social sciences: two-populations mean field theory. Int. J. Mod. Phys. B 22(14), 1–14 (2008)

6. Barra, A., Contucci, P., Gallo, I.: Parameter Evaluation of a Simple Mean-Field Model of Social Interaction. Mathematical Models and Methods in Applied Science 19, 1427–1439 (2009)
7. Barra, A., Contucci, P., Sandell, R., Vernia, C.: Integration indicators in immigration phenomena. A Statistical Mechanics Perspective (2013), http://arxiv.org/abs/1304.4392
8. Heilmann, O.J., Lieb, E.H.: Monomers and dimers. Phys. Rev. Lett. 24(25), 1412 (1970)
9. Watts, D.J., Strogatz, S.H.: Collective dynamics of small world networks. Nature 393, 6684 (1998)
10. McFadden, D.: Economic choices. The Amer. Econ. Rev. 91 (2001)

Searching in Unstructured Overlays Using Local Knowledge and Gossip

Stefano Ferretti

Abstract. This paper analyzes a class of dissemination algorithms for the discovery of distributed contents in Peer-to-Peer unstructured overlay networks. The algorithms are a mix of protocols employing local knowledge of peers' neighborhood and gossip. By tuning the gossip probability and the depth k of the k-neighborhood of which nodes have information, we obtain different dissemination protocols employed in literature over unstructured P2P overlays. The provided analysis and simulation results confirm that, when properly configured, these schemes represent a viable approach to build effective P2P resource discovery in large-scale, dynamic distributed systems.

1 Introduction

This paper deals with resource discovery in large-scale, dynamic Peer-to-Peer (P2P) distributed communication systems. In this context, it has been recognized that an interesting approach consists in exploiting unstructured overlay networks [2, 6, 8], which are alternative to traditional structured solutions [7]. Indeed, there are some clear drawbacks related to unstructured networks, that make structured ones more effective in some distributed systems. In particular, the main weakness of unstructured nets is that links among nodes do not depend on the distribution of the contents. This means that in general it is not possible to provide a bound on the number of nodes that might be involved during the lookup of a resource. On the other hand, the advantages are the easier manageability and the possibility of implementing resource discovery systems based on partial-match and complex queries. Conversely, several structured P2P approaches (e.g. those based on DHTs) strongly limit the expressiveness of the queries to retrieve contents. For these reasons, understanding if, how and when unstructured overlays can support resource and content lookup represents an interesting research topic.

Stefano Ferretti
Department of Computer Science and Engineering, University of Bologna,
Mura Anteo Zamboni 7, Bologna, 40127 Italy
e-mail: s.ferretti@unibo.it

P. Contucci et al. (eds.), *Complex Networks V*,
Studies in Computational Intelligence 549,
DOI: 10.1007/978-3-319-05401-8_7,

A main aspect refers to the algorithm employed to distribute queries among nodes, that strongly influences the performance of the whole system. In this paper, we study a simple class of dissemination algorithms, which are a mix of push-gossip based and informed propagation schemes [4]. Each node has knowledge of its k-neighborhood, i.e. those nodes that are distant at most k hops from it. This information is exploited during the routing of messages in the overlay, i.e. a node sends the message to those 1-neighbors that can relay the message to the k-neighbours that hit the query. Moreover, the node gossips the message to its remaining 1-neighbors. The tuning of the parameters of the algorithm (i.e. gossip probability threshold and depth k of the k-neighborhood) allows to pass, for instance, from pure locally "best neighbor selection" dissemination protocols (gossip probability set equal to 0), e.g. [11], to flooding schemes (gossip probability set equal to 1). Similarly, if the depth k of the k-neighborhood is set $k = 0$, a pure gossip strategy is obtained; when k is set equal to the network diameter, we have a scheme with full-knowledge of the net.

We present an analytical framework that models the described family of communication protocols. A numerical analysis over scale-free network topologies is performed, and it is compared with a simulation of the system. Results confirm that dissemination protocols exploiting the combination of gossip and local knowledge about nodes' neighborhood, are a useful tool to build lookup discovery services over large-scale unstructured P2P systems. Moreover, the framework can be practically exploited to tune the gossip probability at peers and build effective lookup discovery services over P2P unstructured overlays. In many cases, it is sufficient to maintain information on the 2-neighborhood (or even 1-neighborhood, with a higher gossip probability) to have that queries percolate through the overlay, hence obtaining a number of query hits of the order of the number of resources (matching the query) present in the network.

The remainder of this paper is organized as follows. Section 2 presents the system model and the local protocol executed at each node. Section 3 presents the mathematical model. Section 4 outlines results coming from numerical analysis and simulation. Finally, Section 5 provides some concluding remarks.

2 System Model and Protocol

Let consider unstructured overlay networks, with peers that connect each other through a pseudo-random attachment process which shapes the overlay based on a specific network topology, defined through a degree probability distribution. The link creation process does not depend on the placement of contents in the P2P system [5]. We denote with Π^1 the 1-neighborhood of a node n (n's friends); in general Π^k is the k-neighborhood of a node, i.e. nodes at most k hops away from n. Nodes know how to reach all its k-neighbors. We assume the existence of a RELAY(m) procedure that returns the node that n has to contact to reach m. Of course, if m is a 1-neighbor of n, RELAY(m) returns m.

When a peer n holds (removes) from its cache a novel resource item, it informs its k-neighborhood, through some multicast message sent through the overlay. Hence,

Algorithm 1. Query distribution protocol executed at node n

```
Require: Query Q generated at n ∨ Q received in a message relayed by a neighbor peer m
 1: if Q already handled then
 2:     Return
 3: end if
 4: if QUERYHIT(Q) then                                          {local query hit}
 5:     s = ORIGINATOR(Q)
 6:     rp = PROFILEMATCHINGRESOURCE(Q)
 7:     msg = ⟨"available", rp⟩
 8:     SEND(msg, s)
 9: end if
10: DECREASETTL(Q)
11: if TTL(Q) > 0 then                                     {relay to hitting nodes}
12:     R ← {RELAY(i) | i ∈ Π^k ∧ i has an item matching Q} \ m
13:     for all r ∈ R do
14:         SEND(Q, r)
15:     end for
16:     for all i ∈ Π^1 \ {R ∪ m} do                                      {gossip}
17:         if RANDOM() < γ then
18:             SEND(Q, i)
19:         end if
20:     end for
21: end if
```

upon reception at m of a message stating that n holds (deletes) a novel resource item, m adds (removes) a related entry in its neighbor table. This way, each time m receives a query that hits that resource item, m can forward the query towards n. It is clear that the higher the depth k of the neighborhood, the higher the amount of control messages to be transmitted to maintain correct information.

The distribution of a query is based on pure local decisions [4]. We assume that each query contains all the information needed to perform a matching among the requested (type of) item and resources available in the system; in other words, resources are described through a profile (or some metadata). Algorithm 1 shows the pseudo-code of the peer (n) behavior executed to disseminate a query. When n creates or receives a novel query from a neighbor m (which has not be handled already, lines 1–3), first, it checks if there is a query hit locally; in this case, the query originator is contacted directly (lines 4–9).

Then, n multicasts the query to those k-neighbors that own an item that hits the query (lines 12–15). This is accomplished by sending the message to its 1-neighbors that will relay it to the target nodes. However, this is done only if the message has a positive Time-To-Live (TTL) (lines 10–11). (We are assuming that the TTL value allows to cover the whole network; typically, this can be obtained using low values of the order of the logarithm of the network size.) Finally, n gossips the message with a probability $\gamma \leq 1$ to the remaining set of 1-neighbors (lines 16–20) [4].

The considered family of protocols groups together different typical schemes employed over unstructured P2P overlays. Figure 1 shows the protocols we obtain

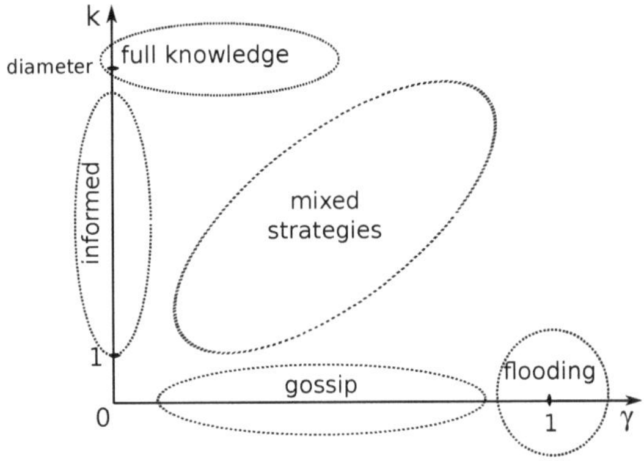

Fig. 1 Discovery protocols obtained through the setting of the depth of the k-neighborhood and the gossip probability γ

depending on the gossip threshold γ and depth of the k-neighborhood. In fact, when $k = 0$ and $\gamma > 0$, we have a gossip protocol, i.e. queries are randomly disseminated. When $\gamma = 1$ we have a flooding protocol, i.e. messages are relayed through all nodes' links. Informed protocols are those where peers have knowledge of their k-neighborhood (without using gossip) [11]; they are thus placed on the k-axis, with $\gamma = 0$. Finally, if we ideally set the k value equal to the network diameter, then we obtain full-knowledge schemes, where the overlay is exploited to route messages.

3 System Analysis

The goal of this analysis is to estimate the average amount of query hits $\langle h \rangle$ that would occur, given an estimate of the resource popularity (i.e. how much resources, that would hit the query, are distributed in the net) and a given degree distribution probability characterizing the unstructured overlay topology.

Each query dissemination process is considered as a standalone, independent task. This is a correct assumption if peers have a buffer cache whose size is sufficiently large to handle simultaneous queries. Otherwise, the model should be extended to consider possible buffer overflows.

We assume to work with very large and dynamical P2P systems. We already mentioned that, for small-sized and stable nets, the use of unstructured overlays can be avoided, since other approaches can be proficiently employed, such as centralized solutions or structured distributed systems (e.g. DHTs). The high number of nodes, together with the random nature of contacts among peers in the overlay, augments the probability of having a low clustering in the network [6, 10]. A consequence of the random nature of the attachment process is that, regardless of the node degree distribution, the probability that a 2-neighbor is also a 1-neighbor of a node, goes as N^{-1}, being N the number of nodes in the overlay. Hence, this situation can be ignored for high N values. This assumption is supported by previous works, asserting that it is undesirable for an unstructured P2P overlay to have high clustering [12].

In fact, clustering reduces the connectivity of a cluster to the rest of the net, increases the probability of partitioning, and it may cause redundant message delivery.

We denote with p_i the probability that a peer has i 1-neighbors (its degree). Let q_i be the excess degree distribution [10], i.e. the probability that, following a link in the overlay, we arrive to a peer m that has other i links (hence the degree of m is $i+1$). Given p_i, we have that $q_i = \frac{(i+1)p_{i+1}}{\sum_j jp_j}$.

Probabilities p_i and q_i represent two similar concepts i.e. the number of contacts of a considered peer (its degree), and the number of contacts obtained following a link of a peer (its excess degree), respectively. In the following, we introduce measures obtained by considering the degree p_i of a node, as well as the excess degree q_i of a link. Hence, with a slight abuse of notation we denote all the probabilities/functions related to the excess degree with the same letter used for the degree, with an arrow on top of it, just to recall that the quantity refers to a link. Thus, for instance, the generating functions for p_i and q_i are denoted as $G(x) = \sum_i p_i x^i$, $\vec{G}(x) = \sum_i q_i x^i$.

We denote with ρ the probability that a node has a resource item matching the considered query, and with γ the gossip probability. If the considered protocol employs the 1-neighborhood Π^1 only, then the probability that a node n does not transmit a query to a neighbor m is $(1-\rho)(1-\gamma)$, i.e. the probability that m does not hit the query, and n decides not to gossip to m. Hence, the probability τ_1 that n transmits the query to a neighbour m, having only knowledge of its 1-neighborhood Π^1 is $\tau_1 = 1-(1-\rho)(1-\gamma)$.

With this in view, the probability that none of the n's 1-neighbours hit the query is $\sum_i p_i(1-\rho)^i = G(1-\rho)$. This result is obtained by considering all the possible cases of n having degree i and its i neighbours do not hit the query. Similarly, the probability that, given a randomly chosen edge of n, we arrive to a node m that does not have any neighbour (apart from the link we considered to arrive to m from n) that hit the query is $\sum_i q_i(1-\rho)^i = \vec{G}(1-\rho)$.

Following this reasoning, it is possible to determine the probability τ_2 of relaying a query to a node m when n has knowledge of its 2-neighborhood Π^2. In fact, such probability is $\tau_2 = 1-(1-\rho)(1-\gamma)\vec{G}(1-\rho)$, i.e. n does not transmit to m if: m does not hit the query (probability $(1-\rho)$); n decides not to gossip m (probability $(1-\gamma)$); and n knows that its 2-neighbours connected through m do not hit the query (probability $\vec{G}(1-\rho)$ measured above).

The approach can be exploited to measure τ_k, with any given value of k. For instance, the probability that following a link we arrive to a node which has no neighbors in its Π^2 that hit the query is $\sum_i q_i(1-\rho)^i[\vec{G}(1-\rho)]^i = \vec{G}((1-\rho)\vec{G}(1-\rho))$. Through this result we might obtain τ_3, and so on.

Now, the probability that n forwards a message to i of its neighbors is

$$f_i = \tau_k^i \sum_{j \geq i} p_j \binom{j}{i} (1-\tau_k)^{j-i}. \qquad (1)$$

f_i considers all the possible cases of n having a degree j, which forwards the query to $i(< j)$ neighbors, while not forwarding the query to its remaining $j-i$ neighbors. Similarly, the probability that following a link we arrive to a node that forwards the query to i other nodes is readily obtained by substituting, in (1) above, p_j with q_j, i.e. $\overrightarrow{f}_i = \tau_k^i \sum_{j\geq i} q_j \binom{j}{i} (1-\tau_k)^{j-i}$.

If we consider the generating function F of the f_i coefficients, we have

$$\begin{aligned} F(x) &= \sum_i f_i x^i = \sum_i \tau_k^i x^i \sum_{j\geq i} p_j \binom{j}{i} (1-\tau_k)^{j-i} \\ &= \sum_j p_j \sum_{i=0}^{j} \binom{j}{i} \tau_k^i x^i (1-\tau_k)^{j-i} \\ &= \sum_j p_j (\tau_k x + 1 - \tau_k)^j \ = \ G\big(\tau_k x + 1 - \tau_k\big). \end{aligned}$$

The average value of coefficients f_i is given by the derivative of F measured at $x=1$, i.e. $F'(1) = \sum_i i f_i$,

$$F'(x)\Big|_{x=1} = \frac{dG}{dx}\big(\tau_k x + 1 - \tau_k\big)\Big|_{x=1} = \tau_k G'(1) = \tau_k \langle p \rangle,$$

where $\langle p \rangle$ is the mean node degree.

Similarly,, $\overrightarrow{F}'(x)\Big|_{x=1} = \tau_k \overrightarrow{G}'(1) = \tau_k \langle q \rangle$, where $\langle q \rangle$ is the mean value of the excess degree, $\langle q \rangle = \sum_i i q_i = \frac{\sum_i i(i+1) p_{i+1}}{\sum_j j p_j} = \frac{\langle p^2 \rangle - \langle p \rangle}{\langle p \rangle}$.

With these measures, it is possible to obtain the whole number of nodes reached by a message starting from a given node, regardless of the number of hops [10]. Let consider the probability r_i that i peers receive a query, starting from a given node and $\overrightarrow{r}_i$ is the probability that i peers are reached starting from a link. $\overrightarrow{r}_i$ can be defined using the following recurrence,

$$\begin{aligned} \overrightarrow{r}_0 &= 0, \\ \overrightarrow{r}_{i+1} &= \sum_{j\geq 0} \overrightarrow{f}_j \sum_{a_1+a_2+\ldots+a_j=i} \overrightarrow{r}_{a_1} \overrightarrow{r}_{a_2} \ldots \overrightarrow{r}_{a_j}. \end{aligned} \tag{2}$$

Equation (2) can be explained as follows. It measures the probability that following a link we disseminate the query to $i+1$ peers. (The case $\overrightarrow{r}_0$ is impossible, since at the end of a link there must be a node.) One peer is that reached at the end of the link itself. Then, we consider the probability that the peer forwards to other j links (varying the value of j). Each link k allows to disseminate the query to a_k peers, and the sum of all these reached peers equals to i.

Similarly, we can calculate r_i as follows

$$\begin{aligned} r_0 &= 0, \\ r_{i+1} &= \sum_{j\geq 0} f_j \sum_{a_1+a_2+\ldots+a_j=i} \overrightarrow{r}_{a_1} \overrightarrow{r}_{a_2} \ldots \overrightarrow{r}_{a_j}. \end{aligned} \tag{3}$$

In this case, we start from the peer itself, considering it forwards to j nodes; and as before, from these j links we can reach i other peers, in total.

The use of generating functions, $R(x) = \sum_i r_i x^i$, $\vec{R}(x) = \sum_i \vec{r}_i x^i$, allow to handle equations (2–3). In fact, after some algebraic manipulation we have

$$\vec{R}(x) = x \sum_{j \geq 0} \vec{f}_j [\vec{R}(x)]^j = x\vec{F}(\vec{R}(x)) \tag{4}$$

and, similarly,

$$R(x) = x \sum_{j \geq 0} f_j [\vec{R}(x)]^j = xF(\vec{R}(x)). \tag{5}$$

From these generating functions, it is possible to measure the average number $\langle r \rangle$ of peers that receive a query through the dissemination protocol, i.e. $\langle r \rangle = \sum_i i r_i = R'(1)$. On the other hand, taking (5) and differentiating

$$R'(1) = \left[F(\vec{R}(x)) + xF'(\vec{R}(x))\vec{R}'(x)\right]_{x=1} = 1 + F'(1)\vec{R}'(1),$$

Similarly, from (4), $\vec{R}'(1) = \left[\vec{F}(\vec{R}(x)) + x\vec{F}'(\vec{R}(x))\vec{R}'(x)\right]_{x=1} = 1 + \vec{F}'(1)\vec{R}'(1)$. Thus, $\vec{R}'(1) = \frac{1}{1-\vec{F}'(1)}$, and final formula for $\langle r \rangle$ is

$$\langle r \rangle = 1 + \frac{F'(1)}{1 - \vec{F}'(1)} = 1 + \frac{\tau_k \langle p \rangle^2}{(1+\tau_k)\langle p \rangle - \tau_k \langle p^2 \rangle}. \tag{6}$$

Now, $\langle r \rangle$ is the number of peers that receive the query, regardless if these nodes have a resource item matching it. To obtain the average number of query hits $\langle h \rangle$, it suffices to multiply $\langle r \rangle$ by the probability ρ that a peer has a resource item matching that query, i.e. $\langle h \rangle = \rho \langle r \rangle$.

Equation (6) has a divergence when $(1+\tau_k)\langle p \rangle = \tau_k \langle p^2 \rangle$, meaning that, under the assumption that the network has an infinite size, the query reaches an infinite number of nodes, i.e. the query percolates through the network. In other words, an amount of nodes of the order of the network size receives the query.

4 Evaluation

This section presents an assessment performed by considering the analytical model and simulation. While during the assessment we tested different network topologies, we will focus here on results concerned with scale-free networks only. These networks are characterized by nodes having a degree following a power law distribution $\sim p^{\alpha}$. They are characterized by the presence of hubs, i.e. nodes with degrees significantly higher than the average, that have an important impact on the net connectivity. The interest on scale-free networks in this work relates to the fact that several real P2P systems are indeed scale-free networks [3, 10].

In this study, we considered not only traditional scale free networks, but also those with an abrupt cutoff c that limits the maximum degree that peers can maintain, so as to bound the workload that hubs in the P2P system must sustain.

4.1 Simulation

We have built a discrete-event simulator mimicking the presented protocol. The simulator was written in C code and it allows testing the behavior of a set of nodes executing the presented dissemination protocol. It is able to generate a random network based on a chosen degree distribution. In particular, once having (randomly) assigned a specific target degree to each node, using the selected degree distribution, a random mapping is made so that links are created until each node has reached its own target degree. During the initialization phase, for each node a random choice was made to place resources; the resource availability was set based on a probability ρ, i.e. for each network node, an item was present with probability ρ.

To build scale-free networks, the construction method was the one proposed in [1]. This algorithm differs from other well known proposals, which build networks with a power law distribution by continuously adding novel nodes, hence having networks that grow in time. Conversely, we build a network of fixed size, characterized by two parameters a, b. More specifically, the number y of nodes which have a degree x satisfies $\log y = a - b \log x$, i.e. $y = \lfloor \frac{e^a}{x^b} \rfloor$. Thus, the total number of nodes $N = \sum_{x=1}^{\lfloor e^{\frac{a}{b}} \rfloor} \frac{e^a}{x^b}$, being $\lfloor e^{\frac{a}{b}} \rfloor$ the maximum possible degree of the network, since it must be that $0 \leq \log y = a - b \log x$. Once the number of nodes and their degrees have been determined, edges are randomly created among nodes until nodes reach their desired degrees. In the reported results, the parameters were set to $a = 6, b = 1$, resulting in networks composed of 2482 nodes.

For each overlay, we varied the values of σ, ρ in a range going from 0.01 up to 0.5, using a step of 0.01. Thus, 2500 simulation scenarios were considered. For each of these settings, we repeated the simulation using a corpus of 20 different randomly generated networks (characterized by the mentioned statistical properties of the target topology). During each simulation execution, we analyzed the dissemination of 400 queries sent by random nodes.

4.2 Results

In a scale free network (without cutoffs) it is known that when $\alpha > -2$ the mean diverges; when $-3 < \alpha < -2$, the mean is finite but the variance and higher moments diverge [10]. Hence, in these cases a query easily percolates through the network and resources are found with high probability. Indeed, results from our assessment confirm this. (We do not show them in charts.)

For this reason we focus, for now, on overlays with a lower value for such exponent, i.e. $\alpha = -3.2$. Figure 2 shows the average amount of query hits in this specific scenario, obtained via the analytical model and simulation, when peers know their

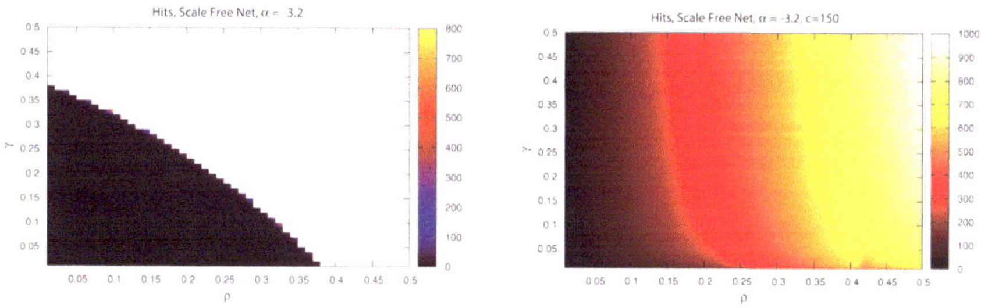

(a) Model, Π^1 (b) Simulation, Π^1

Fig. 2 Average amount of query hits; power law degree distribution with exponent $\alpha = -3.2$. Results are shown for Π^1. When Π^2 is considered, the model returns an ∞ amount of query hits regardless of ρ, σ values (hence not shown in the figure); simulation results confirmed that a high majority of nodes is reached and that queries percolate through the net.

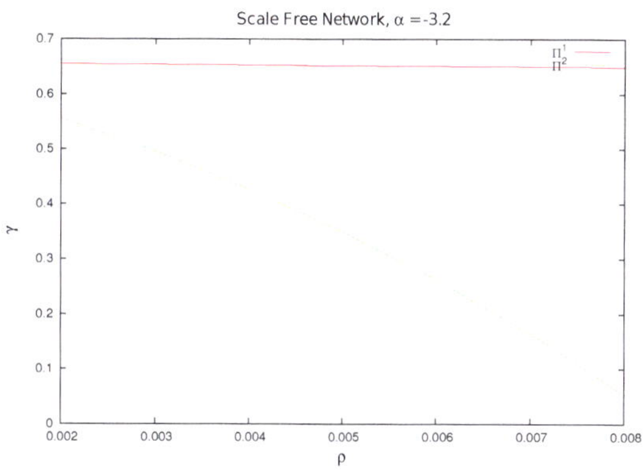

Fig. 3 Minimum γ to find at least one resource; power law degree distribution with exponent $\alpha = -3.2$

1-neighborhood Π^1. (In fact, when peers have knowledge of Π^2, the number of receivers diverges, and thus each query percolates through the network.) It is possible to observe that with lower values of γ, ρ a limited amount of network nodes receive the disseminated queries. Then, by increasing these two values, we reach a transition phase; and after that, the query percolates. One might notice some differences between the two charts referring to the analysis and simulation. Actually, these are perfectly reasonable since the analysis assumes an infinite network size; hence, once a message percolates an infinite amount of nodes is reached. Conversely, simulations employed finite networks; hence, we obtain smoother transitions where a finite (nevertheless significant, when percolation occurs) amount of nodes is reached. With this in view, we can conclude that the two approaches provide similar results.

Figure 3 shows the minimum value of the gossip probability γ, to have that at least one resource is found through a query in a scale free network with $\alpha = -3.2$. The

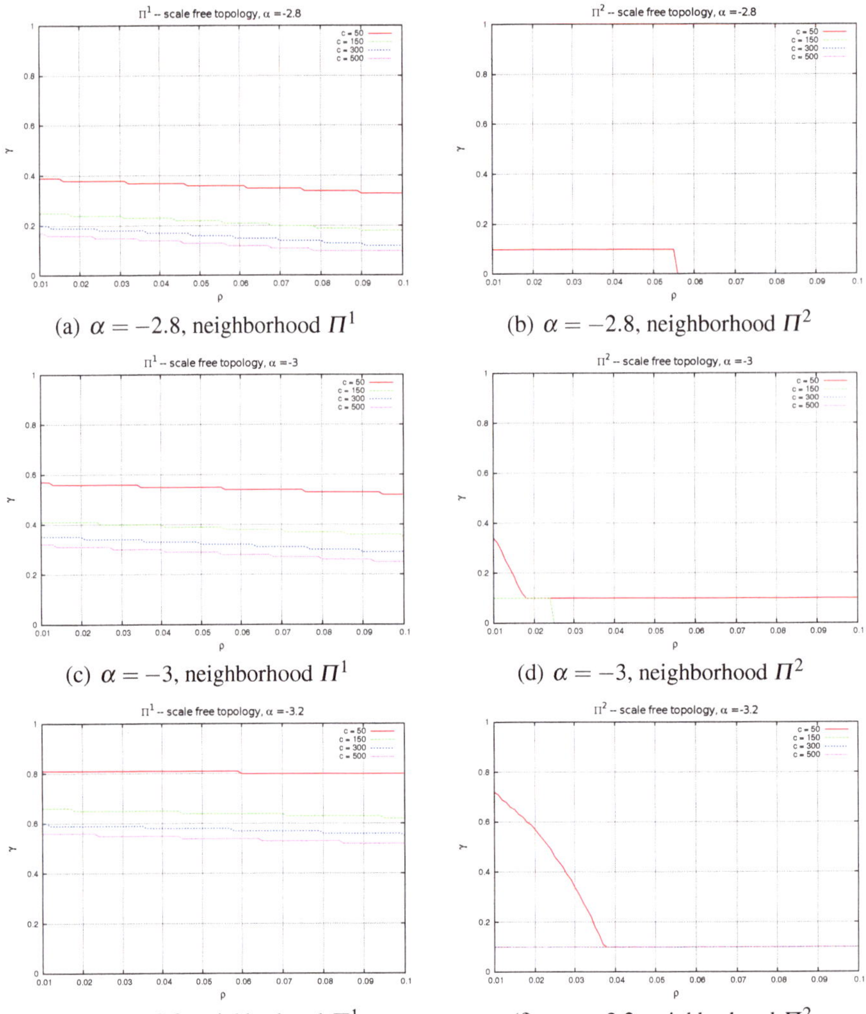

(a) $\alpha = -2.8$, neighborhood Π^1 (b) $\alpha = -2.8$, neighborhood Π^2

(c) $\alpha = -3$, neighborhood Π^1 (d) $\alpha = -3$, neighborhood Π^2

(e) $\alpha = -3.2$, neighborhood Π^1 (f) $\alpha = -3.2$, neighborhood Π^2

Fig. 4 γ value to obtain an infinite amount of query hits; scale-free network topologies with different power law distributions

outcome has been obtained through a numerical analysis exploiting the mathematical model. When peers have knowledge of Π^2, with a resource presence probability $\rho > 0.008$ the gossip probability can be set $\gamma = 0$; hence, a non-negligible threshold for the gossip probability is needed only for rare items. This result is due by the presence of hubs that manage information of a high number of nodes.

It has been already mentioned that scale-free networks are characterized by the presence of hubs; moreover, we already mentioned the importance of introducing a cutoff that limits the maximum amount of contacts a peer may have in the overlay.

Figure 4 shows the percolation transition values (i.e. those values of γ and ρ above which queries do percolate through the net) for different scale-free networks, when varying the exponent α of the degree distribution[1] (different rows in the figure), the depth k of the k-neighborhood (different charts in each row), and different settings for the cutoff c (different curves on each chart). Results are obtained through numerical measurements exploiting the analytical model. In this case, the cutoff has an influence on the ability of nodes to disseminate the query. In fact, the lower the cutoff the lower the number of links leaving from the hubs, and thus the more difficult is to spread the query. An interesting result related to the introduction of the cutoff, in line with what already mentioned, is that the lower the exponent α of the power law distribution, the higher the γ to let queries percolate. This is due to the fact that the presence of the cutoff avoids that the first and second moments of the degree diverge. Moreover, the lower the exponent α the faster the distribution goes to 0, and thus the higher the probability that nodes have low degrees, and thus the lower the connectivity of the network and its ability to spread contents.

Similarly, and as expected, in Figure 4 the higher the cutoff the lower the γ to let queries percolate, since the presence of nodes with higher degrees (hubs) augments the connectivity of the network and its ability to spread contents.

Of course, when nodes have knowledge of 2-neighbors, very small γ values are needed with lower cutoffs (see charts on the right in the figure), while negligible values of γ are necessary for higher settings of the cutoff c.

To sum up, outcomes confirm that lookup operations can be easily built over scale-free unstructured overlays.

5 Conclusions

We analyzed the performance of a class of simple dissemination protocols, employing local knowledge of peers' neighborhood and gossip, to perform resource lookup over P2P unstructured overlays. The provided analytical framework allows to tune the gossip probability to spread queries through the overlay, given a network topology and a resource probability distribution. These network parameters can be estimated using some techniques such as entropy-reduction protocols [9].

We tested our approach over scale-free networks. It turns out that, in certain scenarios, it might be difficult to locate rare items with naive informed schemes without gossip (especially if Π^1 is exploited); this is in accordance with some previous results [11]. However, in most cases very low gossip probabilities are sufficient. Thus, when networks are large in size and with a high level of churn, these solutions represent an interesting alternative to dissemination strategies built on top of costly structured distributed systems.

[1] In this case, the cutoff imposes a limit on the moments of the degrees, that do not diverge; hence, it is interesting to consider networks with values of α higher than those considered above.

References

1. Aiello, W., Chung, F., Lu, L.: A random graph model for power law graphs. Experimental Math. 10, 53–66 (2000)
2. Cholvi, V., Felber, P., Biersack, E.: Efficient search in unstructured peer-to-peer networks. In: Proc. of the 16th ACM Symposium on Parallelism in Algorithms and Architectures, SPAA 2004, pp. 271–272. ACM, New York (2004)
3. D'Angelo, G., Ferretti, S.: Simulation of scale-free networks. In: Simutools 2009: Proc. of the 2nd International Conference on Simulation Tools and Techniques, ICST, pp. 1–10. ICST, Brussels (2009)
4. D'Angelo, G., Ferretti, S., Marzolla, M.: Adaptive event dissemination for peer-to-peer multiplayer online games. In: Proc. of the Int. Conf. on Simulation Tools and Techniques (SIMUTools 2011). ICST (2011)
5. Ferretti, S.: On the degree distribution of faulty peer-to-peer overlay networks. EAI Endorsed Transactions on Complex Systems 12(1), 11 (2012)
6. Ferretti, S.: Publish-subscribe systems via gossip: a study based on complex networks. In: Proc. of the 4th Annual Workshop on Simplifying Complex Networks for Practitioners, SIMPLEX 2012, pp. 7–12. ACM, New York (2012)
7. Hidalgo, N., Rosas, E., Arantes, L., Marin, O., Sens, P., Bonnaire, X.: Dring: A layered scheme for range queries over dhts. In: Proc. of the 2011 IEEE 11th International Conference on Computer and Information Technology, CIT 2011, pp. 29–34. IEEE (2011)
8. Keidar, I., Melamed, R.: Evaluating unstructured peer-to-peer lookup overlays. In: Proceedings of the 2006 ACM Symposium on Applied Computing, SAC 2006, pp. 675–679. ACM, New York (2006)
9. Montresor, A., Jelasity, M., Babaoglu, O.: Robust aggregation protocols for large-scale overlay networks. In: Proc. of the 2004 Int. Conference on Dependable Systems and Networks, DSN 2004, pp. 19–28. IEEE Computer Society, Florence (2004)
10. Newman, M.E.J.: Random graphs as models of networks, pp. 35–68. Wiley-VCH Verlag GmbH and Co. KGaA (2005)
11. Puttaswamy, K.P.N., Sala, A., Zhao, B.Y.: Searching for rare objects using index replication. In: 27th IEEE International Conference on Computer Communications, pp. 1723–1731. IEEE (2008)
12. Voulgaris, S., Gavidia, D., van Steen, M.: Cyclon: Inexpensive membership management for unstructured p2p overlays. Journal of Network and Systems Management 13(2), 197–217 (2005)

Self-organizing Techniques for Knowledge Diffusion in Dynamic Social Networks

Luca Allodi, Luca Chiodi, and Marco Cremonini

Abstract. In this paper, we model a knowledge diffusion process in a dynamic social network and study two different techniques for self-organization aimed at improving the average knowledge owned by agents and the overall knowledge diffusion within the network. One is a weak self-organization technique requiring a system-level central control, while the other is a strong self-organization technique that each agent exploits based on local information only. The two techniques are aimed at increasing the knowledge diffusion by mitigating the hype effect and the network congestion that the system dynamics shows systematically. Results of simulations are analyzed for different configurations, discussing how the improvements in knowledge diffusion are influenced by the emergent network topology and the dynamics produced by interacting agents. Our theoretical results, while preliminary, may have practical implications in contexts where the polarization of interests in a community is critical.

1 Introduction

Dynamic social networks represent a multidisciplinary research strand that has been studied with increasing interest since it emerged from sociology and complex systems research. Social networks exhibit peculiar characteristics with respect to non-social networks, in particular regarding the degree correlation of adjacent nodes

Luca Allodi
DISI - University of Trento,
via Sommarive 5 I-38123 Povo (Tn)
e-mail: luca.allodi@disi.unitn.it

Luca Chiodi · Marco Cremonini
DI - University of Milano,
via Bramante 65 I-26013 Crema (Cr)
e-mail: luca.c.chiodi@gmail.com, marco.cremonini@unimi.it

P. Contucci et al. (eds.), *Complex Networks V*,
Studies in Computational Intelligence 549,
DOI: 10.1007/978-3-319-05401-8_8,

and the clustering [1–3]. With respect to node degree, social networks are typically *assortative*, meaning that the degree correlation of adjacent nodes is positive, i.e., nodes of high degree tend, on average, to be connected with other nodes of high degree. The second peculiar characteristic, *clustering*, has been defined in term of network transitivity, that is, given an edge between a pair of nodes A and B and another edge between nodes A and C, a network is said to be highly transitive if it is likely that there will also be a connection between nodes B and C [2]. For social networks, it has been observed that the clustering coefficient is typically greater, possibly orders of magnitude greater, than in typical random graphs [6, 7].

In the literature, many works have applied models of social networks to real case studies. The email exchange in a community of people, for instance, represents a relevant case study [4], as well as the dynamics showed by individuals joining and leaving groups of interests, which may stem from leisure (e.g. the case of online games) to scientific research or corporate projects [5].

In this paper, we model a social network with a fixed number of nodes (i.e. in the following we will refer to *agents* interacting according to some rules, rather than nodes and edges of the network) and mechanisms for the dynamic creation of connections among them. Our model shows the emergent characteristics of typical social networks, such as *assortativeness* [1, 2], *high clustering coefficients* [6, 7], and transition from several clustered communities of agents to a giant interconnected component [8]. The process of knowledge diffusion consists of agents that know a variable set of *topics*, each one characterized by an *interest* and a *quality*. Agents interact by selecting, first, a topic based on their own interests and then the agent in their neighborhood that owns the best quality associated to the chosen topic. In this sense, knowledge spreads from agents knowing more about a topic to those that know less. At network level, each time an interaction between two agents succeeds a directed link, from the requestor to the respondent agent, is created, if it were the first interaction, or the link's weight is increased with the number of exchanges. This interaction model is similar to important models of network influence based on belief and interest, such as the seminal one by DeGroot [9] and of knowledge diffusion, such as the one discussed by Cowan *et al.* [10]. The aim of this work is to study two self-organization strategies, *weak* and *strong self-organization*, according to the definition introduced in [11], aimed at improving knowledge diffusion, both based on the communication efficiency as the heuristic observed—at network level, in one case, at agent level in the other—to adjust the behavior. The effects of the two strategies are compared to the natural network behavior under different configurations. Our results exhibit interesting similarities with research in different settings and scenarios. In particular, works on organizational learning and parallel problem solving [12, 13] have showed how agents that learn too fast can reduce total system knowledge.

2 Related Work

Works more closely related to ours are those that have investigated the knowledge diffusion in networked environments and that have described self-organizing systems. In particular, Cowan and Jonard, two economists, modelled and analyzed the dynamics of knowledge diffusion in a multi-agent scenario [14]. They showed how network structure affects the dynamics of knowledge diffusion, and they demonstrated that the average knowledge is maximal when the structure is a small-world. Their approach and, qualitatively, many of their results are closely related to our work. Different from them, ours is a dynamic model and we studied the emergent behavior of the system, while in their case a number of static network configurations have been considered. From a different field, but equally important for our research, the work by Brun *et al.* [15] has analyzed the importance of feedback loops in designing self-adaptive systems. Their findings, although not explicitly addressed at modelling dynamic social networks, are relevant in our context. Similarly to them, our self-organizing strategies exploit a control loop, either at system level or at agent level, to adjust the behavior. Walter *et al.* presented a model closely related to ours, although based on a static network and different in the research goal [16]. They considered a model of trust-based recommendation system on a social network, which assumed the transitivity of trust along a chain of relationships connecting agents. Differently from them, we admit only a limited degree of trust transitivity (which is restricted to the best friend-of-friends). Important for the analysis of mixing patterns and community structures in networks is the work by Girvan and Newman [8]. This research analyzed most of the characteristics that our model of social network presents and that we have tested and discussed in this work, from the assortative mixing to the formation of communities, from the relevance of friend-of-friend relationships to the dynamics of the growing network.

3 Model Description

We consider a set of N agents, $n_1, n_2, ..., n_N$, each one characterized by a *Personal state* PS_{n_i} (what n_i knows) and a *Friend state* FS_{n_i} (who n_i knows). The *Personal state* has the form $PS_{n_i} = (\bigcup_{j \in T_i}(topic_j, quality_{i,j}, interest_{i,j}))$, where T is the set of topics that the population knows; each agent n_i knows a variable subset of them, $T_i \subseteq T$. The *Friend state* has the form $FS_{n_i} = (\bigcup_{j \in N_i}(n_j, answers_{i,j}))$, where n_j are the identifiers of agents connected with n_i and $answers_{i,j}$ is a counter to keep track of the number of interactions with each peer. The setup has been defined to be the most neutral, with topics T_i assigned to each agent and associated qualities selected randomly, interests distributed uniformly and no friends. More specifically:

Topics: A random set T_i of *topics* is defined for each agent. The maximum number of topics assigned to the agents can be limited by setting the maximum rate $\lambda_T \in (0, 1]$, so that $|T_i| \leq \lambda_T \cdot |T|$.

Quality and Interest: The *quality* associated to each topic of an agent's Personal state is set to a random value in $[1, 100]$. For the *interest*, the initial value is equally distributed among all topics, and is calculated as $100/|T_i|$.

Friends: Agents start with no friends at setup, therefore, in the early stage of network formation, the local search fails and the selection of peers turns to the random choice.

Topic 0: To permit the selection of new topics, a dummy topic called $topic_0$ is always held by each agent. If $topic_0$ is selected, as for all ordinary topics, the system choose, randomly, another topic *not belonging* to the agent's Personal state. Next, the agent looks for a peer based on the new topic. The $quality$ associated to $topic_0$ is always zero, while the $interest$ is calculated as for the other topics during a simulation.

3.1 Network Construction

The network is dynamically formed according to the following steps:

1. Given agent $n_{i'}$, select a topic ($topic_{j*}$) in the Personal state. The choice of the topic is a weighted random selection with values of the associated interests ($interest_{i,j*}$) as weights, this way topics with higher interest are more likely to be selected;
2. Among $n_{i'}$ friend agents and their "best friend" holding topic ($topic_{j*}$), select agent $n_{i''}$ with maximum value of topic's quality ($quality_{i,j*}$);
3. If $quality_{i'',j*} > quality_{i',j*}$ then the communication takes places and agent $n_{i'}$ increases $quality_{i',j*}$ of $topic_{j*}$;
4. Otherwise, if none holds $topic_{j*}$ or exhibits a topic's quality greater than that of agent $n_{i'}$, then select an agent $n_{i'''}$ randomly among the whole population;
5. if $n_{i'''}$ holds $topic_{j*}$ and $quality_{i''',j*} > quality_{i',j*}$, then the communication takes places and $quality_{i',j*}$ increases, otherwise the communication fails.

Best friend-of-friends. Given agent $n_{i'}$, and a selected $topic_{j*}$, for each of its friends, the "best friend" agent is the one owning $topic_{j*}$ and the higher value of the *answer* attribute. The reason for this setup is that we consider unrealistic in a social context to scan all agents with a distance of 2 from the one selected. The selection based on the *answer* attribute represents a basic form of transitive trust. It is worth noting that the inclusion of "best friends" fosters network transitivity and the formation of triads, two key characteristics of social networks.

Start up. At start up, agents have no connection with others (i.e., Friend state is empty). When, for an agent, the 5-steps algorithm is executed, a topic is selected in *Step1*, then *Step2* and *Step3* fail and in *Step4* a random agent is selected. If *Step5* succeeds, then the connection is established. This mechanism triggers the network formation at start up.

3.2 State Update

After a successful interaction, the agent that started the communication is updated. The *quality* and the *interest* of the topic for which the communication took place increase and the other interests, associated to the other topics owned by the agent, are decreased. We decided that the topic's quality increases with increasing marginal increments, according to the assumption that an agent distrusts another one when they interact for the first time and this distrust progressively diminishes as interactions occur. The discount starts at a given value (i.e. ρ) and goes to zero exponentially. Motivations for this assumption could be found in the literature about information aggregation [17] and collective behavior [18] and refers both to the prevalence of egocentrism in assimilating new information and to trust dynamics. The *quality gain* obtained by agent $n_{i'}$ is:

$$\delta quality_{j*} = \frac{quality_{i'',j*} - quality_{i',j*}}{\gamma + \rho e^{-\frac{x}{\theta}}} \tag{1}$$

with: $\gamma \geq 1$ setting the nominal fraction of $\delta quality$ that $n_{i'}$ could learn from another agent; x the value of the attribute *answers* representing the number of past interactions that agent $n_{i'}$ had with agent $n_{i''}$; ρ the initial discount; θ the factor that controls the rate at which agents increase their trust towards the others.

The dynamics we have assumed for the *interest* associated to the topic for which the interaction took place is similar to that of the quality, but with two important differences: It only depends on the $\delta quality$ value and, accordingly, all other interests on topics owned by the agent decreases (studies in cognitive science have showed the tendency of people to shift their attention and interest, rather than behave incrementally [18]). The function is:

$$\delta interest_{i',j*} = \alpha(1 - e^{-\frac{\delta quality_{i',j*}}{\beta}}) \tag{2}$$

with $\alpha > 1$ and $\beta > 1$ the two parameters that control, respectively, the scale and slope of the interest increase.

Parameter β is key for our following analysis of the knowledge diffusion and the self-organization strategies. Self-organizing mechanisms could tune β to reduce ($\beta \uparrow$) or increase ($\beta \downarrow$) the speed at which the agent's interests change.

Finally, all interests associated to topics different from $topic_{j*}$ are reduced by $\delta interest_{i',j \neq j*}(t_k, t_{k-1}) = \delta interest_{i',j*}(t_k, t_{k-1})/(|T_{i'}| - 1)$, that is the value of the interest gain for $topic_{j*}$ at t_k divided by the number of topics $|T_{i'}|$ minus one.[1]

[1] The interest reduction applies to $topic_0$ as well, which is included in the total number $|T_i|$ of topics known by agent $n_{i'}$.

3.3 Metrics

Three metrics have been defined: *Communication Efficiency (CE)*, *Average Knowledge (AK)* and *Knowledge Diffusion (KD)*. The first one is a heuristic observed either at system-level by a central control process or locally by each agent evaluating its own behavior. *CE* measures how often agents are able to successfully interact with others with respect to the number of requests they made during a simulation. Γ is the number of requests made by all agents:

$$CE = \frac{Total\ No.\ of\ Answers}{Total\ No.\ of\ Requests} = \frac{\sum_{i=1}^{N}\sum_{j\in N_i} answers_{i,j}}{\Gamma} \tag{3}$$

The meaning is that if $CE = 0$, then there has been no communication since every interaction failed; if $CE = 1$ every interaction succeeded.

AK and *KD* metrics are used to evaluate two different characteristics of the knowledge diffusion dynamics and to analyze its efficiency and the benefits of the proposed self-organizing techniques. *AK* is calculated as the average quality with respect to the topics actually owned by agents. *KD*, instead, is the average quality with respect to the case of perfect diffusion of knowledge (i.e., all agents holding all topics). While *AK* is maximised by increasing only the average quality of each agent, regardless of the number of known topics, *KD*, instead, depends from the diffusion of topics among agents.

$$AK = \frac{\sum_{i=1}^{N}\sum_{j=1}^{|PS_i|} quality_{i,j}}{\sum_{i=1}^{N}(|PS_i|-1)}\ ;\ KD = \frac{\sum_{i=1}^{N}\sum_{j=1}^{|PS_i|} quality_{i,j}}{|N|\times(|T|-1)} \tag{4}$$

4 Network Simulations

During simulations, the number of agents ($N = 100$) and duration ($\Gamma = 50000$) have been kept constant. Key parameters that were varied are: the *number of topics in the network* $|T|$, the *maximum rate of topics assigned to agents at setup* λ_T (e.g., $\lambda_T = 0.1$ means that at setup agents know *at most* 10% of topics T), and β defined in Equation 2. Combined, they deeply influence the emergent network structure and the aggregation of agents in communities. In our simulations, the parameter λ_T affects the mean network degree. The typical transition [8] from small communities to the giant one happens for $\lambda_T \approx 0.55$.

Figure 1 shows the results for four typical base configurations. In all cases, *CE* has an initial spike (i.e. the *hype effect*) as a result of polarization of agent interests, which, due to the positive feedbacks of successful interactions and the quality increase, tend to exhibit bursts of interaction with the same peer for the same topic. While this effect greatly increases the performance of the network in the early stages, it also quickly dissipates and *CE* abruptly drops until agents start rebalancing their interests by choosing other topics. The dynamics of the communication has visible effects on *KD*: An inefficient communication, in general, implies a slower diffusion of knowledge within the network, as showed by the values of *KD* (see case

A with *C* and case *B* with *D*). Similar results, although on a different model (based on broadcasting in a static network), have been presented in [10]. With respect to *AK*, instead, the four cases do not exhibit relevant differences. In all cases, at the end of the simulation, the average knowledge owned by agents has reached a level close to 80%, meaning that, limited to the topics they individually own, they have reached good quality. As a final remark to this set of results, we stress the fact that where *AK* and *KD* strongly differ (configurations *A* and *B* are those with the largest differences), the network behavior is inefficient with respect to the goal of diffusing knowledge both in quality *and* in quantity. In those cases agents, on average, have improved on topics they owned at setup, but did not acquire much knowledge about new ones.

5 Weak and Strong Self-organization

The analysis of these configurations has showed that reducing the tendency of agents to polarize, caused by interests on just a few topics that grow too fast, communication and knowledge diffusion generally improve. Self-organization mechanisms could modify at run-time some critical parameters. In particular, we consider the interest function and parameter β. The reason for choosing to change only β is because it has both a intuitive meaning in real-world contexts and a direct reference with previous research. Increasing β means, essentially, to reduce the speed of learning avoiding to concentrate on few issues only [13], or be driven more by our own belief than by information received from others [17].

To decide when β should be adjusted, we adopted a simple heuristic based on the dynamics of the *CE*: when its trend changes from increasing to decreasing, β is modified. With this rationale for our approach to self-organization, two strategies have been designed:

Weak Self-Organization is a system-level strategy that assumes the presence of a central control process able to observe the system dynamics of *CE* and to adjust β, a global parameter, at run-time when communication starts dropping.

Strong Self-Organization is an local strategy where each agent observes its own communication dynamics and acts on its own interest function. To this end, the model has been modified to introduce a *local communication efficiency* CE_{n_i}, for each agent, and a *local interest function*, with the same form of Equation 2, except for β_{n_i}, now specified for each agent, rather than as a global parameter. Another modification has been to set a threshold ω that automatically triggers the change of β_{n_i} for each agent.

The results of the system behavior with *weak self-organizations* are shown in Figure 2 (A-D). Qualitatively, we can see that with this simple technique the initial polarization of the network, with its negative effect, is not reduced. The benefit of the weak self-organization strategy becomes evident when the network has absorbed the excessive polarization following the spike and regains a higher level of efficiency. From that point on, the *CE* remarkably improves in all four cases, with respect to the corresponding ones of Figure 1.

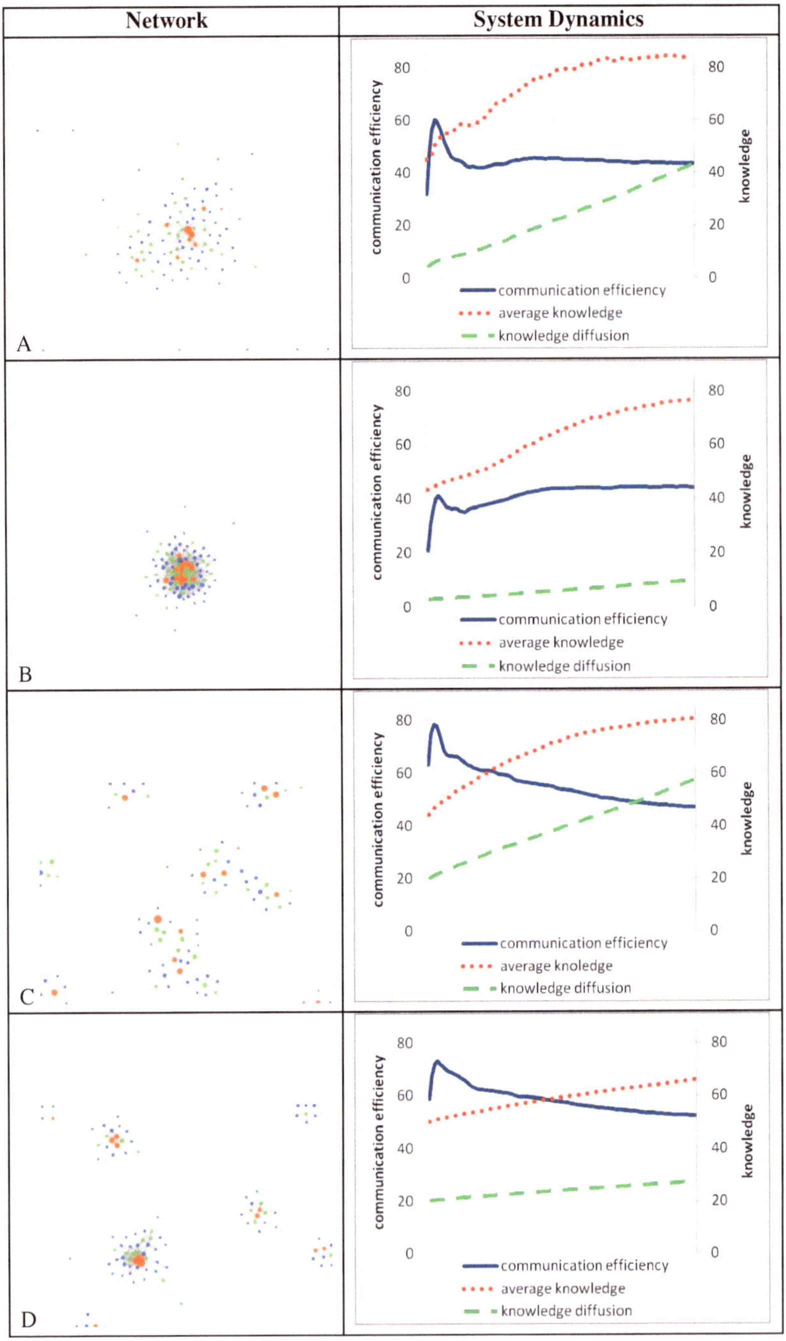

Fig. 1 System dynamics with different configurations. The x-axis represents simulation time (number of ticks Γ); the y-axis is a scale 0-100. Parameters: A) $\lambda_T = 0.1$, $|T| = 20$; B) $\lambda_T = 0.1$, $|T| = 100$; C) $\lambda_T = 1.0$, $|T| = 20$; D) $\lambda_T = 1.0$, $|T| = 100$.

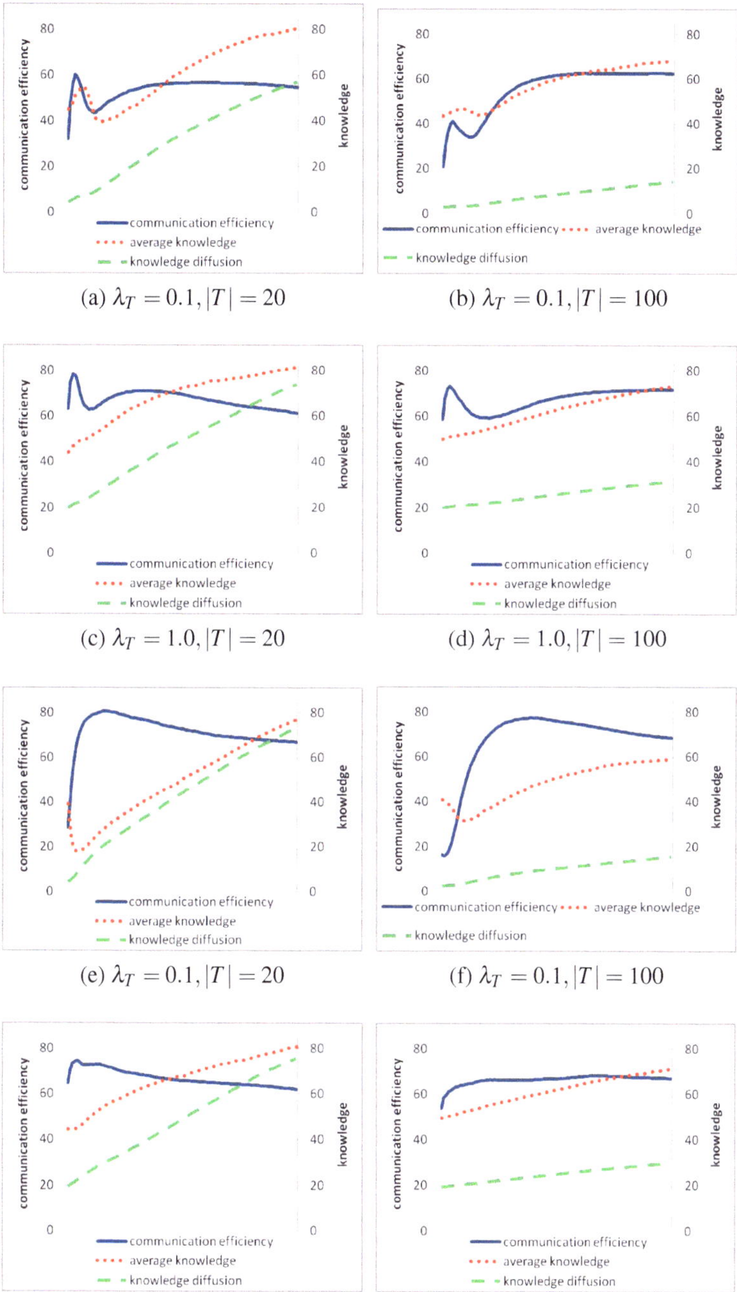

(a) $\lambda_T = 0.1, |T| = 20$ (b) $\lambda_T = 0.1, |T| = 100$

(c) $\lambda_T = 1.0, |T| = 20$ (d) $\lambda_T = 1.0, |T| = 100$

(e) $\lambda_T = 0.1, |T| = 20$ (f) $\lambda_T = 0.1, |T| = 100$

(g) $\lambda_T = 1.0, |T| = 20$ (h) $\lambda_T = 1.0, |T| = 100$

Fig. 2 (A-D): Weak self-organization; (E-H): Strong self-organisation

The worsening of AK is a consequence of the lower clustering and the more frequent random choice of peers (i.e. compare A with C and B with D). This effect is amplified by adjusting β: Agents increment the number of topics, but do not sufficiently interact to improve their AK.

Combined, these results suggest that when knowledge diffusion is adjusted at system level by manipulating the interest dynamics, the number of information in the network $|T|$ with respect to the number of agents N is a critical characteristics, that not only strongly influences the original dynamics, but it also affects the efficacy of the self-organization solution. On the contrary, it could be worthless to force structural modifications, such as inducing the creation of larger communities or splitting giant ones.

Finally, it is remarkable to note how CE always increases of about the same rate in all four cases. This suggests that when self-organization strategies are applied and benefits should be evaluated, an increase in the communication efficiency should not be considered a reliable sign of better knowledge diffusion or a parameter suitable for comparative analyses.

In case of *strong self-organisation*, agents adjust their local β_{n_i} parameter by evaluating their local CE_{n_i}. For the simulations, the threshold, shared by all agents, is set to $\omega = 0.8$, meaning that when an agent sees its own CE_{n_i} dropping below 80%, then it adjusts β_{n_i} to 500; when, instead, it sees CE_{n_i} raising over 80%, it switches back to $\beta_{n_i} = 5$.

The results are showed in Figure 2 (E-H). The more complex dynamics of the strong self-organizing technique has clearly deeper effects on system behavior than the weak self-organization case, which lead to further possibilities to manipulate the evolution of a dynamic social network. By considering the numerical results of Table 1, we observe that reducing the initial spike produces variable benefits. With respect to AK, the network structure is clearly the dominant factor influencing the performances, with highly clustered configurations C and D increasing AK, while lower clustered ones A and B reducing AK. Similar results have been found in [14]. With respect to KD, the solution always produces a gain, but in this case the number of topics is the dominant factor. Configurations A and C with few topics increase KD of more than 16%, while configurations B and D with more topics increase KD of about 4-5%. In more detail, from Table 1, we note that the differences observed for the weak self-organization case between configurations with few topics A and C and configurations with more topics B and D are confirmed in case of strong self-organisation. The same holds for the differences between network structures, A and B having one main component with respect to C and D with small clusters. With respect to KD, there is a tendency of the strong self-organization technique to outperform the weak one, in all configurations except D, meaning that there is a better knowledge diffusion. The opposite holds with respect to AK, due to the reduced tendency of agents to polarize on just few topics. The reasons is that in the weak technique, the global parameter β is adjusted when the *average* communication efficiency of agents drops. This means that some agents could be already extremely polarized and almost unable to interact, while others could be still efficiently interacting. Heterogeneity of agents behavior affects the effectiveness of the weak

Table 1 Results of weak and strong self-organization techniques

	Base System			Weak Self-Organization			Strong Self-Organization		
	CE	*AK*	*KD*	*ΔCE*	*ΔAK*	*ΔKD*	*ΔCE*	*ΔAK*	*ΔKD*
A	39.50	77.44	37.60	**+15.18%**	*-1.07%*	**+16.74%**	**+28.10%**	*-0.32%*	**+35.17%**
B	44.12	76.14	9.68	**+18.45%**	*-8.04%*	**+4.73%**	**+24.41%**	*-17.35%*	**+5.98%**
C	46.96	80.41	57.41	**+14.26%**	**+1.13%**	**+16.42%**	**+14.97%**	**+0.41%**	**+18.24%**
D	52.39	66.11	27.36	**+19.18%**	**+7.08%**	**+4.30%**	**+14.48%**	**+4.98%**	**+2.60%**

self-organization technique. Differently, the strong self-organization techniques operates locally, therefore each agent adjusts its own behavior when needed. This has the effect of preventing extreme polarization.

6 Conclusions

In this paper we presented a model of dynamic social network based on knowledge exchange among agents. The results, although still preliminary, are promising and some strikingly similitudes with previous studies based on different assumptions and different network structures have been found.

In particular, it appears that social network analysis applied to different self-organization strategies could provide important insights for relating one strategy with another. In particular, our focus is on those phenomena that typically exhibit network congestion due to excessive agent polarization, which, as a consequence, exhibit an initial exceptional communication efficiency, followed by a steep decrease. Some well-known examples are: the *hype effect* typically present in the adoption cycle of new technologies [19], the formation of blockbusters in cultural markets [20] or the choice of news published by media [21]. As a final example, in school education, either in traditional classes or through e-learning systems, there is anecdotal evidence that polarization may emerge and could be detrimental for the overall level of knowledge acquired by students. This could be the case of excessive interest of students on few topics only with respect to a more balanced distribution of efforts and time for learning. In all these examples, intuitively, our model and self-organization techniques could be well-suited for describing the effects of manipulating the speed of interest grow.

References

1. Jin, E.M., Girvan, M., Newman, M.E.J.: Structure of growing social networks. Physical Review E 64(4), 046132+ (2001)
2. Newman, M.E.J., Park, J.: Why social networks are different from other types of networks. Physical Review E 68(3) (2003)
3. Skyrms, B., Pemantle, R.: A dynamic model of social network formation. Proceedings of the National Academy of Sciences 97(16), 9340–9346 (2000)

4. Tyler, J.R., Wilkinson, D.M., Huberman, B.A.: Email as spectroscopy: automated discovery of community structure within organizations, pp. 81–96. Kluwer, B.V., The Netherlands (2003)
5. Newman, M.E.J.: Coauthorship networks and patterns of scientific collaboration. Proceedings of the National Academy of Sciences, 5200–5205 (2004)
6. Newman, M.E.J.: The structure and function of complex networks. SIAM Review 45(2), 167–256 (2003)
7. Watts, D.J., Strogatz, S.H.: Collective dynamics of 'small-world' networks. Nature 393, 440–442 (1998)
8. Newman, M.E.J., Girvan, M.: Mixing patterns and community structure in networks. In: Pastor-Satorras, R., Rubi, M., Diaz-Guilera, A. (eds.) Statistical Mechanics of Complex Networks. Lecture Notes in Physics, vol. 625, pp. 66–87. Springer, Heidelberg (2003)
9. DeGroot, M.H.: Reaching a Consensus. Journal of the American Statistical Association 69(345), 118–121 (1974)
10. Cowan, R., Jonard, N.: Knowledge creation, knowledge diffusion and network structure. In: Kirman, A., Zimmermann, J.-B. (eds.) Economies with Heterogeneous Interacting Agents, vol. 503, pp. 327–343. Springer (2001)
11. Di Marzo Serugendo, G., Gleizes, M.P., Karageorgos, A.: Self-organization in multi-agent systems. In: The Knowledge Engineering Review, vol. 20(2), pp. 165–189. Cambridge University Press, United Kingdom (2005)
12. Miller, K.D., Zhao, M., Calantone, R.J.: Adding interpersonal learning and tacit knowledge to March's exploration-exploitation model. Academy of Management Journal 49(4), 709–722 (2006)
13. Lazer, D., Friedman, A.: The Network Structure of Exploration and Exploitation.. Administrative Science Quarterly 52(4), 667–694 (2007)
14. Cowan, R., Jonard, N.: Network structure and the diffusion of knowledge. Journal of Economic Dynamics and Control 28(8), 1557–1575 (2004)
15. Brun, Y., et al.: Engineering self-adaptive systems through feedback loops. In: Cheng, B.H.C., de Lemos, R., Giese, H., Inverardi, P., Magee, J. (eds.) Software Engineering for Self-Adaptive Systems. LNCS, vol. 5525, pp. 48–70. Springer, Heidelberg (2009)
16. Walter, F.E., Battiston, S., Schweitzer, F.: A model of a trust-based recommendation system on a social network. Auton Agent Multi-Agent Syst. 16(1), 57–74 (2008)
17. Bettencourt, L.M.A.: The rules of information aggregation and emergence of collective intelligent behavior. Topics in Cognitive Science 1(4), 598–620 (2009)
18. Goldstone, R.L., Gureckis, T.M.: Collective behavior. Topics in Cognitive Science 1(3), 412–438 (2009)
19. Gartner, Gartner hype cycle, http://www.gartner.com/technology/research/methodologies/hype-cycle.jsp
20. Watts, D.J., Hasker, S.: Marketing in an unpredictable world. Harvard Business Review (September 2006)
21. Media Standards Trust, Shrinking world: The decline of international reporting in the british press, http://mediastandardstrust.org/publications/shrinking-world-the-decline-of-international-reporting-in-the-british-press/

Complex Network Analysis of Ozone Transport

Guoxun Tian and Mehmet Hadi Gunes

Abstract. Ozone transport has an important effect on the local ozone concentrations. In this paper, we model the California State and the Eastern States of America ozone transport as complex networks to understand large-scale dynamics. California has a complex ozone transport due to geomorphology and is divided into 15 air basins to manage air pollution. Unlike California, the ozone transport in the Eastern States of America is mainly affected by wind speed and direction. Through different centralities, we can identify nodes that have higher ozone output, higher ozone income, and the maximum ozone throughput. In general, both networks exhibit similar properties. However, even though Californian network has higher average degree it has smaller clustering than the Eastern State network. Moreover, while Californian counties can be divided into four communities, the Eastern States remain as a single community. Both of these points to the fact that ozone transport within Eastern States is more uniform than between the counties of California.

1 Introduction

In the upper atmosphere, ozone protects the earth from exposure to harmful ultraviolet rays since ozone has a strong absorption of ultraviolet [1]. However, at the ground level, ozone is one of the most important air pollutants. The harms of human health are considerable [2]. In this paper, we establish an *ozone transport network* to analyze large scale characteristics of ozone transfer. To our knowledge, this is the first study to map the ozone transport as a complex network.

Control of ozone emissions is not very effective in reducing ozone concentrations since there are relatively few anthropogenic sources of ozone. Ozone is

Guoxun Tian
Colorado State University, Fort Collins, CO, USA
e-mail: tgxfff@hotmail.com

Mehmet Hadi Gunes
University of Nevada, Reno, NV, USA
e-mail: mgunes@cse.unr.edu

P. Contucci et al. (eds.), *Complex Networks V*,
Studies in Computational Intelligence 549,
DOI: 10.1007/978-3-319-05401-8_9,

primarily formed in the air. After the formation, ozone is cumulated in the air at day time with sunlight and is transported to other places, propagating the air pollution, affecting the people in a downwind area which may be far away from the pollution sources. Transport and formation are the two main sources of ozone pollution.

Long-range transport of ozone and its precursors has become a question in the West since they are isolated to a local area because of the topography. The United States Environmental Protection Agency (US EPA) use a new ozone standard based on the average of ozone concentrations. Areas failing to meet these standards set by the EPA were defined as a "non-attainment areas" for ozone [3].

There were 4 states in the West with areas that do not comply with new standard, including California, Nevada, Arizona, and Colorado. Among which California was the worst. Several factors have to be taken into account to figure out why California has the highest ozone level and so many non-attainment areas. The first one is the long distance pollutant transport from Eastern Asia. Also, the ozone and its precursor can be transported from Eastern Asia to America. Several studies show that the concentration of air pollutants in Eastern Asia significantly affects the background levels of ozone in Western North America [4]. The second factor we have to consider is the ozone transport between different areas in California. To better manage air ozone transport, California is divided into 15 air basins based on its topography features or political boundary [5]. There are 68 counties are included in these 15 air basins. In this study, a complex network is built for these 68 counties to analyze the transport of ozone between them so the responsibility can be better understood.

In the Eastern States of America, control programs throughout the northeast and southeast was established due to the formation and transport of ozone over long distances. These control programs are often more regional in scope. Through these programs, Eastern States and local governments have found that pollutants which are transported from the large industrial combustion sources in the Midwest and Southeast contribute to non-attainment classifications in these areas. The speed and direction of wind is the main factor of this long-range transport. It was found that moderate to high wind speeds have a moderate to high potential for a contribution from transport [6]. That is high pollution episodes from the Midwest and Southeast, which has heavy NOx pollution, can be taken by airflow streams to the Northeast. This long distance transport has been displayed using back trajectory analysis. In this study, the complex network of ozone transport between Eastern States is also constructed to better understand the ozone transport.

2 Californian County Network

2.1 Ozone Transport Classification between Air Basins

California State is divided to 15 air basins for the ozone transport research. In this paper, we divide it into 68 counties for a finer granularity analysis using complex network metrics. In our network, the nodes represent counties and the edges represent the ozone transport between them. Each county is treated as a sub-region of the corresponding air basin. The transport between air basins is estimated based on the assessment of the impacts of transported pollutants on ozone concentrations.

California Environmental Protection Agency Air Resources Board (CEPAARB) estimate the transported pollutants on ozone concentrations, identify upwind air basins and the downwind air basins, and make an assessment of the relative contributions of upwind emissions to downwind ozone concentrations [7]. Classification of aerosol transport between different regions made by CEPAARB is [5]: *Inconsequential*, *Significant*, and *Overwhelmed*.

Using statistical analysis, we can estimate the transport effect by comparing the diurnal profile of hourly ozone concentration averaged over all potential transport days with similar plots for days not in the potential transport [8]. In this network, the weight of edges is determined based on the ozone transport classification between air basins using data from [7]. We assign weights of 0.1 to *inconsequential*, 0.6 to *significant*, and 1 to *overwhelmed*. The average value of the combinations represents the weight of the edges. As [7] only shows the ozone transport between air basins, refinement of the air basins is needed. Thereby, we estimate the ozone transport between different counties using following assumptions.

- **Assumption 1:** Ozone transport only exists between two adjacent counties. If these two counties are at different air basins, the edge between them has the same weight as the ozone transport between the two adjacent air basins.

This assumption is used to estimate the ozone transport between two counties which are included in different air basins. For example, air transport from San Joaquin Valley to North Central Coast has weight of 0.6. Fresno County is in San Joaquin Valley and adjacent to Mono County which is in North Central Coast. So the ozone transport from Fresno County to Mono County also has weight 0.6.

- **Assumption 2:** Air pollution can move freely within an air basin. In the same air basin, the maximum transport weight is set to 1. The actual weight value is directly proportional to the NOx emission of the air basin since it plays an important role in the formation of ozone. The quantification rule is:
  a) Within the air basin which has the maximum NOx emission (MNE) which is equal to 34% of state total, the ozone transport has a weight of 1.
  b) Within other air basins, the weight of the ozone transport is directly proportional to the air basin NOx emission (NE) using the equation $Weight = \frac{NE}{MNE}$.

For example, in San Francisco Bay Area air basin, there is a heavy concentration of industrial facilities, several airports, and a dense freeway and surface street network. The NOx emission has 16% of state total. San Francisco County and San Mateo County are adjacent, so the weight between two of them is quantified as 0.16/0.34=0.47. However, in North Coast air basin, the northern part is sparsely populated and the air is very clean. So the ozone transport in that area is neglected.

- **Assumption 3:** Ozone transport between non-adjacent counties is ignored.

For example, ozone transport from San Francisco County to Mountain Counties and South Central Coast is neglected since they are not adjacent. In reality, ozone transport does exist between non-adjacent counties. It is neglected in this study because the ozone transport through aloft flow is much less than the direct wind transport, and each county always has a lot of non-adjacent neighbors.

2.2 Analysis of the Californian County Network

In this section, we analyze different measures of the ozone transport network of Californian counties.

Closeness Centrality: In a graph, the farness of a node is defined as the sum of its distances to all other nodes, and its closeness is defined as the inverse of the farness, i.e., $closeness(i) = \frac{1}{\sum_j d_{ij}}$ where i is the focal node, j is any node in the network and d_{ij} is the shortest distance between them. As a consequence, a node with higher closeness has lower total distance to all other nodes. For a network, closeness is used to estimate how long it will take to spread information from one node to other nodes sequentially.

In our case, instead of information spreading, ozone transport is considered. *A node that has higher closeness can transport ozone to other nodes in a shorter time*. The closeness centrality value give us a hint about which nodes has the higher ozone output. The average closeness is 0.164 and the top 10 counties have closeness centrality higher or equal to 0.237.

Compared to other air basins, San Francisco Bay Area Air Basin has the most counties in the top 10 in terms of centrality. This is consistent with the history since this air basin has violated the state and federal health-based standards many times over the years, and has contributed to air pollution problems in all of the surrounding air basins.

Betweenness Centrality: Betweenness centrality is a measure of a node within a network. It was introduced as a measure for quantifying the control of a node on the communication between other nodes. It is defined as $C(n) = \sum_{i,j} \frac{\sigma_{i,j}(n)}{\sigma_{i,j}}$ where $\sigma_{i,j}$ is the total number of shortest paths between nodes i and j and $\sigma_{i,j}(n)$ is the number of shortest paths between nodes i and j that pass node n. Nodes of high betweenness centrality are important since communication or transport is more efficient along the shortest paths.

For this ozone transport network, *the node with higher betweenness has the higher ozone throughput*. The average betweenness centrality of this network is 0.010 and the top 10 counties have betweenness higher or equal to 0.044. Despite the nodes with highest betweenness, compared to other air basins, San Joaquin Valley Air Basin includes the biggest portion, five counties, among the top 10 counties. This air basin is at the center of California State and has the second highest NOx emission. The location and high NOx emission give this air basin the ability to affect other areas.

PageRank: The PageRank value indicates an importance of a particular page. A hyperlink to a page counts as a vote of support. A page with high PageRank means it receives a high rank and it is linked "incoming link" to by many pages. It is defined as $PR(p_i) = \frac{1-d}{N} + d\sum_{p_j \in M(p_i)} \frac{PR(p_j)}{L(p_j)}$ where N is the total number of pages, $p_1, p_2, \ldots\ldots, p_N$ are the pages under consideration, $L(p_j)$ is the number of out bound links on page p_j, and $M(p_i)$ is the set of pages that link to p_i.

In this application, instead of a page, a node of the ozone transport is considered. *A node that has higher PageRank is easier to be contaminated by incoming ozone transport from other nodes.* The average PageRank of this network is 0.011 and the top 10 counties have PageRank greater or equal to 0.026. Mojave Desert Air Basin, which is heavily impacted by other air basins, includes three of them.

Clustering: Clustering coefficient has been defined as $C = \frac{3\times\text{number of triangles}}{\text{number of connected triples of nodes}}$ or $C = \frac{number\ of\ closed\ triplets}{number\ of\ connected\ triples\ of\ nodes}$. The average clustering coefficients of the Californian County network is 0.25. The clustering coefficients are shown in Figure 1.

Assortativity: Assortativity coefficient indicates how nodes of different types are connected amongst themselves, and is defined by $\mathrm{r} = \frac{\sum_i e_{ii} - \sum_i a_i b_i}{1 - \sum_i a_i b_i}$ where $a_i = \sum_j e_{ij}$ and $b_j = \sum_i e_{ij}$ and e_{ij} is the fraction of edges from node of type i to a node of type j. Assortativity coefficient of the ozone transport network is 0.18, which means this network is slightly assortative, i.e., nodes of high or low degrees are a bit likely to affect similar degree areas.

Average Degree: The average degree of this network is 5.06. Degree distribution in Figure indicates that the network has an exponential degree distribution. Different from typical complex networks that have power law degree distribution, each node has approximately the same number of nodes connected to them.

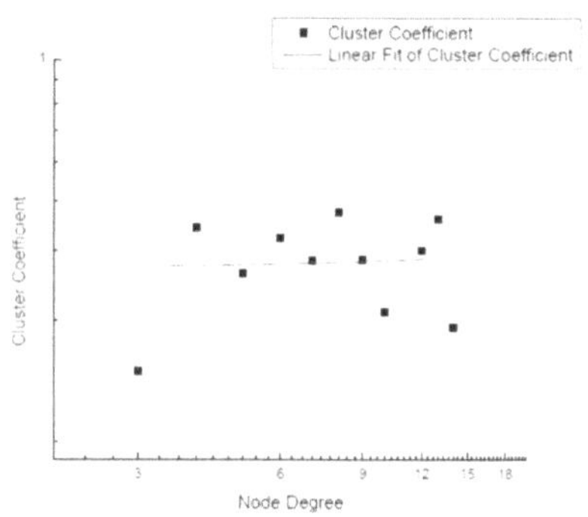

Fig. 1 Clustering coefficients distribution of Californian County network (0 clustering and 0 degree nodes are ignored)

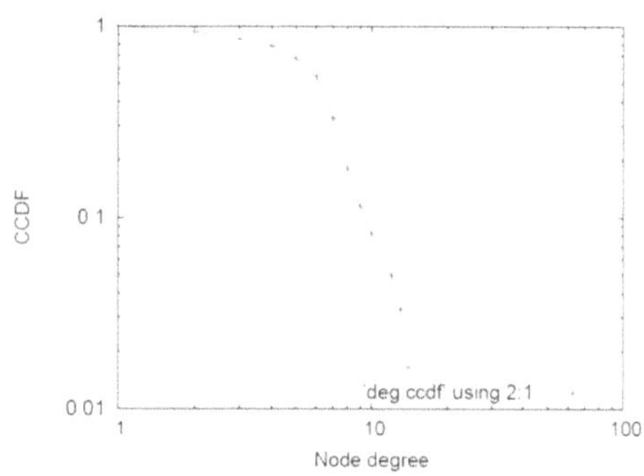

Fig. 2 Degree distribution of Californian County network (0 degree nodes are ignored)

Community Structure: Community detection mechanisms try to identify strongly connected communities within a given network. We use simulated annealing approach which tries to minimize the Hamiltonian $H(\{\sigma\}) = -\sum_{i \neq j}(A_{ij} - \gamma p_{ij})(\sigma_i, \sigma_j)$ where p_{ij} is the probability of vertices i and j being connected. It can be shown that minimizing this Hamiltonian, with $\gamma = 1$ is equivalent to maximizing Newman's modularity. By increasing the parameter γ, it's possible also to find subcommunities. As shown in Figure 3, the network can be divided to 4 groups, i.e., network consists of four regions that are more densely interrelated.

Fig. 3 Communities of the Californian ozone transport network. Different color represents different group.

3 Eastern State Network

In this section, we establish a non-weighted Eastern State network. Transport of ozone and precursors between the Eastern States of America has no boundaries and it is highly related to the wind speed and direction. Ozone can travel across states and provinces easily. At different wind directions and speeds, the ozone concentration pattern is consistent with an atmospheric ozone lifetime of about one day. High ozone concentrations are typically located downwind of areas with the highest emissions with high wind speed (i.e., >6m/s).

Transport during high and low ozone days is investigated in [9]. Transport conditions were established for regionally high (90%) and low (10%) daily maximum 1-hour ozone concentrations. Since the absolute ozone transport between different states is difficult to quantify, a non-weighted ozone transport network is constructed based on the wind directions. In the graph, if wind blow air from state A to B, a non-weighted edge from A to B in the transport network is established. For example, as shown in Figure 5, wind vector shows that wind blow air from Texas to Oklahoma, so an edge from Texas (nodes 0) to Oklahoma (nodes 1) is established.

3.1 Analysis of the Eastern State Network

In this section, we analyze some network metrics of the ozone transport network of the Eastern States.

Closeness Centrality: Compared to the Californian County network, the ozone transport network of the Eastern States of America has a higher average closeness centrality. The average value is 0.278 and Illinois and Kentucky have highest closeness with value 0.367.

Betweenness Centrality: The average betweenness value of this network is 0.025, which is close to the Californian County network. New York has highest betweenness with value 0.13.

PageRank: The average PageRank of this network is 0.014 and is little higher than the Californian County network. New York State has highest PageRank with value 0.45.

Clustering: The average clustering coefficient of the ozone transport network is 0.33, which is higher than the Californian County network. This indicates higher dependence between different regions.

Assortativity: Assortativity coefficient of this ozone transport network is 0.26, which is higher than the Californian County network and shows the network is a bit more assortative. That is states of higher degree are more likely to affect or be affected by other high degree regions.

Average Degree: The average degree of this network is 3.6, which is lower than the Californian County network. Degree distribution is shown in Figure 4. Similar to Californian County network, this figure also shows that the network has an exponential degree distribution.

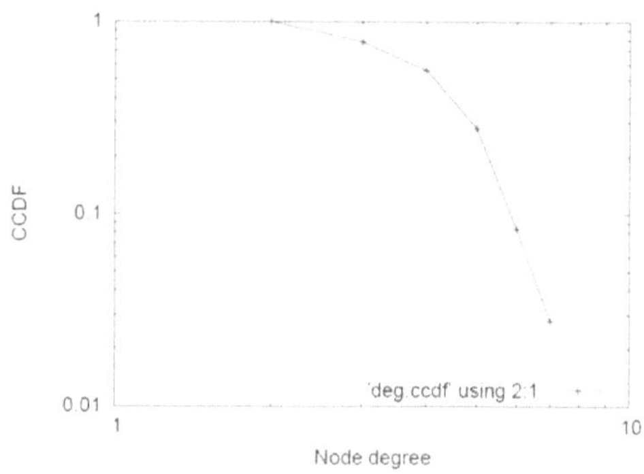

Fig. 4 Degree distribution of the ozone transport of Eastern States

Community Structure: The community structure is shown in Figure 5 where the network is divided into 4 groups. Compared with Californian County network, this network has higher average clustering coefficient, which means this network is more connected. The group barrier, however, is not clear as three of the communities can be accepted as a single group for a total of two groups.

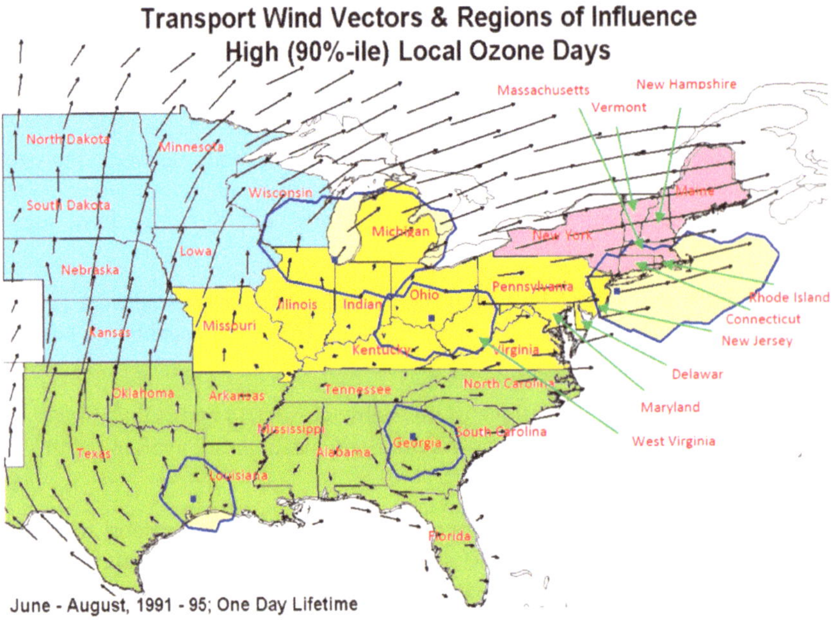

Fig. 5 The communities of ozone transport in Eastern States (modified from [9])

4 Discussion

This paper presents a new way to analyze the ozone transport. We explore the properties of the ozone transport from the complex network perspective. We establish two ozone transport networks, one for Californian Counties and another one for the Eastern States of the United States of America. Some complex network properties of these networks are different as shown in Table 1. Compared to the Californian County network, the Eastern State network has a higher average closeness centrality, a higher clustering coefficient, a higher assortativity coefficient, and a lower average degree. The PageRank centrality, betweenness centrality and degree distribution are similar. Both networks have exponential degree distribution since the ozone transport between non-adjacent areas is ignored.

Generally, the explored properties of the networks are consistent with the existing research. For example, San Francisco Bay Area Air Basin, which is heavy polluted and has contributed to air pollution problems in all of the surrounding air basins, has the biggest portion among the top 10 counties in terms of closeness centrality. Similarly, in terms of closeness centrality, 3 of the top 10 counties are included in San Francisco Bay Area Air Basin. The other 7 counties, which are distributed in several air basins, give us some details of the ozone transport. Those counties may also be playing or will potentially play important roles in the ozone transport.

Table 1 Properties of Californian County network and Eastern State network

	Californian Counties	Eastern States
Average closeness centrality	0.164	0.278
Average betweenness centrality	0.010	0.025
Average PageRank	0.013	0.014
Average clustering coefficients	0.25	0.33
Assortativity coefficient	0.18	0.26
Average degree	5.06	3.6
Degree distribution	Exponential	Exponential
Community structure	4 groups	2 or 4 groups

Among the Eastern States of America, Michigan, one of the biggest source areas of ozone transport plays an important role in the ozone transport network. It has the 3rd highest closeness centrality, 2nd highest betweenness centrality and 2nd highest PageRank. New York, which has the highest closeness centrality and PageRank, may be playing or will potentially play an important role in the ozone transport as well. The community structure of Eastern State network also shows that New York is a key node is the only conjunction node of the two groups.

As we can see from Table 1, these two networks have similar properties. Despite the fact that the average degree of Californian county network is higher than Eastern State network, the average clustering coefficient of Californian county network is lower since the geographic characteristics of the Californian counties. These facts point to the geomorphology effect for ozone transport and the fact that ozone transport within Eastern States is much more uniform than between the Californian counties.

Acknowledgement: This material is based in part upon work supported by the National Science Foundation under grant number EPS- IIA-1301726

Disclaimer: Any opinions, findings, and conclusions or recommendations expressed in this material are those of the author(s) and do not necessarily reflect the views of the National Science Foundation.

References

[1] Seinfeld, J.H., Pandis, S.N.: Atmospheric Chemistry and Physics: From Air Pollution to Climate Change. John Wiley and Sons, Inc. (1998)

[2] Avol, E.L., Linn, W.S., Venet, T.G., Shamoo, D.A., Hackney, J.D.: Comparative respiratory effects of ozone and ambient oxidant pollution exposure during heavy exercise. J. Air Pollut. Control Assoc. 34, 804–809 (1984)

[3] Wierman, S.S.: Will the 8-hr ozone standard finally get off the ground. EM, 20–28 (September 2003)
[4] Schwarzhoff, P.: Long-range Transport of Pollutants to the West Coast of North America. Georgia Basin/Puget Sound Research Conference (2003)
[5] Garcia, C., Gouze, S., Wright, W.: Assessment of the Impacts of Transported Pollutants on Ozone Concentrations in California. California Air Resources Board (2001)
[6] Douglas, S.G., Hudischewskyj, A.B., Lolk, N.K., Guo, Z.: Analysis of Southern California Wind Profiler and Aircraft Data. SAI Report (1997)
[7] Austin, J., Gouze, S.: Ozone Transport: 2001 Review. California Environmental Protection Agency Air Resources Board (2001)
[8] Tian, G.: Complex Network Analysis of Ozone Transport. MS Thesis, University of Nevada, Reno (2012)
[9] Schichtel, B.A., Husar, R.B.: Eastern north america transprot climatology during average, high and low ozone days (1999)

The Small World of Seismic Events

Douglas S.R. Ferreira, Andrés R.R. Papa, and Ronaldo Menezes

Abstract. The understanding of long-distance relations between seismic activities has for long been of interest to seismologists and geologists. In this paper we have used data from the world-wide earthquake catalog for the period between 1972 and 2011, to generate a network of sites around the world for earthquakes with magnitude $m \geq 4.5$ on the Richter scale. After the network construction, we have analyzed the results under two viewpoints. Firstly, in contrast to previous works, which have considered just small areas, we showed that the best fitting for networks of seismic events is not a pure power law, but a power law with an exponential cutoff. We also have found that the global network presents small-world properties. The implications of our results are discussed.

1 Introduction

The general belief in seismic theory is that the relationship between events that are located far apart is hard to be understood. However we live today in a world where data is being collected on most aspects of our lives and better yet, computer power is cheaply available for analyzing the data. The work on seismic data analysis is no

Douglas S.R. Ferreira
Instituto Federal de Educação, Ciência e Tecnologia do Rio de Janeiro, Brazil
Observatório Nacional, Rio de Janeiro, Brazil
e-mail: douglas.ferreira@ifrj.edu.br

Andrés R.R. Papa
Geophysics Department, Observatório Nacional, Rio de Janeiro, Brazil
Instituto de Física, Universidade do Estado do Rio de Janeiro, Rio de Janeiro, Brazil
e-mail: papa@on.br

Ronaldo Menezes
BioComplex Laboratory, Department of Computer Sciences,
Florida Institute of Technology, Melbourne, Florida, USA
e-mail: rmenezes@cs.fit.edu

P. Contucci et al. (eds.), *Complex Networks V*,
Studies in Computational Intelligence 549,
DOI: 10.1007/978-3-319-05401-8_10,

different; we have now large collections of millions of seismic events from around the world which deserves analysis. In this paper we have found some evidence that point to small-world characteristics in the existent data on seismic events. An event in a particular geographical site appears to be related to many other sites around the world and not only to other events at nearby sites.

Since the work from Barabási and Albert [5] researchers have turned their attention not on mining the data itself but rather organizing the data in a network which captures relationships between pieces of data and only then mining the network structure and hence the relations between pieces of data. The network can review information hard to see from mining the raw pieces of data. The use of networks as a framework for the understanding of natural phenomena is nowadays called *Network Science*.

Through the analysis of a model using successive earthquakes, recent studies [1, 2] have applied concepts of complex networks to study the relationship between seismic events. In these studies, networks of geographical sites are constructed by choosing a region of the world (e.g. Iran, California) and its respective earthquake catalog. The region is then divided into small cubic cells, where a cell will become a node of the network if an earthquake occurred therein. Two different cells will be connected by a directed edge when two successive earthquakes occurred in these respective cells. If two earthquakes occur in the same cell we have a loop, i.e., the cell is connected to itself. This method of describing the complexity of seismic phenomena has found that, at least for some regions, the common features of complex networks (e.g. scale-free, small-world) are present. However, in spite of the importance of the results which show that seismic networks for some specific regions present small-world effects, these results are somewhat expected since it makes sense for areas located geographically close to each other to be correlated.

In this paper we have used data from the world-wide earthquake catalog for the period between 1972 and 2011, to generate a network of sites around the world. Since only seismic events with $m \geq 4.5$ are recorded for all locations around the world, we then consider them *significant events* and used this set in our analysis.

2 Theoretical Background

2.1 Complex Networks Features

Scale-free networks are defined as those in which the degree distribution of nodes (or vertices) follows a power law, that is, the probability that a network will have nodes of degree k, denoted by $P(k)$ is given by $P(k) \sim k^{-\gamma}$, where γ is a positive constant. This equation states that scale-free networks have a very small number of highly-connected nodes (called hubs) and a large number of nodes with low connectivity. These networks exist in contrast with general random networks with a very large number of nodes in which the probability distribution follows a Poisson distribution. Random networks have nice properties but the truth of the matter is that most real networks are not random.

One of the best approaches for defining Small-world Networks is based on the work of Watts [13] which states that in small-world networks, every node is "close" to every other node in the network. It is generally agreed that "close" refers to the average path length in the network, ℓ, which has the same order of magnitude as the logarithm of the number of nodes, i.e., $\ell \sim \ln N$. In addition, and what makes small-world networks even more interesting, is the fact that these networks have a high degree of clustering representing a transitivity in the relation of nodes; if a node i has two connections, the theory argues that the two connections are also likely to "know" each other. More formally, the clustering coefficient, C_i, of that node is given by:

$$C_i = \frac{\triangle(i)}{\triangle_{all}(i)} \quad (1)$$

where, $\triangle(i)$ is the number of the directed triangles formed by i with its neighbors and $\triangle_{all}(i)$ is the number of all possible triangles that i could form with its neighbors; the clustering coefficient of the entire network, C, is just the average of all C_i over the number of nodes in the network, N. In random networks the clustering coefficient can be estimated using the closed form $C_{\text{rand}} = \langle k \rangle / N$, where $\langle k \rangle$ is the average degree in the random network.

In the context of networks of seismic events, if it contains hubs, one can argue that the distribution of earthquakes should also follow a power law. On the other hand, if the network has small-world properties one can argue that there is some indication of long-range relations between far-apart earthquake sites.

3 A Geographical Network from Seismic Events

The use of networks to understand phenomena associated with geographical locations has been used in many instances in science including diseases [9], scientific collaborations [8, 10], and organ transplantation [12] to mention just a few. Seismic activity is intrinsically linked to geography because todays instruments can pinpoint with great accuracy the location in the globe where each seismic event take place.

It is important to precisely locate the geographical location of a seismic event but if we want to understand relations between events we should concentrate on creating a network in which locations are linked based on an acceptable criteria. In this paper we use the same procedure employed by [1] in their studies of earthquakes in specific regions of the world. The construction of the network is as follows. We first have to decide on what should represent the nodes. Obviously our first choice are the sites where the earthquake took place. The problem of doing this is that an earthquake epicenter is rarely located exactly in the same location and given the accuracy of today's instruments we would have an infinitely large number of possible sites. We decided instead to define nodes representing a larger region of the world we here call it cell. A cell will become a node of the network if an earthquake has its epicenter therein. The creation of edges follows a temporal order of seismic events. For instance, if an earthquake occurs in a cell C_1 and the next earthquake in a cell C_2, we assume a relation between C_1 and C_2 and we represent the event by a directed edge in the network. The process continues linking cells according to the temporal order.

The degree of each node (the total number of its connections) is not affected by the direction of the network. The nature of the way the network is constructed means that for each node in the network, its in-degree is equal to its out-degree (the exceptions are only the first and last sites in the sequence of seismic events but for all practical purposes we can disregard this small difference).

Although the use of temporal ordering of events is not new in our paper, there are two main differences between our study and others. Firstly and most importantly, the region considered in our investigation is the entire globe, instead of just some specific geographical subarea of the globe; this is, to the best our knowledge, the first worldwide study of seismic events using networks and consequently the first one to investigate the possibility of long-range links between seismic events. Secondly, we have used a two dimensional model in which the depth dimension of the earthquake epicenter is not considered, since we are interested in looking for spatial connections between different regions around the world and besides 82% of the earthquakes, in our dataset, have their hypocenters in a depth less than or equal to 100 km.

Before we divide the globe into cells, we need to choose the size of such cells particularly because we are dealing with the entire globe; if the cells are too small we will not have any useful information in the network, if the cells are large we lose information due to the grouping of events into a single network node. There are no rules to define the sizes. Therefore we have taken three different sizes, the same sizes used in previous studies [1, 11], where the authors conducted studies about earthquake networks using data from California, Chile and Japan. The quadratic cells have, 5 km $\times$ 5 km, 10 km $\times$ 10 km and 20 km $\times$ 20 km. To set up cells around the globe, we have used the latitude and longitude coordinates of each epicenter in relation to the origin of the coordinates, i.e., where both latitude and longitude are equal to zero (we have chosen the referential at the origin for simplicity). So, if a seismic event occurs with epicenter E with location (θ_E, ϕ_E), where θ_E and ϕ_E are the values of latitude and longitude in radians of the epicenter, we are able to find the distances north–south and east–west between this point and the origin. These distances can be calculated, considering the spherical approximation for the Earth, by:

$$\begin{aligned} S_E^{ns} &= R.\theta_E \\ S_E^{ew} &= R.\phi_E.\cos\theta_E, \end{aligned} \tag{2}$$

where S_E^{ns} and S_E^{ew} are, the north-south and east-west distances for the earthquake E, respectively, and R is the Earth radius, considered equal to 6.371×10^3 km. With this computation we can identify the cell in the lattice for each event using the values of S_E^{ns} and S_E^{ew}.

Note that the distances between different cells are irrelevant for the present part of our study. For now we are just interested in the connectivity of nodes. However, from the sequence there are important consequences to be obtained which we present below.

The seismic data used to build our network was taken from the Global Earthquake Catalog, provided by Advanced National Seismic System[1], which records events from the entire globe. The data spans all seismic events between the period from January 1, 1972 to December 31, 2011. This catalog has a limitation because it is not consistent in all regions of the world; it includes events of all magnitudes for the United States of America but only events with $m \geq 4.5$ (in the Richter scale) for the rest of the world (unless they received specific information that the event was felt or caused damage). Therefore, in order to obtain a more homogeneous distribution of data through the world, we have analyzed only events with $m \geq 4.5$ and we also excluded data that represent artificial seismic events ("quarry blasts"). In the end, we were left with 185 747 events, where 82% of them happen near the surface of the world (depth $\leq$ 100 km).

4 Results

Given the network build as described in the previous section, we have performed a few experiments to understand this structure. Following the procedure explained earlier, the 185 747 events yielded three different networks depending on the size used for the cells: 20 km $\times$ 20 km with 65 355 nodes, 10 km $\times$ 10 km with 104 516 nodes, and 5 km $\times$ 5 km with 144 974 nodes.

4.1 Scale-Free Property of the Seismic Network

It has been shown recently [3, 11] that seismic networks for specific regions of the globe (e.g. California) appear to have scale-free properties, or in other words that the construction of the network employs preferential attachment as described by [5] insofar that a node added to the network has a higher probability to be connected to an existing node that already has a large number of connections. This is somewhat trivial to understand because active sites in the world will tend to appear in the temporal sequence of seismic events many times. The preferential attachment states that the probability P that a new node i will be linked to an existing node j, depends on the degree $\deg(j)$ of the node j, that is, $P(i \rightarrow j) = \deg(j)/\sum_u \deg(u)$. This rule generates a scale-free behavior whose connectivity distribution follows a power-law with a negative exponent as shown in Section 2.1.

In [3, 11], earthquake networks were built for some specific regions (California, Chile and Japan), and their connectivity distributions were found to follow power-laws. However, if we look carefully to the connectivity distribution and plot its cumulative probability, instead of its probability density, we can observe that the power-law distributions that emerge from these network are truncated. According to [4], there are at least two classes of factors that may affect the preferential attachment and consequently the scale-free degree distribution: the aging of the nodes and the cost of adding links to the nodes (or the limited capacity of a node). The

[1] `http://quake.geo.berkeley.edu/anss`

aging effect means that even a highly connected node may, eventually, stop receiving new links. The presence of an aging-like effect in our work could be expected from the fact that relaxation times for tectonics are much longer that the time interval under study. Some cells can stop of receiving new connections during a period of time comparable to our own time window by a temporal quiet period due to a transitory stress accumulation. The second factor that affects the preferential attachment occurs when the number of possible links attaching to a node is limited by physical factors or when this node has, for any reason, a limited capacity to receive connections, like in a network of world airports. We have not found a suitable parallel to this factor in the case of earthquakes. When any of these factors is present, the distribution is better represented with a power law with an exponential cutoff, $P(k) \sim k^{-\alpha} e^{-k/k_c}$, where α and k_c are constants.

In Fig. 1, we plot the cumulative probability distribution for the earthquake network built for the Southern California (32°N – 37°N and 114°W – 122°W), using the data catalog provided by Advanced National Seismic System, where we considered all seisms with magnitude $m > 0$ for the period between January 1, 2002 and December 31, 2011. The total number of events is 147 435. It is possible to observe in this plot that the data is better fitted to a power-law with exponential cutoff than a pure power law which is a good fit only for small values of k with an exponent $\gamma - 1 = 0.513$, which is consistent with the value $\gamma = 1.5$ reported in [3] for the probability density function. These results apply for a network built using cell sizes of 5 km × 5 km. It is noteworthy that in a probability density plot, the cutoff does not seem to exist, because the fluctuations are higher than in a cumulative probability plot.

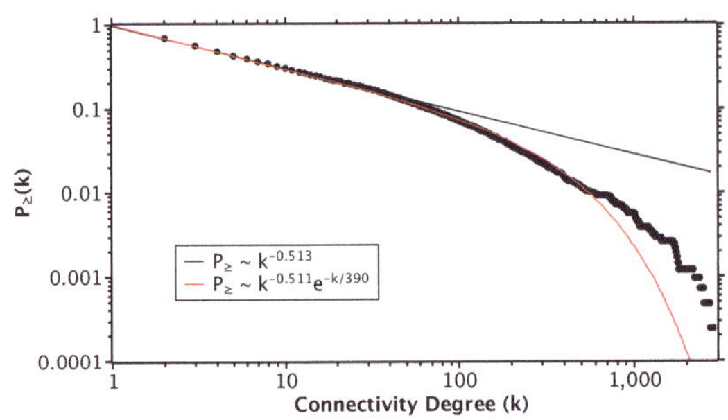

Fig. 1 Cumulative probability distribution of connectivity for the earthquake network in California using cell size 5 km × 5 km. The solid lines represent two possible fittings: a power-law (black) and a power-law with exponential cutoff (red). There are 4 187 nodes in this network.

Looking at the world earthquake network constructed using the data from the Global Earthquake Catalog, we note that the aging-cost effect are visibly stronger in the connectivity distribution; the exponential cutoff is clearly visible in both the degree distribution and the cumulative degree distribution presented in Fig. 2.

Fig. 2(left) represents the connectivity distribution for the global networks using the three different cell sizes for the global lattice. It is interesting to note that, comparing these plots, we observe that the behavior is the same in all three cases (in the sense that they present a power law with an exponential cutoff), which indicates that the cell size does not change the complex features behind the global seismic phenomena.

In Fig. 2(right), we have the same plot of Fig. 2(left), but using the cumulative probability only for cell size 20 km × 20 km. Note that the cumulative probability plot for the global network shows the same exponential cutoff behavior than for local network, as shown in Fig. 1.

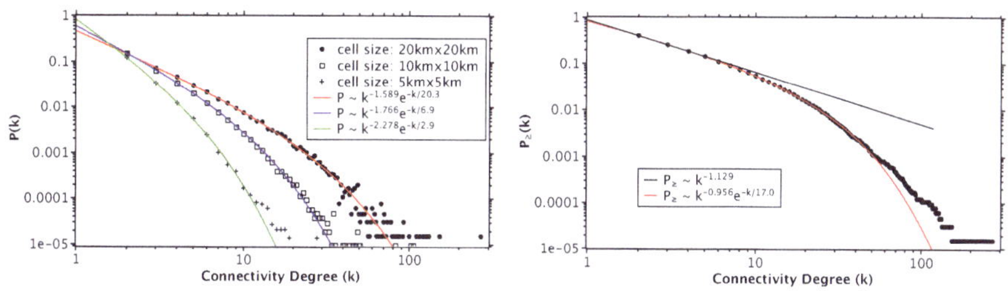

Fig. 2 Connectivity distributions in the global earthquake network. Plot for the cell sizes 20 km × 20 km (solid circles), 10 km × 10 km (squares) and 5 km × 5 km (cross), where the solid lines represent the best fit using power-law with exponential cutoff (on the left). Cumulative probability for the cell size 20 km × 20 km. The solid lines represent a standard power-law (black) and a power-law with exponential cutoff (red) (on the right).

4.2 *Small-World Property of the Global Seismic Network*

Small-world networks [13] have the general characteristic that they contain groups of near-cliques (dense areas of connectivity) but long jumps between these areas. These two properties lead to a network in which the *average shortest path* is very small and the *clustering coefficient* very high. It is important to note that the term *average shortest path* does not refer to a spatial distance but the number of "steps" on the network to move from a node to another.

Here we would like to test if the global seismic network has small-world properties. The consequence of such a finding would be an indicative that seismic events around the world are correlated and not independent. To study these properties we need to introduce slight changes to our original network. The first is that the loops have to be removed, since we are looking for correlations between nodes and it only makes sense when these nodes are different. The second change is, that we move from a network with multi-graph characteristics to a weighted network. That is, if two nodes are linked by w edges in the original network, they will be linked by a single edge with weight w in the new version of the network.

We have analyzed the seismic network for the entire world under two viewpoints: directed and undirected. The cell size used in this construction was 20 km × 20 km. The data were the same used to construct the Fig. 2. Table 1 shows the results obtained for the clustering coefficient (C) [6] and the average path length (ℓ) [7].

Table 1 Results for the clustering coefficient (C) compared to the clustering coefficient of a random network of the same size (C_{rand}) and the average path length (ℓ) compared to the $\ln N$, where N is 65 355 in network with cell size 20 km × 20 km

Network	C	C_{rand}	ℓ	$\ln N$
Directed	7.0×10^{-3}	4.2×10^{-5}	17.19	11.08
Undirected	4.2×10^{-2}	4.2×10^{-5}	12.24	11.08

From Table 1, we note that both versions of the earthquake network has small-world properties; the clustering coefficient is much higher than an equivalent for a random network, and the average path length has the same order of magnitude as the logarithm of the number of nodes. It is worth noticing that the regional earthquake networks built for California, Japan and Chile also are small-world [3, 11] although the significance of small world at the global level is higher because with these world-wide results we have an indicative of long-range relations between different places around the world.

5 Conclusions

The use of networks to model and study relationships between seismic events has been used in the past for small areas of the globe. Here we demonstrate that similar techniques could also be used at the global level. More importantly, many of the techniques used in complex network analysis were used here to show that there seem to exist long-distance relations between seismic events.

We argued in favor of the long-distance relation hypothesis by showing that the network has small-world characteristics. Given the small-world characteristics of high clustering and low average path length, we were able to argue that seisms around the world appear not to be independent of each other.

Another interesting approach we intend to do in the future relates to using community analysis or community detection to understand how seismic locations are grouped.

Acknowledgements. Douglas S. R. Ferreira thanks the Capes Foundation, Ministry of Education of Brazil, for the scholarship under process BEX 13748/12-2. Andrés R. R. Papa thanks CNPq (Brazilian Science Foundation) for his productivity fellowship.

References

1. Abe, S., Suzuki, N.: Scale-free network of earthquakes. Europhys. Lett. 65(4), 581–586 (2004)
2. Abe, S., Suzuki, N.: Small-world structure of earthquake network. Physica A 337(1), 357–362 (2004)

3. Abe, S., Suzuki, N.: Complex-network description of seismicity. Nonlin. Processes Geophys. 13(2), 145–150 (2006)
4. Amaral, L.A., Scala, A., Barthelemy, M., Stanley, H.E.: Classes of small-world networks. Proc. Natl. Acad. Sci. 97(21), 11149–11152 (2000)
5. Barabási, A.-L., Albert, R.: Emergence of scaling in random networks. Science 286(5439), 509–512 (1999)
6. Barrat, A., Barthelemy, M., Pastor-Satorras, R., Vespignani, A.: The architecture of complex weighted networks. Proceedings of the National Academy of Sciences of the United States of America 101(11), 3747–3752 (2004)
7. Brandes, U.: A faster algorithm for betweenness centrality. Journal of Mathematical Sociology 25(2), 163–177 (2001)
8. Divakarmurthy, P., Biswas, P., Menezes, R.: A temporal analysis of geographical distances in computer science collaborations. In: IEEE Third International Conference on Privacy, Security, Risk and Trust (Passat) and 2011 IEEE Third International Conference on Social Computing (SocialCom), pp. 657–660. IEEE (2011)
9. Newman, M.E.J.: Spread of epidemic disease on networks. Physical Review E 66(1), 016128 (2002)
10. Pan, R.K., Kaski, K., Fortunato, S.: World citation and collaboration networks: uncovering the role of geography in science. Scientific Reports 2 (2012)
11. Pasten, D., Abe, S., Munoz, V., Suzuki, N.: Scale-free and small-world properties of earthquake network in chile. arXiv preprint arXiv:1005.5548 (2010)
12. Venugopal, S., Stoner, E., Cadeiras, M., Menezes, R.: Understanding organ transplantation in the usa using geographical social networks. Social Network Analysis and Mining, 1–17 (2012)
13. Watts, D.J.: Networks, dynamics, and the small-world phenomenon. American Journal of Sociology 105(2), 493–527 (1999)

Discovering Colored Network Motifs

Pedro Ribeiro and Fernando Silva

Abstract. Network motifs are small overrepresented patterns that have been used successfully to characterize complex networks. Current algorithmic approaches focus essentially on pure topology and disregard node and edge nature. However, it is often the case that nodes and edges can also be classified and separated into different classes. This kind of networks can be modeled by colored (or labeled) graphs. Here we present a definition of colored motifs and an algorithm for efficiently discovering them. We use g-tries, a specialized data-structure created for finding sets of subgraphs. G-Tries encapsulate common sub-structure, and with the aid of symmetry breaking conditions and a customized canonization methodology, we are able to efficiently search for several colored patterns at the same time. We apply our algorithm to a set of representative complex networks, showing that it can find colored motifs and outperform previous methods.

1 Introduction

Network motifs are small overrepresented subgraphs appearing more frequently than what would be expected in randomized networks with similar topological characteristics [1]. They have been extensively used in may domains, such as biological [2] or social [3] networks. The vast majority of past research deals essentially with pure structural motifs, with nodes and edges being of the same type, i.e., with no associated *color* or *label*. This is a restriction that limits the information we can obtain, because it is often the case that nodes or edges are of different nature. For example, in metabolic networks, we can distinguish between two sets of nodes: reactions and chemical compounds [4]. By ignoring these labels we may be missing important patterns, and it has been shown that by associating different colors to the nodes, their information content is richer [5]. The same can be said about edges, and

Pedro Ribeiro · Fernando Silva
CRACS & INESC-TEC, DCC-FCUP, Universidade do Porto, Portugal
e-mail: {pribeiro,fds}@dcc.fc.up.pt

P. Contucci et al. (eds.), *Complex Networks V*,
Studies in Computational Intelligence 549,
DOI: 10.1007/978-3-319-05401-8_11,

previous experiments have shown that we would gain if it was possible to distinguish between different types of connections in biological networks [6].

Discovering network motifs is a computationally hard task, intimately connected to the subgraph isomorphism problem. Current methods essentially rely on computing the frequency of subgraphs on both the original graph and on an ensemble of randomized similar graphs. This is traditionally done in one of two distinct approaches: either we enumerate all subgraphs of a certain size and then compute their isomorphisms, or we query individually for one subgraph at a time. The first *network-centric* approach demands that we always look for all possible subgraphs, with no option for looking for specific types. The second *subgraph-centric* approach, considers one subgraph type at a time, not reusing information from previous searches. We recently introduced the g-trie data-structure [7], allowing a *set-centric* approach, in which we can search specifically for a certain set of subgraphs. The algorithms developed show substantial efficiency gains on pure structural motifs, when comparing to previous approaches [8].

In this paper we present an efficient g-tries based algorithm able to discover colored motifs. Our main contribution is two-fold. First, we give a clear definition of what a colored motif can be, including how we can compute its statistical significance. Secondly, we provide an extension of the g-trie data structure and associated algorithms in order to allow the usage of color information on both nodes and edges. This allows for richer searches that do not discard information about the nature of the network constituents. We empirically evaluate our algorithms in a set of representative graphs, showing the feasibility of our approach and how we can outperform past methods.

The remainder of the paper is organized as follows. Section 2 establishes a common graph terminology, explains the problem being tackled, and overviews related work. Section 3 describes the g-trie data structure and details how it can be used for discovering colored motifs. Section 4 shows the results of an empirical evaluation of the developed algorithms, when applied to a set of complex networks, and compares it to the performance obtained by a competing algorithm. Finally, section 5 concludes the paper.

2 Preliminaries

2.1 Graph Notation

We briefly review the main concepts and notation that we use. A *graph* G is composed of a set $V(G)$ of *vertices* or *nodes* and a set $E(G)$ of *edges* or *connections*. A k-graph is a graph of size k, i.e., with k nodes. Every edge is composed of a pair (u,v) of two *endpoints* in the set of vertices. A graph is said to be *colored* if we associate colors, or labels, to each of its vertices and/or edges. Note that non-colored graphs can be seen as a simpler case with only one color for nodes and edges. The *degree* of a vertex u is the number of connections it has to other nodes, and its *neighbourhood*, denoted as $N(u)$, is composed by the set of vertices $v \in V(G)$ such that v

and u share an edge. The *exclusive neighborhood* of a vertex v of graph G relative to a subgraph G_k is defined as $N_{exclusive}(v,G_k) = \{u : u \in N(v) \wedge u \notin N(G_k) \wedge u \notin G_k\}$.

A *subgraph* G_k of a graph G is a graph of size k in which $V(G_k) \subseteq V(G)$ and $E(G_k) \subseteq E(G)$. This subgraph is said to be *induced* when $\forall u,v \in V(G_K)$, $(u,v) \in E(G_k)$ if and only if $(u,v) \in E(G)$. The neighborhood of a subgraph G_k, denoted by $N(G_k)$ is the union of $N(u)$, $\forall u \in V(G_k)$. Two graphs G and H are said to be *isomorphic*, if there is a one-to-one mapping between the vertices of both graphs (including colors) and there is an edge of color c between two vertices of G if and only if their corresponding vertices in H also have an edge of color c.

2.2 Colored Motifs

Milo et al. provided the first definition of network motifs as *"patterns of interconnections occurring in complex networks in numbers that are significantly higher than those in similar randomized networks"* [1]. Here we introduce a similar definition for colored motifs, but with the necessary adaptions. To put it to practice, we need to establish two different concepts: what is the *frequency* of a subgraph and what are *similar randomized networks*.

For the first concept, we resort to the standard definition in the motif detection realm, considering only induced subgraphs, and with two occurrences being different if they have at least one node or edge not shared. As to the second concept, the idea is to test subgraph significance by comparing the frequency of the subgraph in the original network with its frequency on a large number of similar random networks. To be sure that the results are specific to a particular network one should use randomized networks as close as possible to the original one. Milo et al. suggested to keep single-node properties, namely its degree. Adapting this to the colored case, we maintain all color related information. The randomized networks have the same set of nodes and colors, each node keeps the same amount of edges in each color, and each edge connects endpoints of the same colors it was connecting in the original network. We can say that we keep the *colored degree sequence*.

Figure 1 shows an example colored motif following our definition, appearing 4 times on the original network. Different occurrences (indicated with thick lines) may share some of the nodes (e.g., $\{1,6,7\}$ and $\{1,7,11\}$ share nodes 1 and 7). Colors enrich the information and allow us to distinguish between different types of triangle subgraphs. Note how the randomized networks respect the colored degree sequence of the original, with each node keeping the exact same type of colored connections. For instance, in the original network node 1 has two dashed connections to light nodes, one dashed connection to a black node and one continuous connection to another black node. The same happens at each of the similar randomized networks. However, the subgraph we are considering is much less frequent on these networks than in the original one, and hence we consider it a motif, i.e., the subgraph is overrepresented.

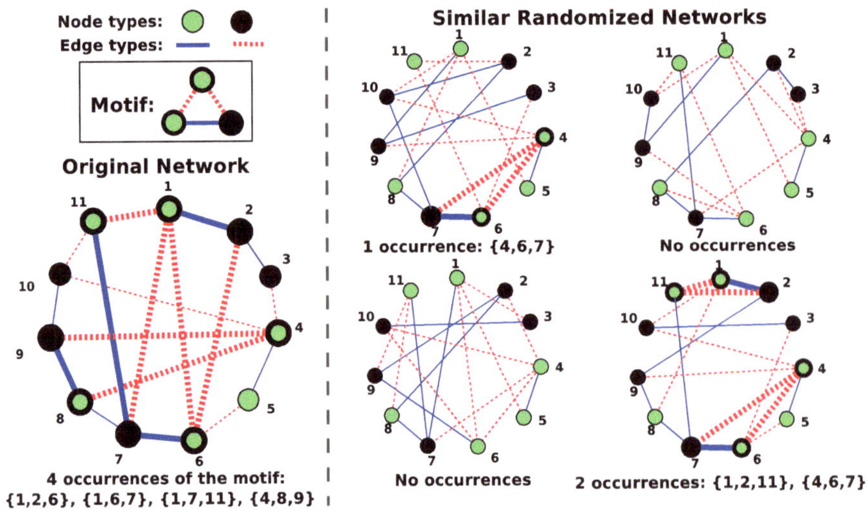

Fig. 1 An example colored network motif with 3 nodes

2.3 Related Work

There has been some work on motif detection using colors only in nodes [9], but not with colored edges, even if it was acknowledged that incorporating connection types would lead to richer results [6]. The work by Adami et al. [5] considered the case of colors only in nodes and uses an entropy-based measurement to determine significance. Quian et al. [10] also use colors solely in nodes and swap colors in the network, that is, the randomized networks are topological equivalent to the original, but with a different permutation of the colors in the nodes. Schbath et al. [11] also incorporate color in motif detection, but they only consider sets of connected colors, ignoring the exact connections between the respective nodes.

The only work we are aware of that directly supports motif finding with colors both in nodes and edges is the FanMod tool [12], which implements the ESU algorithm ESU [13]. Nevertheless, a clear formal definition of colored motif is not given. ESU was initially created for non-colored topological motif detection, a problem for which there are several possible methodologies. Kavosh [14], FaSE [15] and ESU itself are examples of network-centric approaches, where all subgraphs of a given size are counted by an enumeration followed by isomorphism tests to determine the subgraph class of each found occurrence. On the other hand, algorithms such as the one by Grochow and Kellis [16] only search for one subgraph at a time, with no re-usage of any kind of information between the computation of different subgraphs.

Here we use a set-centric approach, extending our own g-trie data structure [7], which previously did not account for color information. This methodology has been shown to be very competitive with running times that can outperform previous existing methods in the case of uncolored motifs [8].

In this paper we concentrate on exact algorithms, that are able to compute exact frequencies. There are however approximation techniques that trade accuracy for better running times. They are based on methodologies such as sampling [13, 17, 18]. The work we show in this paper can be further extended in the future to support such an approach.

3 Finding Colored Motifs with G-Tries

3.1 The Colored G-Trie Data Structure

A g-trie is a multiway tree (with a variable number of children per node) able to store and describe a set of graphs. To avoid further ambiguities, from now on we will use the term node for referring to g-trie tree nodes, and the term vertex to refer to the actual graph vertices stored in the g-trie. Each node contains information about a single vertex and connections to ancestor vertices. In order to support colors, we include in this information the color of the respective node and the color of each edge. A path from the root to a leaf node defines a subgraph.

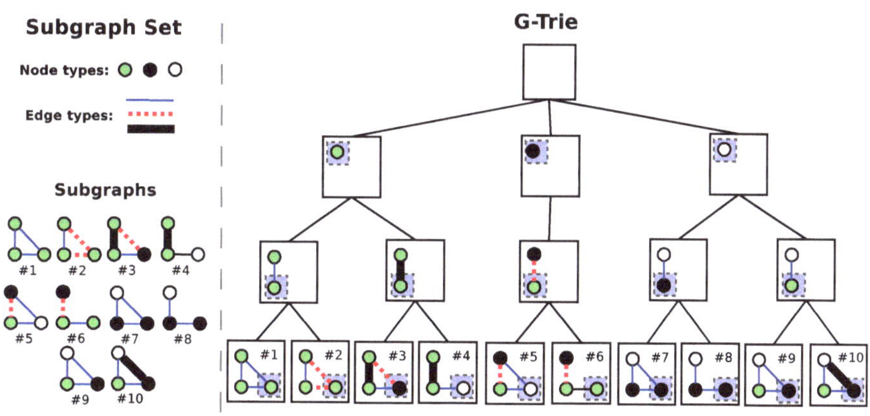

Fig. 2 An example colored g-trie containing 10 different subgraphs

Figure 2 exemplifies the concept, with a g-trie storing 10 colored subgraphs, with 3 different vertex colors and 3 different edge colors. Each g-trie node adds a new graph vertex (inside the small square) to the already existing ones in the ancestor nodes. Note how descendants of a node share a common colored subtopology.

3.2 Motif Discovery

Our methodology has two input parameters: a network to analyze and k, the size (in vertices) of the motifs we are looking for. The flow of the algorithm is, in its essence, the same as what was done originally by Milo et al [1]. First, compute the frequency of all k-sized subgraphs of the original network, i.e., a *subgraph census*. Secondly, compute the frequency of these subgraphs in the set of similar randomized networks. In the following sections we describe how these two steps are made.

3.3 Census in Original Network

The original g-trie algorithm for computing subgraph frequencies needs as input a set of subgraphs to store. When no colors are used and the size k is small, it is possible to use as input the set of all possible k-subgraphs. With colors, the number of different possibilities is much larger (due to the possible permutations between colors) and this option becomes unfeasible even for a relatively small k.

Given this, we opted to follow a network-centric approach for this initial step, i.e., we search for all occurrences of k connected nodes and identify their topological type. At the core, we need an algorithm for enumerating the sets of connected nodes. We opted to use ESU [13] as the base algorithm but in order to avoid a large number of isomorphism tests we incorporate the g-trie in the enumeration process, in a similar way to what was done with the FaSE algorithm [15].

Algorithm 1. Census of subgraphs with k vertices in the original network G.

1: T := new empty g-trie
2: **for all** $v \in V(G)$ **do**
3:     EXTENDSUBGRAPH($\{v\}, \{u \in N(v) : u > v\}, v, T$)
4: **for all** $n \in T.leaves()$ **do**
5:     $frequency[CanonicalLabel(n.Graph)]$ += $n.count$

6: **procedure** EXTENDSUBGRAPH(V_{Subg}, V_{Ext}, v, T)
7:     c := last vertex in V_{Subg} and its connections to previous nodes
8:     **if** T.notHasChild(c) **then** T.CREATECHILD(c)
9:     T := T.child(c)
10:     **if** $|V_{Subg}| = k$ **then** $T.count += 1$
11:     **else**
12:         **while** $V_{Ext} \neq \emptyset$ **do**
13:             remove any $w \in V_{Ext}$
14:             $V'_{ext} := V_{ext} \cup \{u \in N_{exclusive}(w, V_{subg}) : u > v\}$
15:             EXTENDSUBGRAPH($V_{Subg} \cup \{w\}, V'_{ext}, v, T$)

Algorithm 1 details our approach. At the core we are using the ESU algorithm (lines 2 to 3 and 10 to 15), which enumerates all k-subgraphs of a network once and only once. It works by maintaining two vertex lists: V_{Subg} and V_{ext}. The first one represents the current partial subgraph being constructed (set of connected vertices) and the latter a list of all neighbor vertices that can be added. Initially, it chooses each vertex v in the original graph G to be a possible starting point, and its neighbors as possible extensions (lines 2 and 3). Then it recursively removes each element of the extension list (line 13) and creates a new list of possible extensions (line 14). The usage of the exclusive neighborhood, along with the condition $u > v$, breaks symmetries and guarantees that each occurrence is only found once. When the size of V_{Subg} reaches k it means a new occurrence of a k-subgraph is found.

ESU execution naturally creates an implicit recursive search tree. At each time, one new node is added to the current partial subgraph, and it corresponds to a ramification in the g-trie. If that ramification already existis, we just update our current position in the g-trie (line 9). If not, we first create the ramification (line 8) and then follow that path. In the end, a g-trie path from the root to any leaf corresponds to a different node permutation of a certain graph type. We compute a canonical labeling for each leaf in order to identify its subgraph type (lines 4 and 5) but we avoid the need for that computation on all other occurrences of the same type whose node permutation corresponds to an automorphism, i.e., corresponds to the same path in the g-trie. Note that two leaves may be isomorphic, because the order in which vertices are traversed may implicitly define different paths. Ideally, one wants a single canon path in the g-trie for the same subgraph type but in here we trade memory for better running time, postponing the canonization for the randomized networks. The actual canonical labeling algorithm used is explained in section 3.5.

3.4 Census in the Randomized Networks

The bulk of the computation work is to discover the subgraphs frequency in the randomized networks. Typically one uses around 100 random networks [1], with each taking an amount of time similar to the time needed for the subgraph census in the original network. Since we know which subgraph types appeared in the original network, we limit our search to those. Our approach is to build a new g-trie containing only a single representation of each subgraph we are interested in. For that, we take all subgraphs found in the original network, apply the canonical labeling described in section 3.5, and insert such canonical representation in the g-trie.

Henceforth, we can use the new g-trie (with a single copy of each subgraph type) to greatly constrain the frequency calculation in the random networks. The core idea is to backtrack through all possible connected sets of nodes, and at the same time only follow the possibilities that map exactly to a g-trie path. We take advantage of common substructures identified in the g-trie in the sense that at a given time we have a partial isomorphic match for several different candidate subgraphs (all the descendants in the g-trie). We also use symmetry breaking conditions to further constrain the search and avoid redundant computation, in a similar way to what we have done with uncolored motifs [8].

Algorithm 2 details our method to count all occurrences of the g-trie colored subgraphs on a single randomized network. The idea is to find matches for all possible g-tries paths, i.e., all possible subgraphs. In the beginning we follow all g-trie root children and start with an empty partial match (lines 1 and 2). We then find all

Algorithm 2. Census of subgraphs of G-Trie T in graph G

1: **for all** children c of $T.root$ **do**
2:     MATCH(c, $\emptyset$)
3: **procedure** MATCH(T, V_{used})
4:     V := MATCHINGVERTICES(T, V_{used})
5:     **for all** node v of V **do**
6:         **if** isLeaf(T) **then** FOUNDMATCH($V_{used} \cup \{v\}$)
7:         **for all** children c of T **do** MATCH($c, V_{used} \cup \{v\}$)
8: **function** MATCHINGVERTICES(T, V_{used})
9:     **if** $V_{used} = \emptyset$ **then** $V_{cand} := \{v \in V(G) : v$ respects color of g-trie node$\}$
10:     **else** $V_{cand} := \{v \in N(V_{used}) : v$ respects color and sym. conditions$\}$
11:     $Vertices := \emptyset$
12:     **for all** $v \in V_{cand}$ **do**
13:         **if** $V_{used} \cup v$ respects connections of T **then** $Vertices := Vertices \cup \{v\}$
14:     **return** $Vertices$

candidate vertices to fill the position of that g-trie node (line 4). If we are at a g-trie leaf, we found a complete match to a subgraph and we can increment its frequency (line 6). If not, we continue as before, recursively following all possible g-trie paths from there, i.e., all subgraphs that may start with the current partial match (line 7). In order to find the candidate vertices (lines 8 to 14) we have a look at the neighbors of the current partial match (line 10) and only use those that respect the symmetry breaking conditions (line 10) and that at the same time respect the color connections with the vertices already in the partial match (line 13).

3.5 Canonical Labeling of a Colored Subgraph

For both previous algorithms we need to be able to produce a canonical labeling of a colored subgraph. This allows us to identify the subgraph type at each leaf in the g-trie that is dynamically constructed as we enumerate the subgraphs in the original network. At the same time, this canonical representation is used for choosing the path that will individually represent each isomorphic type in the g-trie that is used for searching in the randomized networks. The choice of this labeling will therefore directly interfere with the g-trie shape for this second part, including how much common-substructure is found, as explained in [8].

We opted to use the `GTCanon` labeling, which was created precisely for usage with g-tries [8]. At the core, this labeling uses `nauty`, a very efficient graph isomorphism program [19]. However, natively, `nauty` supports colors only in nodes and therefore we had to adapt it so that edge colors are also supported. In practice, for the labeling part, we convert each subgraph to an equivalent subgraph that uses colors only in vertices. The idea is to use several layers as is, for instance, exemplified in [20]. If we have k colors in the edges, we can substitute each vertex $v_j \in V(G)$ in the graph by a connected graph of k vertices $v_j^0, v_j^1, \ldots, v_j^k$ (we use cycles of size k). If there is an edge of color i between vertices v_j and $v_{j'}$, we add an edge between v_j^i and $v_{j'}^i$. This would imply a new graph with $|V(G)| \times k$ nodes, that is, with k layers. We reduce this to $|V(G)| \times log(k)$ by using the binary expansion of each color number and letting each layer represent a single bit in that binary representation.

4 Experimental Evaluation

In order to evaluate our approach, we implemented the described algorithms in C++. All experiments were run on an Intel Core 2 6600 (2.40GHz) with 2GB of memory. We use four different networks, with varied topological features, as summarized in Table 1. All networks are undirected and simple, i.e., with no self-loops or multiple edges between the same pair of nodes.

Table 1 The set of four networks used for experimentation

Name	Nodes	Edges	Avg. Degree	Nr of colors in: nodes	edges	Brief Description	Ref
`blogs`	1,490	16,715	22.4	2	1	links between political blogs	[21]
`dblp`	2,878	11,324	7.9	3	3	co-authorship of papers	[22]
`flights`	7,976	15,677	3.9	1	2	flights between airports	[23]
`elections`	8.297	100.753	24.3	2	2	elections for wikipedia admin	[24]

Given the nature of this paper, we always apply some kind of coloring in the networks used, even if the original dataset did not natively come with that color assignment. In `blogs`, all edges are of the same type and there are 2 colors in the nodes, indicating political leaning (left/liberal or right/conservative). By contrast, in `flights` all nodes are of the same type and there are 2 colors for the connections, indicating domestic and international flights. `dblp` is a co-authorship network in which we created 3 categories for the nodes (bottom 10%, top 1% and the rest) to differentiate the number of different co-authors each one has (the same was done for edges and number of co-authorships). Finally, `elections` is a signed network where there are 2 node colors (users being voted on and user casting votes) and 2 edge colors (supporting or opposing).

The only publicly available competing algorithm that performs a similar task is ESU, through its Fanmod tool [12], and therefore we directly compare our work with it, turning on colors on both nodes and connections. We experimented to compute motifs as we increase the size k of the subgraphs. We consider a typical usage of 100 randomized networks, generated with a markov chain process that keeps the same colored degree sequence of the original network, both on ESU and g-tries. We measured the execution time for each subgraph census, as shown in Table 2. For practical reasons, we limited the size k so that it would be feasible to run an entire motif computation with Fanmod.

Table 2 Experimental results

network	k	Execution Time (seconds)						Speedup
		ESU (via Fanmod)			G-Tries			G-Tries
		Original	Avg.Random	Total	Original	Avg.Random	Total	vs ESU
blogs	3	2.1	2.1	209.06	0.73	0.29	29.73	7.0x
	4	232.10	263.45	26,577.10	53.04	15.10	1,563.04	17.0x
dblp	3	0.50	0.25	25.50	0.15	0.02	2.15	11.9x
	4	8.11	11.80	1,188.11	1.90	0.17	18.90	62.9x
	5	276.03	479.57	48,233.03	70.02	5.50	620.02	77.8x
flights	3	1.59	1.63	164.59	0.48	0.05	5.48	30.0x
	4	139.36	187.00	18,839.36	35.01	4.23	458.01	41.1x
elections	3	23.02	33.55	3,378.02	7.51	1.70	177.51	19.0x
	4	6,987.34	7,434.25	750,412.02	800.86	256.68	26,468.85	28.4x

We can observe that our algorithm consistently outperforms ESU. The exact speedup is not easy to predict and highly depends on the topology of the analyzed network. Nevertheless, the results show a tendency to achieve better speedups as k increases, being one to two orders of magnitude faster than ESU. Given the nature of the associated computational task, the time to compute grows exponentially as k increases. However, even an increase of just one node in the size of the motifs that we are able to calculate may be very important, because patterns previously unknown may become "visible". Furthermore, faster execution times may allow the analysis of larger networks. Regarding the memory usage, the census on the original network is more expensive on our method than with ESU, since as explained in section 3.3 we opted to trade memory for better running time. However, it is possible to use any other algorithm for this initial census, including ESU itself. Moreover, the task that dominates the total running time is the census on the ensemble of random networks and in that part g-tries are not more costly in terms of memory than ESU, because they natively provide a compact representation of the set of subgraphs in which we can store their respective frequencies.

The main focus of this paper is the algorithm in itself, but we feel it is important to show the usefulness of colored motifs and figure 3 shows some of the motifs we found. The significance of a subgraph is computed as in the original Milo et al. paper [1]. For instance, a subgraph is overrepresented if the probability that it appears more often in the randomized networks than in the original network is smaller than a certain threshold. Here we are not worried about defining what the exact value of the threshold should be, but rather we want to show some general trends and our goal is to give the reader a feel of the kind of information colored motifs provide. For instance, without colors on both nodes and edges, the 5 patterns indicated for `dblp` would be indistinguishable. However, we can see that some of them are overrepresented (appearing more often than expected), some are underrepresented (appearing less than expected) and others are neutral (appearing in expected frequencies). The same happens on the other networks. Finding explanations for the depicted

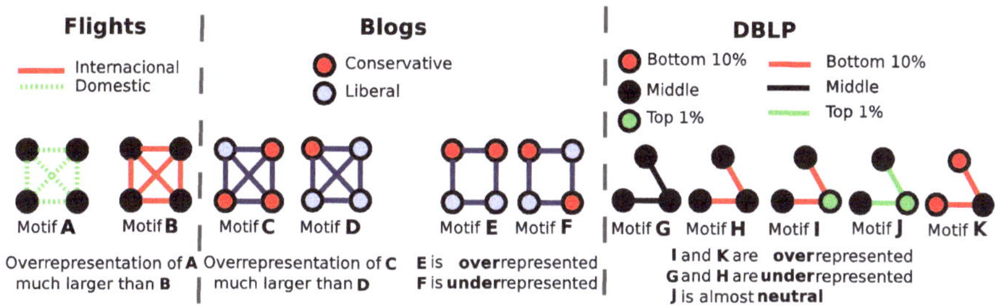

Fig. 3 Some examples of motifs found on the networks used for experimenting

frequencies would be another (interesting) task, but what remains is that by considering colors we have a richer mining environment, able to expose patterns that would be invisible when only using the traditional uncolored analysis.

5 Conclusions

In this paper we gave a definition of colored network motifs and described a complete methodology to find such characteristic patterns. The main computational problem we tackled was calculation of subgraphs frequency and at the core or our algorithms lies the g-trie data structure, responsible for representing sets of subgraphs. We adapted existing algorithms and combined them in order to produce a solution that we have shown to be not only feasible, but also substantially faster than a direct competing algorithm. We also tried to show some of the potential in the usage of colored motifs.

In the near future we plan to publicly release the software created for the described algorithms and to explore extensions of these algorithms such as an approximation version with sampling capabilities or a parallel approach. We also intend to apply the concept of colored motifs on several real-world networks, and to experiment with different types of randomized networks

Acknowledgements. Pedro Ribeiro is funded by an FCT Research Grant(SFRH/BPD/81695/2011). This work is partly funded by ERDF through the COMPETE Programme and by National Funds through FCT within project FCOMP-01-0124-FEDER-037281.

References

1. Milo, R., Shen-Orr, S., Itzkovitz, S., Kashtan, N., Chklovskii, D., Alon, U.: Network motifs: simple building blocks of complex networks. Science 298(5594), 824–827 (2002)
2. Albert, I., Albert, R.: Conserved network motifs allow protein-protein interaction prediction. Bioinformatics 20(18), 3346–3352 (2004)
3. Wu, G., Harrigan, M., Cunningham, P.: Characterizing wikipedia pages using edit network motif profiles. In: 3rd Int. Workshop on Search and Mining User-Generated Contents (SMUC), pp. 45–52. ACM, New York (2011)

4. Schbath, S., Lacroix, V., Sagot, M.: Assessing the exceptionality of coloured motifs in networks. EURASIP Journal on Bioinformatics and Systems Biology (2008)
5. Adami, C., Qian, J., Rupp, M., Hintze, A.: Information content of colored motifs in complex networks. Artificial Life 17(4), 375–390 (2011)
6. Yeger-Lotem, E., Sattath, S., Kashtan, N., Itzkovitz, S., Milo, R., Pinter, R.Y., Alon, U., Margalit, H.: Network motifs in integrated cellular networks of transcription-regulation and protein-protein interaction. Proc. of the National Academy of Sciences of the United States of America 101(16), 5934–5939 (2004)
7. Ribeiro, P., Silva, F.: G-tries: an efficient data structure for discovering network motifs. In: 25th ACM Symposium on Applied Computing (SAC), pp. 1559–1566. ACM (March 2010)
8. Ribeiro, P., Silva, F.: G-tries: a data structure for storing and finding subgraphs. Data Mining and Knowledge Discovery (2013)
9. Bruno, F., Palopoli, L., Rombo, S.E.: New trends in graph mining: Structural and node-colored network motifs. IJKDB 1(1), 81–99 (2010)
10. Qian, J., Hintze, A., Adami, C.: Colored Motifs Reveal Computational Building Blocks in the C. elegans Brain. PLoS ONE 6(3), e17013+ (2011)
11. Schbath, S., Lacroix, V., Sagot, M.F.: Assessing the exceptionality of coloured motifs in networks. EURASIP J. Bioinformatics and Systems Biology 2009 (2009)
12. Wernicke, S., Rasche, F.: Fanmod: a tool for fast network motif detection. Bioinformatics 22(9), 1152–1153 (2006)
13. Wernicke, S.: Efficient detection of network motifs. IEEE/ACM Transactions on Computational Biology and Bioinformatics 3(4), 347–359 (2006)
14. Kashani, Z., Ahrabian, H., Elahi, E., Nowzari-Dalini, A., Ansari, E., Asadi, S., Mohammadi, S., Schreiber, F., Masoudi-Nejad, A.: Kavosh: a new algorithm for finding network motifs. BMC Bioinformatics 10(1), 318 (2009)
15. Paredes, P., Ribeiro, P.: Towards a faster network-centric subgraph census. In: International Conference on Advances in Social Networks Analysis and Mining, pp. 264–271. IEEE (2013)
16. Grochow, J.A., Kellis, M.: Network motif discovery using subgraph enumeration and symmetry-breaking. In: Speed, T., Huang, H. (eds.) RECOMB 2007. LNCS (LNBI), vol. 4453, pp. 92–106. Springer, Heidelberg (2007)
17. Ribeiro, P., Silva, F.: Efficient subgraph frequency estimation with G-tries. In: Moulton, V., Singh, M. (eds.) WABI 2010. LNCS, vol. 6293, pp. 238–249. Springer, Heidelberg (2010)
18. Kashtan, N., Itzkovitz, S., Milo, R., Alon, U.: Efficient sampling algorithm for estimating subgraph concentrations and detecting network motifs. Bioinformatics 20(11), 1746–1758 (2004)
19. McKay, B.D., Piperno, A.: Practical graph isomorphism, {II}. Journal of Symbolic Computation 60, 94–112 (2013)
20. Kao, M.Y. (ed.): Encyclopedia of Algorithms. Springer (2008)
21. Adamic, L.A., Glance, N.: The political blogosphere and the 2004 U.S. election: divided they blog. In: 3rd International Workshop on Link Discovery (LinkKDD), pp. 36–43. ACM, New York (2005)
22. Kang, U., Papadimitriou, S., Sun, J., Tong, H.: Centralities in large networks: Algorithms and observations. In: SIAM International Conference on Data Mining, pp. 119–130 (2011)
23. Opsahl, T.: Why anchorage is not (that) important: Binary ties and sample selection (August 2011), `http://toreopsahl.com/2011/08/12/`
24. Leskovec, J., Huttenlocher, D., Kleinberg, J.: Signed networks in social media. In: SIGCHI Conference on Human Factors in Computing Systems, pp. 1361–1370. ACM (2010)

Structure Comparison of Binary and Weighted Niche-Overlap Graphs

Nayla Sokhn, Richard Baltensperger, Louis-Felix Bersier,
Ulrich Ultes-Nitsche, and Jean Hennebert

Abstract. In ecological networks, niche-overlap graphs are considered as complex systems. They represent the competition between two predators that share common resources. The purpose of this paper is to investigate the structural properties of these graphs considered as weighted networks and compare their measures with the ones calculated for the binary networks. To conduct this study, we select four classical network measures : the degree of nodes, the clustering coefficient, the assortativity, and the betweenness centrality. These measures were used to analyse different type of networks such as social networks, biological networks, world wide web, etc. Interestingly, we identify significant differences between the structure of the binary and the weighted niche-overlap graphs. This study indicates that weight information reveals different features that may provide other implications on the dynamics of these networks.

Keywords: Network Measures, Weighted Networks, Food-webs, Niche-Overlap Graphs.

1 Introduction

Complex systems have recently gained much interest. Many analyses have been conducted to understand the structure of these systems and to uncover their unique patterns [1, 2, 3]. Networks have emerged across many fields including biology, ecology, social networks [4, 5, 6] and many others. All these different networks were found to have a special architecture and a

Nayla Sokhn · Louis-Felix Bersier · Ulrich Ultes-Nitsche · Jean Hennebert
University of Fribourg, CH 1700 Fribourg, Switzerland

Nayla Sokhn · Richard Baltensperger · Jean Hennebert
University of Applied Sciences of Western Switzerland,
CH 1700 Fribourg, Switzerland

P. Contucci et al. (eds.), *Complex Networks V*,
Studies in Computational Intelligence 549,
DOI: 10.1007/978-3-319-05401-8_12,

particular behavior. It was shown that social networks belong to the small word property [7], known as the «six degrees of separation» phenomena. Food-webs and niche-overlap graphs turned out to follow a single scale exponential distribution [8, 9] while other networks such as the biology cells and the World Wide Web were found to follow a scale-free power law distribution [10]. In these real systems, two entities are connected if there is a relationship between them. For instance, in a social network, the relationship would be «being a friend with», in a food-web «feeding on a species», in a niche-overlap graph «competition between species». However, in order to have a better understanding of these networks, it is important to quantify the relationship between nodes. This is done by giving a weight to the links of the network. For example, in the scientific collaborator network, the weight is equal to the number of coauthored papers between two authors. For the world wide web network, the weight is defined by the load of data transferred between two hosts [11]. For niche-overlap graph, the weight is characterized by the number of common prey between two predators. In order to analyse weighted networks, researchers generalized some network measures by considering the weight of the links [12, 13]. Here, our aim is to first investigate the structure of weighted niche-overlap graph using four classical metrics: node degree, clustering coefficient, assortativity and betweenness centrality. We then compare the results with the ones obtained by analysing the binary niche-overlap graphs. To our knowledge, this is the first study that considers niche-overlap graphs as weighted networks and conducts an analysis to reveal their structure. The rest of the paper is organized as follows. Section 2 describes the food-webs and the niche-overlap graphs. Section 3 presents the structural properties used to inspect the binary and weighted networks. Section 4 illustrates and discusses the results. Finally, Section 5 concludes.

2 Ecological Networks: Food-Webs and Niche-Overlap Graphs

Food-webs are examples of ecological networks. They describe the interactions between consumers and resources. These complex systems are illustrated by a directed network. Nodes characterize species and directed links map the feeding connections between them. Other networks, namely niche-overlap graphs, are also examples of ecological networks. These graphs depict the competition between consumers. Two predators (consumer) are linked if they share at least one prey (resource). Niche-overlap graphs are drawn considering the information (who eat whom) retrieved from the food-webs. There are two different ways of using this information: (1) searching only for the common prey for each predator or (2) taking in consideration the number of common prey for each predator. In the second case, the weight $\omega_{i,j}$ assigned to each edge will be defined using the Jaccard index [14]:

$$\omega_{i,j} = \frac{|\text{prey}_i \bigcap \text{prey}_j|}{|\text{prey}_i \bigcup \text{prey}_j|}, \tag{1}$$

where prey_i and prey_j are the prey of predator i and j respectively.

These weights provide important information on the competition between predators. Two nodes might have the same number of links. However the strength of their links might be different.

3 Datasets and Network Measures

3.1 Datasets

We selected a collection of 15 real food-webs and built their corresponding niche-overlap graphs (Table 1). Weighted niche-overlap graphs were also generated to assess the comparison with the binary ones.

Table 1 Empirical food-webs and their associated niche-overlap graphs are presented by their name, order and size (number of links)

	Food-web		Niche-overlap			Food-web		Niche-overlap	
Graph	Order	Size	Order	Size	Graph	Order	Size	Order	Size
Chesapeake	33	71	27	95	Mangrove	90	1151	84	2148
Cypdry	68	468	53	855	LRL North Spring 2	144	2095	111	2520
Cypress	64	437	50	827	LRL North Summer	165	2706	121	3064
Cypwet	68	459	53	854	LRL North Winter	109	1257	86	1501
Everglades	63	617	58	1214	LRL South Winter	102	1328	83	1418
Gramdry	66	664	60	1267	LRL South Spring 1	151	2399	112	2965
Saint Martin	44	218	38	312	LRL South Summer	173	2901	119	3652
Mangrovedry	94	1210	86	2315					

3.2 Network Measures

In order to assess a comparison between the architecture of these binary and weighted graphs, we selected four classical network measures that were used to analyse different networks such as social networks, biological networks, world wide web networks and others [15, 16, 17]. These measures are presented below:

Degree: The degree D_v is the number of links that a node v has.

Weighted Degree: The node strength D_v^W is the sum of the weights of the links that a node v has.

By taking into consideration the strength of each link, we obtain additional information on the importance of the competition that a predator has. D_v^W is certainly lower than D_v since the weights $\omega_{i,j}$ are in the interval $[0,1]$.

Clustering Coefficient: The clustering coefficient C_v measures the tendency that the neighbors of a node v are linked to each other's. It is given by:

$$C_v = \frac{2E_v}{D_v(D_v-1)} = \frac{\sum_{j,h} a_{vj}a_{vh}a_{jh}}{D_v(D_v-1)}, \tag{2}$$

where a_{vj} is 1 if species v and j are connected (i.e. in competition) and 0 otherwise. The factor $\frac{D_v(D_v-1)}{2}$ is the potential number of links among the neighbors. E_v is the number of links among the neighbors of v i.e. the actual number of triangles in which node v participates: $\frac{1}{2}\sum_{j,h} a_{vj}a_{jk}a_{kv}$. The clustering coefficient of the whole network, is the average clustering coefficient C over all the nodes.

Weighted Clustering Coefficient: Many definitions of the weighted clustering coefficient have been proposed in the literature [13, 18, 19, 20, 21]. In this paper, we restrict our analysis on the following two definitions : the one proposed by Barrat et al. [13] which reflects how much of node strength is associated with adjacent triangle edges and the one proposed by Onnela et al. [21] which shows how large triangle weights are compared to network maximum.

Barrat et al. take into account only two links of the triangle:

$$C_v^{W(B)} = \frac{1}{s_v(D_v-1)} \sum_{j,h} \frac{(w_{vj}+w_{vh})}{2} a_{vj}a_{vh}a_{jh}, \tag{3}$$

where s_v accounts for the strength of node v:

$$s_v = \sum_j a_{vj}w_{vj}.$$

The factor $s_v(D_v-1)$ is the normalization factor to ensure that the weighted clustering is in the interval $[0,1]$ and $\frac{w_{vj}+w_{vh}}{2}$ is the weights' average of the links between node v and its neighbors j and h.

If $C^{W(B)} > C$, this shows that the interconnected triples are more likely to be created by the links with larger weights. If $C^{W(B)} < C$, this indicates that these triples are formed by the links with lower weights [22].

Onnela et al. consider all the three link weights of a triangle:

$$C^{W(O)} = \frac{2}{D_v(D_v-1)} \sum_{j,h} (\hat{w}_{vj}\hat{w}_{jh}\hat{w}_{hv})^{\frac{1}{3}} = C_v \bar{I}_v, \tag{4}$$

where $\hat{w}_{vj}$ is equal to $w_{vj}/\max(w)$. The actual number of triangles in which node v participates is replaced by the average intensity $\bar{I}_v$ of the triangle, which is the geometric mean of the links' weights $(\hat{w}_{vj}\hat{w}_{jh}\hat{w}_{hv})^{\frac{1}{3}}$ [11].

Betweenness Centrality: The betweenness centrality of a node v introduced by Freeman [23] identifies the number of shortest paths that passes through node v (denoted by $\sigma_{st}(v)$) among all the shortest paths (σ_{st}) in the network. This measure is given by:

$$BC_v = \sum_{s,t \neq v} \frac{\sigma_{st}(v)}{\sigma_{st}}. \tag{5}$$

If the betweenness centrality BC of a node v is equal to 0, it belongs to *only one complete subgraph* (a clique) of a graph G [24].

Weighted betweenness Centrality: The weighted betweenness centrality of a node v is calculated by taking into consideration the weights of the links in the network when finding the shortest path that passes through v, $\sigma^w_{st}(v)$ and the ones among the network σ^w_{st} :

$$BC^W_v = \sum_{s,t \neq v} \frac{\sigma^w_{st}(v)}{\sigma^w_{st}}. \tag{6}$$

In the weighted version of the betweenness centrality, if a species has a BC^W equal to 0, this does not ensure that it belongs to one unique clique.

Assortativity Coefficient: The assortativity coefficient R of a graph measures the tendency of degree correlation. It is calculated using the correlation coefficient of Pearson applied to the degrees of each node in the network.

It is defined as:

$$R = \frac{\frac{1}{M}\sum_{\Phi}(\prod_{v \in F(\Phi)} D_v) - (\frac{1}{2M}\sum_{\Phi}(\sum_{v \in F(\Phi)} D_v))^2}{\frac{1}{2M}\sum_{\Phi}(\sum_{v \in F(\Phi)} D_v^2) - (\frac{1}{2M}\sum_{\Phi}(\sum_{v \in F(\Phi)} D_v))^2}, \tag{7}$$

where M is the total number of links in the network, $F(\Phi)$ denotes the set of two nodes linked by the Φth link [15].

Weighted Assortativity Coefficient: The weighted assortativity coefficient R^W suggested by Leung et al. [25] is given by :

$$R^W = \frac{\frac{1}{H}\sum_{\Phi}(w_\Phi \prod_{v \in F(\Phi)} D_v) - (\frac{1}{2H}\sum_{\Phi}(w_\Phi \sum_{v \in F(\Phi)} D_v))^2}{\frac{1}{2H}\sum_{\Phi}(w_\Phi \sum_{v \in F(\Phi)} D_v^2) - (\frac{1}{2H}\sum_{\Phi}(w_\Phi \sum_{v \in F(\Phi)} D_v))^2}, \tag{8}$$

where H is the total weight of all links in the network and w_Φ denotes the weight of the Φth link.

If $R^W > R$, this implies that the links with a larger weights are pointing to the neighbors with larger degree. If $R^W < R$, this shows that the links with a larger weights are pointing to the neighbors with smaller degree [26].

4 Results and Discussion

The average degree of species of the 15 weighted niche-overlap graphs was significantly lower compared to the binary ones (Fig. 1 (a)). This indicates that even though species compete with many other species, they actually share few resources between them, thus providing weak links.

The distribution of the weighted clustering coefficient proposed by Barrat et al. $C^{W(B)}$ is slightly higher than the one for the binary clustering coefficient C whereas the one suggested by Onnela et al. $C^{W(O)}$ is considerably lower among the others (Fig. 1 (b)). The differences between both definitions comes from the fact that Onnela et al. take into account the weights between neighbors of node v and the weights of the edges between neighbors. On the other hand, Barrat et al. consider only the weights of the triangle forming the edges linked to node v but not the edges connecting the neighbors of v. Both weighted clustering coefficient ($C^{W(B)}$ and $C^{W(O)}$) provide us with complementary information. $C^{W(B)}$ being close the C yields to two conclusions : (1) the absence of correlation (randomized network), (2) the network is divided in two sets, one where triples are constituted by larger weights and others by smaller weights. $C^{W(O)}$ being significantly lower is due to the weight normalization by the global $\max(w)$ and to a broad distribution of weights in networks [27].

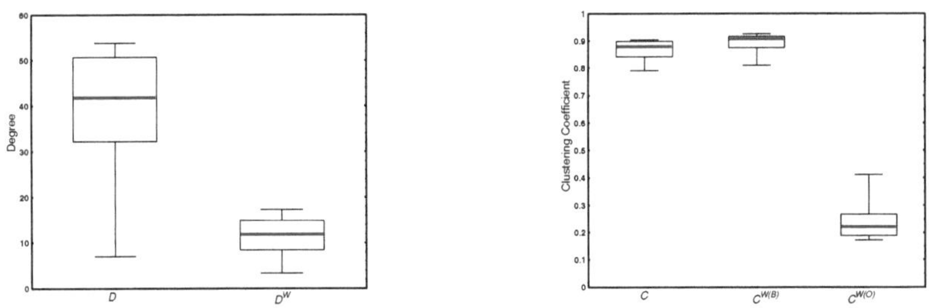

(a) (b)

Fig. 1 Box plots (minimum, quartiles and maximum) illustrating the distribution of degree and clustering coefficient respectively of the 15 niche-overlap graphs. Medians are indicated by red lines. D and D^W correspond to the binary and weighted degree respectively. C denotes the binary clustering coefficient. $C^{W(B)}$ and $C^{W(O)}$ the one proposed by Barrat et al. and Onnela et al. respectively.

The percentage of species with a betweenness centrality equal to 0 differed between the binary and the weighted niche-overlap graphs (Fig. 2 (a)). A higher number of species with a $BC^W = 0$ was detected in the weighted version. This points out that some species have a stronger competition (a high number of shared prey) among the others.

Interestingly, the assortativity coefficient for the weighted networks was positive whereas for the binary ones it was close to 0 and slightly negative (Fig. 2 (b)). This points out that by considering the strength of links, niche-overlap graphs reveal a fairly tendency to be assortative. This expresses that

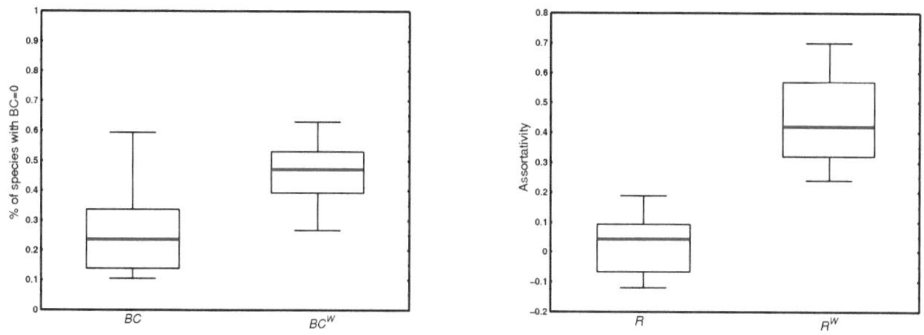

(a) (b)

Fig. 2 Box plots (minimum, quartiles and maximum) illustrating the distribution of betweenness centrality and assortativity respectively of the 15 niche-overlap graphs. Medians are indicated by red lines. BC and BC^W correspond to the binary and weighted betweenness centrality respectively. R and R^W correspond to the binary and weighted assortativity respectively.

predators with a high number of common prey tend to be connected with predators who also have a high number of common prey. Nevertheless, by considering simply the presence or absence of links (ignoring the weights), highlights a different assemblage of predators, indeed for binary niche-overlap graphs, predators tend to be linked randomly.

5 Conclusion

In this work, a set of 15 real networks was considered to conduct a comparison between the structure of the binary and weighted niche-overlap graphs. Our analysis showed significant differences between both structures indicating the influence of the weights on the architecture and on the assemblage of species. We believe that our study provides new insights and additional topological information on the structure of niche-overlap graphs studied in the context of foodwebs.

References

1. Albert, R., Barabási, A.L.: Statistical mechanics of complex networks. Reviews of Modern Physics 74(1), 47 (2002)
2. Newman, M.E.: The structure and function of complex networks. SIAM Review 45(2), 167–256 (2003)
3. Newman, M.E.: Fast algorithm for detecting community structure in networks. Physical Review E 69(6), 066133 (2004)
4. Little, J., Shepley, D., Wert, D.: Robustness of a gene regulatory circuit. The EMBO Journal 18(15), 4299–4307 (1999)

5. Dunne, J.A., Williams, R.J., Martinez, N.D.: Network structure and biodiversity loss in food webs: robustness increases with connectance. Ecology Letters 5(4), 558–567 (2002)
6. Girvan, M., Newman, M.E.: Community structure in social and biological networks. Proceedings of the National Academy of Sciences 99(12), 7821–7826 (2002)
7. Watts, D.J.: Small worlds: the dynamics of networks between order and randomness. Princeton University Press (1999)
8. Dunne, J.A., Williams, R.J., Martinez, N.D.: Small networks but not small worlds: unique aspects of food web structure. Proc. Nat. Acad. Sci. (2002)
9. Sokhn, N., Baltensperger, R., Hennebert, J., Ultes-Nitsche, U., Bersier, L.F.: Structure analysis of niche-overlap graphs. In: NetSci 2013 (2013)
10. Barabási, A.L., Albert, R., Jeong, H.: Scale-free characteristics of random networks: the topology of the world-wide web. Physica A: Statistical Mechanics and its Applications 281(1), 69–77 (2000)
11. Ioannis, A., Eleni, T.: Statistical analysis of weighted networks. arXiv preprint arXiv:0704.0686 (2007)
12. Newman, M.E.: Analysis of weighted networks. Physical Review E 70(5), 056131 (2004)
13. Barrat, A., Barthelemy, M., Pastor-Satorras, R., Vespignani, A.: The architecture of complex weighted networks. Proceedings of the National Academy of Sciences of the United States of America 101(11), 3747–3752 (2004)
14. Jaccard, P.: Distribution de la flore alpine dans la bassin de dranses et dans quelques regions voisines. Bulletin de la Societe Vaudoise des Sciences Naturelles 37, 241–272 (1901)
15. Newman, M.E.: Assortative mixing in networks. Physical Review Letters 89(20), 208701 (2002)
16. Adamic, L.A.: The small world web. In: Abiteboul, S., Vercoustre, A.-M. (eds.) ECDL 1999. LNCS, vol. 1696, pp. 443–452. Springer, Heidelberg (1999)
17. Freeman, L.C.: Centrality in social networks conceptual clarification. Social Networks 1(3), 215–239 (1979)
18. Zhang, B., Horvath, S., et al.: A general framework for weighted gene co-expression network analysis. Statistical Applications in Genetics and Molecular Biology 4(1), 1128 (2005)
19. Kalna, G., Higham, D.J.: Clustering coefficients for weighted networks. In: Symposium on Network Analysis in Natural Sciences and Engineering, p. 45 (2006)
20. Lopez-Fernandez, L., Robles, G., Gonzalez-B, J.M.: Applying social network analysis to the information in cvs repositories (2004)
21. Onnela, J.P., Saramäki, J., Kertész, J., Kaski, K.: Intensity and coherence of motifs in weighted complex networks. Physical Review E 71(6), 065103 (2005)
22. Boccaletti, S., Latora, V., Moreno, Y., Chavez, M., Hwang, D.U.: Complex networks: Structure and dynamics. Physics Reports 424(4), 175–308 (2006)
23. Freeman, L.C.: A set of measures of centrality based on betweenness. Sociometry, 35–41 (1977)
24. Grassi, R., Scapellato, R., Stefani, S., Torriero, A.: Betweenness centrality: extremal values and structural properties. In: Networks, Topology and Dynamics, pp. 161–175. Springer (2009)
25. Leung, C.C., Chau, H.F.: Weighted assortative and disassortative networks model. Physica A: Statistical Mechanics and its Applications 378(2), 591–602 (2007)

26. Barthélemy, M., Barrat, A., Pastor-Satorras, R., Vespignani, A.: Characterization and modeling of weighted networks. Physica A: Statistical Mechanics and its Applications 346(1), 34–43 (2005)
27. Saramäki, J., Kivelä, M., Onnela, J.P., Kaski, K., Kertesz, J.: Generalizations of the clustering coefficient to weighted complex networks. Physical Review E 75(2), 027105 (2007)

Core Decomposition in Directed Networks: Kernelization and Strong Connectivity

Vincent Levorato

Abstract. In this paper, we propose a method allowing decomposition of directed networks into cores, which final objective is the detection of communities. We based our approach on the fact that a community should be composed of elements having communication in both directions. Therefore, we propose a method based on digraph kernelization and strongly p-connected components. By identifying cores, one can use based-centers clustering methods to generate full communities. Some experiments have been made on three real-world networks, and have been evaluated using the V-Measure, having a more precise analysis through its two sub-measures: homogeneity and completeness. Our work proposes different directions about the use of kernelization into structure analysis, and strong connectivity concept as an alternative to modularity optimization.

1 Introduction

Complex networks appear in many applications, including social networks analysis on the Web, which is a topical research subject. These networks carry non-trivial topological properties that characterize their connectivity, and affect the dynamics of their behavior. The analysis of complex networks often leads to the analysis of the roles of elements, or groups of elements, composing a network. Communities detection belongs to this research field, and can be very useful to better understand how networks are structured. In this article, we focus on the problem of finding communities in networks, and more specifically finding cores in *directed networks*. Dealing with methods of community detection for directed networks is a difficult task, and few methods exist compared to methods used in the undirected

Vincent Levorato
CESI, 959 rue de la Bergeresse, 45160 OLIVET, FR
e-mail: vlevorato@cesi.fr

Université d'Orléans, LIFO, Bat. IIIA rue Léonard de Vinci, 45067 ORLEANS, FR
e-mail: vincent.levorato@univ-orleans.fr

P. Contucci et al. (eds.), *Complex Networks V*,
Studies in Computational Intelligence 549,
DOI: 10.1007/978-3-319-05401-8_13,

case. Here are some of the most known works in the literature [9, 17, 13] dedicated to directed networks, or which can be adapted to work with directiveness: *Clauset et al.* method [5], *CFinder* based on the Clique Percolation Method (CPM) [21], *Louvain* method [3], *InfoMap* [23], *Simulated Annealing* for modularity [11], *Wu-Huberman* method [31], *MarkovCluster* algorithm [29], *Multistep Greedy* algorithm [24] and *EM* method (Expectation-Maximization) [18]. More recently, others methods concerning directed networks have been proposed, with more or less good results [32, 15, 14].

Generally, these methods are most of the time designed for undirected networks, and adjusted to work in the directed case: they are *not initially dedicated to the directed case*. There is also a significant amount of methods using modularity optimization. However, this kind of approach has its limits, and can *"miss important substructures of a network"* [10]. Some recent work discuss about *reciprocated interaction* [4], that two people should communicate in both directions, the first person expecting messages from the second person, and vice versa. Our approach is based on this simple idea: in a directed network, *a community should be composed of nodes which can communicate with every nodes in the community, in both directions*. Interesting results in our previous exploratory work [19] encourage us to continue in this direction. Usually represented by graphs in undirected networks, this kind of representation can be modeled by the *connected component* concept, and more restrictively by the *clique* concept. Except that for the directed case, it can be represented by the concept of *strongly connected component*. Finding these components should be equivalent to find *cores* to which other elements of the network will be assigned. This work gives the key concepts of this approach, focusing on the cores finding.

Our paper is structured as such: the first section gives the definitions of graph theory and formal concepts needed to understand our method. Then, the second part exposes the different steps of our approach, followed by some experimental results on real networks. The paper ends by a conclusion which opens discussion on future work directions.

2 Graph Theory Notions

2.1 Graph Definitions

In this article, we consider only *directed graphs* (also noted *digraph*). We give here a short reminder of graph theory notions. Formally, a digraph $G = (V,A)$ is the pair composed of [2]:

- a set $V = \{x_1, x_2, ..., x_n\}$ named *vertices* or *nodes*.
- a family $A = (a_1, a_2, ..., a_n)$ of elements of the Cartesian product $V \times V = \{(x,y)/x \in V, y \in V\}$ named *arcs*.

The amount of vertices is noted n (also noted $|V(G)|$) and the amount of arcs is noted m (also noted $|A(G)|$). A *path* P is composed of k arcs such as $P = (a_1, a_2, ..., a_i, ..., a_k)$ where for every arc a_i the terminal end coincides with the initial end of a_{i+1}. Several equivalent notations can be used: $P = ((x_1,x_2),(x_2,x_3),...) = [x_1,x_2,...,x_k,x_{k+1}] = P[x_1,x_{k+1}]$. A *chain* is, like a path, an alternating sequence of vertices and edges, where an edge is an arc without orientation. A *circuit* is a path such that the first node of the path corresponds to the last. It can be viewed as an oriented cycle.

2.2 Connected Components

Here are the different types of connected components we could have in a directed graph [12]:

- a *weakly connected component* WCC of a digraph is a subgraph where: $\forall x,y \in WCC$, there is a chain between x and y.
- an *unilaterally connected component* UCC of a digraph is a subgraph where: $\forall x,y \in UCC$, there is a path between x and y OR there is a path between y and x.
- a *strongly connected component* SCC of a digraph is a subgraph where: $\forall x,y \in SCC$, there is a path between x and y AND there is a path between y and x.

To discover cores in a network, we use a special case of strongly connected component named *strongly p-connected component* by [30] which is related to l-edge-connectivity [7] (we use *p-connected* notation instead of *n-connected* to avoid confusion):

- a *strongly p-connected component* p-SCC of a digraph is a subgraph where: $\forall x,y \in p$-SCC, there is a path of length p or less between x and y, and there is a path of length p or less between y and x, with $p \geq 2$.

3 Core detection

3.1 Related Work

Finding cores in order to find communities is a method that can be related to *pattern* identification [13]. This consists in finding maximal subsets which implies separation between them. Clique finding is one of these methods, but is also very restrictive, because each node must have a direct connection to other nodes. This approach has been relaxed by the n-clique definition where each node is connected to others by at least one path which length is at most n, but that can be outside of the n-clique. The n-clan concept fixes the connectedness issue of the n-clique [20].

Our approach is related to these works, but we don't put strong constraint on the size (triads), and we don't want to avoid circuits (directed k-clique), as it is specifically the configuration we are looking for: strongly p-connected components.

3.2 Searching for Cores Using p-SCC Concept

Our community definition refers to a group containing elements that can communicate with all other elements of the group, corresponding in digraphs, to strongly connected components (SCC) concept. Tarjan based his algorithm on the search of circuits [27]. To our knowledge, no work has considered the SCCs in the case of researching communities. By simply applying Tarjan's algorithm to directed graph generated through LFR benchmark [16], some communities can be found, but SCCs are often oversized. To refine the process, our approach proposes to find p-SCCs (fig. 1). The problem is that in a digraph, the number of circuits may be exponential in the number of vertices [26]. Therefore, processing all circuits of a graph is not relevant, especially if the graph has a significant number of nodes like in large real-world networks.

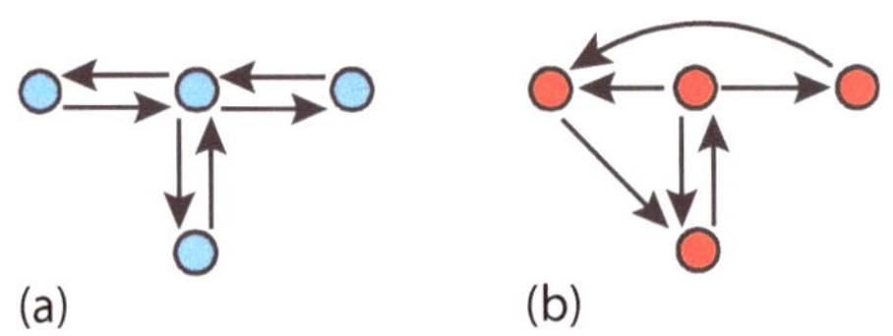

Fig. 1 Examples of p-SCCs: *(a)*: nodes are connected by paths of length at most 2 (2-SCC) *(b)*: nodes are connected by paths of length at most 3 (3-SCC)

Starting from a node s, we look for p-SCC by searching for circuits, but circuits with a given size. Trivially, the length of the path p is bounded by the length of the circuits found into the p-SCC :

$$p \leq 2 \times (c-1)$$

where c is the size of circuits we are searching for.

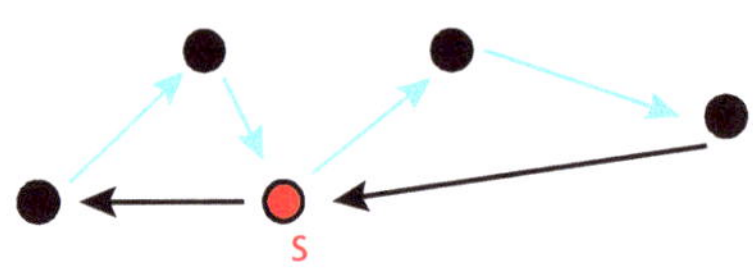

Fig. 2 Searching p-SCC is similar to search circuits from a starting node s. In this case, searching for 4-SCC means searching for circuits of length at most 3. The highlighted path length is 4, the maximal path length which can be found.

For instance, searching for circuits of length 3 starting from a node means searching for at most 4-SCCs (fig. 2). We propose an algorithm which returns a p-SCC starting from a given node (alg. 1). As the algorithm is searching for circuits, and considering that p parameter sets the circuit length, we can only find p-SCCs with p being an even number. It is written in a non-recursive way, but time complexity should be approximatively the same as Tarjan's algorithm, which is $O(n+m)$, with two differences: we don't always need to pass through every arc (depends on path length), but we should pass through nodes several times.

Input: *G*: digraph, *s*:starting node, *p*:path length (even integer)
Data: *astack*: stack of arcs, *vpath*: stack of nodes, *c*: integer (circuit size)
Remark: *Aout(k)* represents set of outing arcs of the node *k*.
source(a) represents the source node of the arc *a*, *dest(a)* represents the destination node of the arc *a*.
Result: *C*:set of nodes (*p*-SCC)

```
C ← ∅; C ← C ∪ {s};
vpath.push(s); c ← (p+2)/2;
foreach a in Aout(s) do
    astack.push(a);
end
while astack ≠ ∅ do
    a ← astack.pop();
    w ← vpath.peek();
    if source(a) ≠ w then
        while source(a) ≠ w do
            w ← vpath.pop();
        end
        vpath.push(w);
    end
    z ← dest(a);
    if z = s then
        C ← C ∪ vpath;
    end
    else
        if |vpath| < c then
            foreach b in Aout(z) do
                astack.push(b);
            end
            vpath.push(z);
        end
    end
end
return C;
```

Algorithm 1. Algorithm for extracting the p-SCC.

3.3 *Digraph Kernelization*

Before describing the whole method based on core detection, we show how we can optimize the search for p-SCCs by "cleaning" the digraph, meaning excluding nodes and arcs which should never belong to a circuit: this brings us to the notion of *graph kernelization*. We are interested in the kernelization which is used in the FVS problem (Feedback Vertex Set) [28]. As we work in the directed case, we used the kernelization technique applied to the directed FVS [8], following the four first rules of the method (it removes self-loop, multiple arcs, isolated nodes, and recursively chained nodes with only outgoing or incoming arcs).

4 Digraph Cores Decomposition Method

This section describes our method for core detection in directed networks. Let use the following notations: G is the input digraph (network), and $\mathscr{K}$ is the set of output cores. The method follows these steps, considering a given p:

1. Kernelize G.
2. For each node of G as the starting node, process p-SCC.
3. Sort p-SCCs by size. Starting from the biggest one, put them one by one in $\mathscr{K}$ if it doesn't intersect existing cores already inserted into $\mathscr{K}$. In case of cores having the same size, take the most connected one (biggest amount of arcs).
4. (optionnal) Remove p-SCCs with size inferior to a given threshold K_{min}.

Illustration: The figure 3 gives an illustration of our method, step by step, with $p = 4$ (meaning we search for circuits of length at most 3). Let take a digraph (a), and apply the first step which is kernelization (b). Some nodes are ignored, and won't be considered. The second step processes 4-SCCs, node by node: in the example (c), only five iterations are represented (nodes with labels 4 ,5 ,8 ,10 ,13), and for each node, a 4-SCC is computed, which can be the same for several nodes (nodes 5 and 8 produce the same 4-SCC, same thing for nodes 4 and 13). The last step (d) extracts the biggest and non-intersecting 4-SCCs giving the final result with 3 cores.

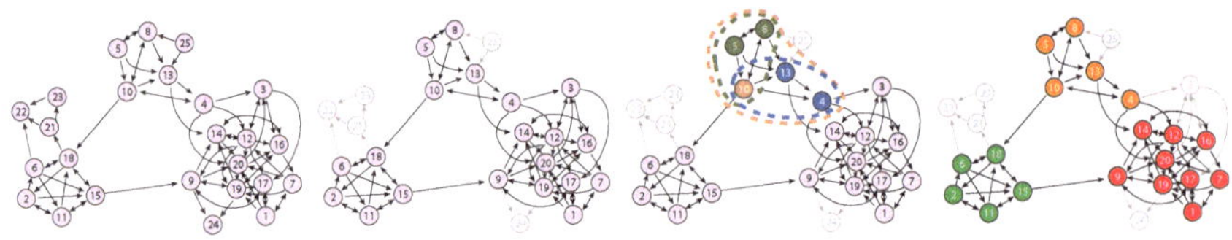

(a) Input digraph (b) Kernelization (c) Processing the 4−SCCs (d) Cores result

Fig. 3 Illustration of the decomposition method of a digraph into cores

This method returns a set of cores which can be used to cluster the remaining nodes of the network to have a complete clustering. As some clustering methods like k-means algorithm, the number of communities is set by the number of cores. In this article, as we don't focus on clustering methods, our experiments use a simple aggregative method like center-based clustering methods, and assign each node to the community having the nearest core, in term of *chain* length. The K_{min} value can be useful to avoid too small cores that shouldn't be considered.

5 Experiments

Validating communities structures corresponds to validate a clustering method. The difficulty is to find an objective measure of quality of clusters. For our experiments, we use *V-Measure* which is an alternative to F-Measure, and *Normalized Mutual Information* from the information theory field to compare the clustering obtained by our method to the reference classes. We based our experiments on real data networks, as some experiments have already been done on generated networks (LFR Benchmarks [16]) in our previous work with good results [19]. Moreover, there is an issue with the LFR Benchmark, as *it produces graphs already kernelized*, which puts a strong constraint on generated graphs. On the contrary, in the experiment part, we observe that the real-world networks we used are strongly kernelizable.

5.1 Clustering Evaluation

The entropy notion is used to express the used measures, and is noted as follow, with X and Y two discrete random variables: $H(X)$ and $H(Y)$ for the marginal entropies, $H(X|Y)$ and $H(Y|X)$ for the conditional entropies, $H(X,Y)$ for the joint entropy. Measures give results between 0 (worst matching) and 1 (best matching). Here are the two evaluation measures definitions:

- **V-Measure** [22] is an entropy based-evaluation measure, composed of two concepts: *completeness* and *homogeneity*. With C a set of classes (reference), K a set of clusters (unsupervised method), the homogeneity is defined as:

$$h = \begin{cases} 1 & if\ H(C,K) = 0 \\ 1 - \frac{H(C|K)}{H(C)} & else \end{cases}$$

  and the completeness is defined as:

$$c = \begin{cases} 1 & if\ H(K,C) = 0 \\ 1 - \frac{H(K|C)}{H(K)} & else \end{cases}$$

  A clustering result satisfies homogeneity if all of its clusters contain only elements which are members of a single class, and a clustering result satisfies completeness if all the elements that are members of a given class are elements of the same cluster. The V-Measure is based on homogeneity and completeness scores such as:

$$V_\beta = \frac{(1+\beta)\cdot h \cdot c}{(\beta \cdot h) + c}$$

  with a β parameter which can be used to weight homogeneity and completeness scores. In our experiments, we set $\beta = 1$ (balanced weights).

- **Normalized Mutual Information** (NMI) [6] Mutual information is a measure used in information theory domain, giving the amount of information that one random variable contains about another. This measure is defined between the cluster assignments K and a pre-existing labeling set of classes C normalized by:

$$NMI(K,C) = \frac{I(K,C)}{\sqrt{H(K)H(C)}}$$

with $I(K,C)$ the mutual information of K and C such that $I(K,C) = H(K) - H(K|C)$.

5.2 Results

In order to test our approach, we used directed network datasets already known in the literature [1, 25]. Three networks have been used (see tab. 1). Several characteristics of the network are given like density, degree information, communities maximum and minimum sizes, but also the mixing parameter μ [17] and the directed modularity Q_d [18].

Table 1 Network datasets

Directed Network	*Political Blog*	*Cora*	*Citeseer*
$\|V\|$	1,222	2,485	2,120
$\|A\|$	19,024	5,209	3,768
Classes	2	7	6
Density	1.27%	0.08%	0.08%
Degrees	$k_{mean} = 31$ $k_{min} = 1$ $k_{max} = 467$	$k_{mean} = 4$ $k_{min} = 1$ $k_{max} = 169$	$k_{mean} = 4$ $k_{min} = 1$ $k_{max} = 100$
Communities size	$\|C\|_{min} = 588$ $\|C\|_{max} = 636$	$\|C\|_{min} = 131$ $\|C\|_{max} = 726$	$\|C\|_{min} = 115$ $\|C\|_{max} = 532$
μ	0.09	0.18	0.28
Q_d	0.41	0.63	0.51

5.2.1 Core Decomposition

In tab. 3, we compare the obtained cores with the reference classes, using the p parameter which corresponds to the path length of a p-SCC, and the K_{min} parameter which is the minimum core size (only relevant results are shown). In most cases, cores have a good completeness score, meaning that we succeed in having nodes which belong to a single class in only one core. On the other hand, the homogeneity score tends to be better when the threshold of the minimum core size is increased (Political Blog and Cora networks), having nodes of a same core belonging to a single class. The interpretation that can be made from these results is that the more the graph is compressed, the less the K_{min} gets an high value. When too many nodes are available to build cores, the K_{min} threshold has to be high to remove some eventual noise, giving less nodes usable in the core creation. In the results of tab. 3, we first consider completeness value, and then the homogeneity value. We give more importance to the completeness score, as it gives better results in the final process communities detection.

Table 2 Network kernel sizes

Directed Network	*Political Blog*	*Cora*	*Citeseer*
$\|V\|_K$	811	2,485	2,120
$\|A\|_K$	15,833	5,209	3,768
Compression rate (nodes)	33%	84%	97%

Table 3 Cores detection on real-world networks

(a) Political Blog

		V-Measure				
p	K min size	h	c	V	Nb Nodes	Nb Cores
2	2	0.35316	0.95295	0.51534	329	20
2	3	0.40471	0.94876	0.56739	308	13
2	4	0.44344	0.94601	0.60383	296	10
2	5	0.52053	0.96137	0.67538	281	7
2	6	0.59364	0.97468	0.73787	269	5
2	7	0.7381	**1.0**	0.84932	255	3
2	**16**	**1.0**	**1.0**	1.0	239	2
4	2	0.60926	0.98967	0.75421	401	12
4	3	0.79398	0.98896	0.88081	380	5
4	4	0.90979	**1.0**	0.95276	372	3
4	**5**	**1.0**	**1.0**	1.0	367	2

(b) Cora

		V-Measure				
p	K min size	h	c	V	Nb Nodes	Nb Cores
4	2	**0.41019**	**0.9412**	0.57137	98	28
4	3	0.57836	0.9352	0.71471	47	11
6	2	0.41481	0.93729	0.5751	103	28
6	3	0.58141	0.92778	0.71485	52	11
6	4	0.70183	0.92268	0.79724	32	6

(c) Citeseer

		V-Measure				
p	K min size	h	c	V	Nb Nodes	Nb Cores
2	1	0.49039	0.93415	0.64315	54	26
2	2	0.19087	0.29364	0.23136	6	2
4	**1**	**0.48994**	**0.93454**	0.64285	58	26
4	2	0.5907	0.77251	0.66948	12	3

5.2.2 Communities Detection

Using an aggregative method, the cores first absorb nodes which are in the kernel but not in the cores, giving pre-built communities. Then, the nodes outside the kernel are absorbed by the pre-built communities to give a final clustering of communities. Clusters are not strongly connected, but unilaterally connected. The results in tab. 4 show that even with a naive method of clustering, the communities structures remain "acceptable". The cores used in the final clustering process are the cores having the best completeness scores in the core decomposition operation. Observing the results, we can make the assumption that the compression of graphs impacts the quality of cores, and therefore the detection of communities. With a small amount of nodes in the kernel, the choice to make between the nodes to build the cores is important, as it determines the final process of communities detection. Also, having only big cores means setting a too high K_{min} threshold value, which can have a negative impact on the cores detection, and some communities cannot be found in the process. For instance, in fig. 4, the central community has been found by our method using the parameters $p = 4, K_{min} = 2$. If we set $K_{min} = 3$, cores of size 2 are excluded, and this central community is not detected.

Table 4 Real-world networks communities detection results based on core decomposition

Network	Measures				Amount of Communities
	h	c	V	NMI	
Political Blog	0.70385	0.69929	0.70156	0.70116	2
Cora	0.35335	0.46349	0.40099	0.40469	28
Citeseer	0.28162	0.38734	0.32613	0.32742	26

(a) Clustering using an aggregative method based on core decomposition.

(b) Reference classes.

Fig. 4 Communities detection comparison on the Cora citation network ($p = 4, K_{min} = 2$)

6 Conclusion

In this article, we focused on an approach dedicated to directed networks, and we gave a method allowing the decomposition of these networks into cores. These cores can be used by any clustering method based on centers to detect communities. Our various contributions can be presented as follows:

- We provide a simple and efficient algorithm to generate these p-SCCs in a digraph. This approach can be classified in the *pattern identification* category that we can find in some method classification, while being flexible enough.
- The interest of using kernelization process has been highlighted : it reduces the core detection process, and can give some information on the network structure.
- An important thing about these results is that we didn't take into account the modularity concept in our approach. As a large part of the communities detection algorithms are dedicated to modularity optimization [13], we want to stress the point that we can have interesting results in communities detection without this concept.

Several options can be considered for the continuation of this work. As we said, we have to apply our method to others real-world datasets. We should also study how to increase the quality of the core detection, and it could be interesting to have the possibility to automatically fix the K_{min} threshold value. Testing other based-centers clustering methods should be done too. Also, the case of overlapping communities should be considered, as our approach could be quickly adaptable with p-SCCs which naturally overlap each other. In our opinion, our work points out that no clear or unanimous consensus about the definition of communities exists, and provides a new point of view on the detection of communities into directed networks, being omnipresent in the Web nowadays.

Acknowledgements. We would like to thank Pr. Ioan Todinca, Mathieu Liedloff, Anthony Perez and the GAMoC team from the LIFO laboratory at University of Orléans (FR) for their listening and advices. Special thanks to Guillaume Cleuziou from the Constraints and Machine Learning team.

References

1. Adamic, L.A., Glance, N.: The political blogosphere and the 2004 u.s. election: divided they blog. In: Proceedings of the 3rd International Workshop on Link Discovery (LinkKDD 2005), pp. 36–43. ACM, New York (2005)
2. Berge, C.: Graphes et Hypergraphes. Dunod, Paris (1970)
3. Blondel, V.D., Guillaume, J.-L., Lambiotte, R., Lefebvre, E.: Fast unfolding of communities in large networks. Journal of Statistical Mechanics (10) (2008)
4. Cheng, J., Romero, D.M., Meeder, B., Kleinberg, J.M.: Predicting reciprocity in social networks. In: SocialCom/PASSAT, pp. 49–56 (2011)
5. Clauset, A., Newman, M.E.J., Moore, C.: Finding community structure in very large networks. Phys. Rev. E 70(6), 066111 (2004)
6. Danon, L., Díaz-Guilera, A., Duch, J., Arenas, A.: Comparing community structure identification. Journal of Statistical Mechanics: Theory and Experiment (9), P09008–P09008 (2005)
7. Diestel, R.: Graph Theory, 4th revised edn. Springer (July 2010)
8. Fleischer, R., Wu, X., Yuan, L.: Experimental study of fpt algorithms for the directed feedback vertex set problem. In: Fiat, A., Sanders, P. (eds.) ESA 2009. LNCS, vol. 5757, pp. 611–622. Springer, Heidelberg (2009)
9. Fortunato, S.: Community detection in graphs. Physics Reports 486(3-5), 74–174 (2010)
10. Fortunato, S., Barthélemy, M.: Resolution limit in community detection. Proceedings of the National Academy of Science 104(1), 36–41 (2006)
11. Guimerà, R., Nunes Amaral, L.A.: Functional cartography of complex metabolic networks. Nature 433, 895–900 (2005)
12. Harary, F.: Graph Theory. Addison-Wesley, Reading (1969)
13. Labatut, V., Balasque, J.-M.: Detection and interpretation of communities in complex networks: Practical methods and application. In: Computational Social Networks: Tools, Perspectives and Applications, pp. 81–113 (2012)
14. Lai, D., Lu, H., Nardini, C.: Finding communities in directed networks by pagerank random walk induced network embedding. Physica A: Statistical Mechanics and its Applications 389(12), 2443–2454 (2010)

15. Lambiotte, R., Sinatra, R., Delvenne, J.-C., Evans, T.S., Barahona, M., Latora, V.: Flow graphs: Interweaving dynamics and structure. Phys. Rev. E 84, 017102 (2011)
16. Lancichinetti, A., Fortunato, S.: Benchmarks for testing community detection algorithms on directed and weighted graphs with overlapping communities. Physical Review E 80(1), 1–8 (2009)
17. Lancichinetti, A., Fortunato, S.: Community detection algorithms: A comparative analysis. Phys. Rev. E 80, 056117 (2009)
18. Leicht, E.A., Newman, M.E.J.: Community structure in directed networks. Phys. Rev. Lett. 100(11), 118703 (2008)
19. Levorato, V., Petermann, C.: Detection of communities in directed networks based on strongly p-connected components. In: IEEE International Conference on Computational Aspects of Social Networks (CASoN), pp. 211–216 (October 2011)
20. Mokken, R.J.: Cliques, clubs and clans. Quality & Quantity 13(2), 161–173 (1979)
21. Palla, G., Derényi, I., Farkas, I., Vicsek, T.: Uncovering the overlapping community structure of complex networks in nature and society. Nature 435, 814–818 (2005)
22. Rosenberg, A., Hirschberg, J.: V-measure: A conditional entropy-based external cluster evaluation measure. In: Joint Conference on Empirical Methods in Natural Language Processing and Computational Natural Language Learning (EMNLP-CoNLL), pp. 410–420 (2007)
23. Rosvall, M., Bergstrom, C.T.: Maps of random walks on complex networks reveal community structure. In: Proceedings of the National Academy of Sciences USA, pp. 1118–1123 (2008)
24. Schuetz, P., Caflisch, A.: Efficient modularity optimization by multistep greedy algorithm and vertex mover refinement. Phys. Rev. E 77, 046112 (2008)
25. Sen, P., Namata, G.M., Bilgic, M., Getoor, L., Gallagher, B., Eliassi-Rad, T.: Collective classification in network data. AI Magazine 29(3), 93–106 (2008)
26. Tarjan, R.: Enumeration of the Elementary Circuits of a Directed Graph. SIAM Journal on Computing 2(3), 211–216 (1973)
27. Tarjan, R.E.: Depth-first search and linear graph algorithms. SIAM Journal on Computing 1(2), 146–160 (1972)
28. Thomassé, S.: A quadratic kernel for feedback vertex set. In: Proceedings of the Twentieth Annual ACM-SIAM Symposium on Discrete Algorithms (SODA 2009), Philadelphia, PA, USA, pp. 115–119. Society for Industrial and Applied Mathematics (2009)
29. Van Dongen, S.: Graph clustering via a discrete uncoupling process. SIAM. J. Matrix Anal. & Appl. 30(1), 121–141 (2008)
30. Wasserman, S., Faust, K.: Social Network Analysis: Methods and Applications. Cambridge University Press (1994)
31. Wu, F., Huberman, B.A.: Finding communities in linear time: A physics approach. The European Physical Journal B Condensed Matter 38(2), 331–338 (2003)
32. Yang, T., Chi, Y., Zhu, S., Jin, R.: Directed network community detection: A popularity and productivity link model. In: SDM 2010: Proceedings of the 2010 SIAM International Conference on Data Mining (2010)

Network Disruption and Recovery: Co-Evolution of Defender and Attacker in a Dynamic Game*

Holly Arnold, David Masad, Giuliano Andrea Pagani, Johannes Schmidt, and Elena Stepanova

Abstract. The evolution of interactions between individuals or organizations are a central theme of complexity research. We aim at modeling a dynamic game on a network where an attacker and a defender compete in disrupting and reconnecting a network. The choices of how to attack and defend the network are governed by a Genetic Algorithm (GA) which is used to dynamically choose among a set of available strategies. Our analysis shows that the choice of strategy is particularly important if the resources available to the defender are slightly higher than the attackers'. The best strategies found through GAs by the attackers and defenders are based on betweenness centrality. Our results agree with previous literature assessing strategies for network attack and defense in a static context. However, our paper is one of the first ones to show how a GA approach can be applied in a dynamic

Holly Arnold
University of Oregon, Eugene, Oregon, USA
e-mail: arnold3@uoregon.edu

David Masad
George Mason University, Fairfax, Virginia, USA
e-mail: dmasad@gmu.edu

Giuliano Andrea Pagani
University of Groningen, Groningen, The Netherlands
e-mail: g.a.pagani@rug.nl

Johannes Schmidt
University of Natural Resources and Life Sciences, Vienna, Austria
e-mail: johannes.schmidt@boku.ac.at

Elena Stepanova
Santa Anna School of Advanced Studies, Pisa, Italy
e-mail: e.stepanova@sssup.it

* The authors have evenly contributed to the work presented in the paper.

P. Contucci et al. (eds.), *Complex Networks V*, 141
Studies in Computational Intelligence 549,
DOI: 10.1007/978-3-319-05401-8_14,

game on a network. This research provides a starting-point to further explore strategies as we currently apply a limited set of strategies only.

1 Introduction

Networks have been used to elegantly model systems with many interacting elements in many different disciplines [16] including biology [10], linguistics and social sciences [18], epidemics [4], infrastructures [17], and banking [3]. A central question in network science is to understand the robustness of a network if nodes or edges fail or come under attack [9, 2]. The study of network robustness has many different applications, such as assessing the vulnerability of power grids [1], subway networks [13], and airline transportation networks [7]. Additionally, social networks of interest are covert networks such as criminal or terrorist organizations [12]. For example, targeting one individual over another by police force might have more effect on the communication capability of the network depending on network topology. Analogously, technical networks of interest are computer networks, where the maintainers of computer networks might attempt to identify the best strategy to defend against cyber attacks or random failures.

Network topology plays a large role in how effective an attack is, and how the network is able to defend itself. Albert *et al.* [2] demonstrated that scale-free networks, unlike random networks, are very robust to random failure but vulnerable to targeted attacks. This is due to the fact that most nodes in their scale-free model had few connections, so the probability of randomly targeting a highly connected and central node was low. The targeted attack, however, was able to remove the small percentage of highly connected nodes rapidly, thereby crippling the network connectivity much faster than random attacks. Several researchers have addressed the issue of network robustness using iterative attack and defense games on networks where attackers and the defending network employ static attack and defense strategies against one another [14, 8, 5]. Holme *et al.* [8] considered static attack strategies on edges as opposed to nodes, and suggested edge betweenness as a more effective target of an attacker than attacks on high degree nodes. Nagaraja and Anderson [14] extend Holme's approach by considering both static attack and defense strategies. The network is allowed to defend, or rewire its connections to become less vulnerable to attack via a set of predefined defense strategies. Likewise attacks on the network are performed with a predefined strategy, where attacks based on node centrality were found to perform best on disconnecting the network. Like Nagaraja and Anderson, Domingo-Ferrer *et al.* [5] allow for iterated attack and defense rounds, and show that the attacker's knowledge of the network is also an important factor in the effectiveness of an attack.

But while previous literature on iterated attack and defense has considered many different attack and defense strategies, to date, no research has been done to allow the attacker (or defender) to dynamically change strategies during the course of the game. We extend previous approaches by allowing the attacker and defender to operate with a set of strategies in each time step and to make decisions based on mixing strategies. This allows not only for the possibility that a single strategy could go to fixation, but also cyclical pattern of attack and defense strategies to emerge. A second possibility is that it could simply be advantageous to attack (or defend) based on mixing strategies during attack and defense rounds. Or, it could be that attack and defense strategies simply reach an equilibrium, where no further improvement of attack (or defense) strategy is found by the participants. We examine attacker strategies which identify network nodes to maximize the damage to the network defender. Contrary, network defenders identify the best way to rewire the network following the attack. The choice of strategies is dynamically determined by a genetic algorithm (GA) for both attackers and defenders, and thus representing coevolution between attacker and defender, or a coevolutionary 'game'.

2 The Model

In our model we have three fundamental entities that we deal with:

1. A **network** composed by a set of n nodes and m edges.
2. An **attacker** attempting to disrupt the network.
3. A **defender** attempting to repair the network after an attack to guarantee its continuing functionality.

An attacker disrupts the network by removing a node and all its associated edges. The defender, on the other hand, is allowed to reintroduce a node that has been previously disconnected as a consequence of an attack by reconnecting it to the network. The defender also adds edges to the network if he has enough resources to spend. In fact, the attacker and defender each have an assigned set of resources that they can use in their attack or defense process. The resources for the attacker correspond to the number of nodes that he can remove, whereas defender resources correspond to the number of edges that can be added to the network following an attack. We assume that attackers and defenders have complete knowledge of the network topology and that they perform their actions one after the other beginning with an attack followed by a defense.

A particular simulation starts by generating an initial (first generation) population of an equal number of attackers and defenders. Their genes are initialized randomly, and attackers and defenders are randomly paired up. Each attacker-defender pair is assigned a network of n vertices and no edges. Based on the rules defined by their genomes (which are explained in detail in section 3), each defender adds new edges to the network, up to a total number of m edges. So we start with a set of disconnected nodes and start to build the network from scratch, not fixing any specific network topology at the start. However, fixing the defender and attacker rules will create networks that are similar in topology.

After the network is initially built, the attacker removes k nodes in the network, k being the amount of resources assigned to the attacker, which are the same for all attackers. The choice of the nodes to remove depends on the attacker genome. Once the attack phase is completed, the defender is allowed to add a total of w edges to the network, w being the amount of resources assigned to the defender, which are the same for all defenders. First, the nodes removed by the attacker in this round are re-connected to the network. The nodes to which they will be connected depends on the defender genome. If defender resources allow additional edges to be inserted into the network, those edges are added to the network by the following rule: the starting point for the edges is a random node from the list of nodes which lost edges in the previous attack. The end point is determined by the genetic algorithm. If there are still resources left after reconnecting each of the nodes that have lost an edge in the previous attack, random nodes in the network are picked as starting points. Again, the end points of the new edges are determined by the genetic algorithm.

This process of attack and defense on the network is repeated for r rounds. In summary, a round is an execution of the game with iterative attacks each based on the k resources for the attacker and a (re-)wiring process consisting of w resources for the defender. In our simulations r is equal to 20, i.e. a total of 20 attack-defense rounds is played in each generation of the genetic algorithm.

After each round, the fitness (see Section 3 for the thorough fitness description) of the attackers and defenders is calculated and a final average fitness after r rounds is computed for each individual in the population. Recombination of individuals and mutations which are necessary to generate a new generation of attackers and defenders are discussed in the next section. We are interested to track over generations the evolution of the fitness function for both, attacker and defender as a measurement of their performance in the game. We track over generations the change in genomes as well, because we are interested to identify prevailing strategies.

3 Genetic Algorithm

The GA is used to evolve the strategies applied by the attackers and defenders and thus, allows for a dynamic development of the strategies that are applied by the two groups. A strategy is a mechanism for both the attacker and the defender to decide which node to attack or edge to create/rewire based on some rules, measures or indicators on the network. First, we define the fitness function, then we discuss the genomes of attackers and defenders, and finally we present recombination and mutation strategies.

3.1 The Fitness Function

We define the fitness of the defender to be the number of nodes of the Largest Connected Component (LCC) divided by the total number of nodes in the initial network n, i.e.

$$f^{def} = \frac{LCC}{n} \tag{1}$$

The attacker's fitness is the opposite, i.e.

$$f^{att} = 1 - f^{def} \tag{2}$$

The size of the LCC is a good proxy of the resilience of the network, its ability to keep its structure connected and thus allow interaction between the nodes. The same metric has been used in previous studies [11, 15], allowing our results to be compared to previously-published ones. However, depending on the application of our model, different fitness functions may be appropriate. In section 6 we discuss this aspect in more detail.

3.2 Attacker Genome

A set of strategies is available to the attacker indexed by $j = \{1, 2, 3\}$ – these strategies have been developed previously in the literature [11, 15, 5]:

1. High-degree removal: nodes are prioritized for removal in decreasing order with respect to their degree.
2. High-centrality removal: nodes are prioritized for removal in decreasing order with respect to their betweenness centrality, which is known to be more related to connectivity than other centrality measures.
3. Random removal: nodes are prioritized randomly.

Each gene G_j corresponds to a weight on one of the strategies, and its value varies from 0 to 100. Each strategy calculates a specific network metric (e.g. degree or betweenness centrality) for every node i. The metric is normalized to the interval $[0, 1]$. Thus, to each node i in the network, a value N_{ij} in the interval $[0, 1]$ is assigned by each strategy. In combination with the importance of the strategy as defined by the genome, this represents the removal ranking of a node i. For each node in the network, the attacker's genome assigns a number

$$TotalN_i = \sum_j G_j N_{ij} \tag{3}$$

which is a linear combination of all available strategies weighted by the attacker genome. The probability of a node i to be attacked Pr_i is $TotalN_i$ divided by the sum over $TotalN_i$ for all network nodes, i.e.

$$Pr_i = \frac{TotalN_i}{\sum_i TotalN_i} \quad (4)$$

A node is removed from the network based on its probability Pr_i.

3.3 Defender Genome

The strategies of the defender are similar to the attacker strategies as they are based on the same weighting algorithm. The starting point of an edge that is added to the network is not determined by this weighting algorithm, but by a sequence of rules as outlined in the previous section. Only the endpoint of the new edge is determined by the defender's genome.

The following strategies are available to the defender indexed by $j = \{1, 2, 3\}$ - these strategies have been developed previously in the literature [11, 15, 5]:

1. Preferential replenishment: nodes are ranked in decreasing order with respect to their degree.
2. Balanced replenishment: nodes are ranked in increasing order with respect to their betweenness centrality.
3. Random replenishment: nodes are ranked randomly.

The weighting of nodes is performed similar to the attacker, i.e. the genome determines how the value of a certain metric for the nodes is weighted. See the description of the attacker genome above for details.

3.4 Genome Reproduction Process

The indexed set of genes G_j, $j = \{1, 2, 3\}$ representing the attacker and the defender genome are initially randomly sampled from a uniform distribution in the range $[0, 100]$. Reproduction consists of gene recombination: two attackers or defenders from the current population are randomly chosen from the current generation. The mechanism of selection follows the principle of genetic algorithms known as *roulette wheel selection* [6]: the probability of being picked is not uniform, but is proportional to the fitness of the agent. A random position in the genome is chosen for crossover. At this position, the two individuals will exchange their genetic material, taking the first part from the first parent and the second part from the second parent[1], as shown in

[1] As we have only 3 genes in the genome, there are only two possibilities: the offspring will inherit the first gene from his first parent and second and third genes from his second parent, or he will inherit the two first genes from the first parent and the third gene from the second parent.

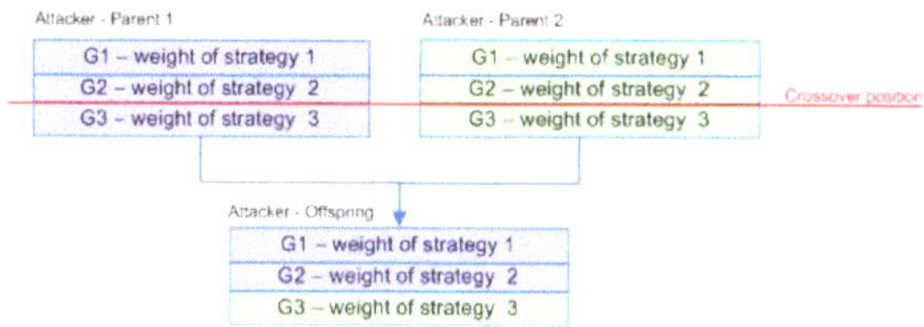

Fig. 1 Example gene crossover

Figure 1. The offspring replaces the previous generation (i.e., parents), thus providing the new base of the genetic material for the following evolution step.

A mutation process occurs with a fixed 5% probability. The mutation in a gene is obtained by sampling a value from a Gaussian distribution with the mean equal to the current value of the gene and a standard deviation of 5.

4 Scenarios

We are interested in the following research problems: first, how does an attacker applying a genetic algorithm perform against a static defender, i.e. a defender with only one, fixed defense strategy. We next look at the inverted scenario, i.e. how a static attacker performs against an evolving defender. Finally, we allow both the attacker and defender to co-evolve against each other. For the purpose of comparison, we also run each static attacker strategy against each static defender strategy. Both defender and attacker have 3 different strategies each. This implies that there are 16 different scenarios to assess in total.

In the base run, we start with a population of 200 attackers and defenders, operating on a network of 100 nodes and 150 edges, and run the GA for 500 generations. Attackers are allowed to remove 3 nodes while defenders rewire 5 edges. In a sensitivity analysis we test different defender budgets of 3,7, or 9 edges. The whole simulation is driven by random choices of attackers and defenders and by a random (although directed) process of selection of individuals in the genetic algorithm. That implies, that a different run of the same simulation may show a different dynamical outcome. At the current moment, we did not run the simulations for several times to analyze the variance of results due to time constraints with the exception of the co-evolution case which was run 25 times. Further runs are left to be presented in future versions of this paper.

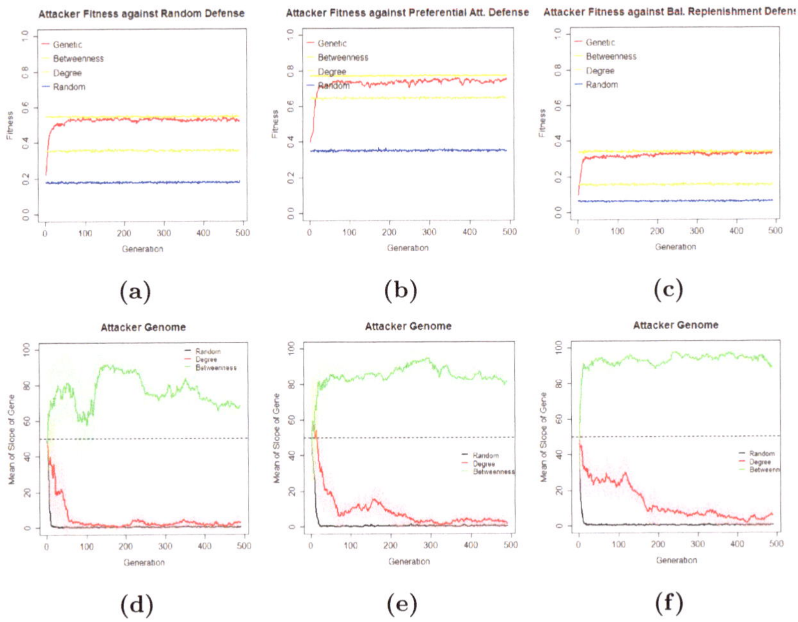

Fig. 2 Top: evolution of the mean of fitness in the attacker population when attackers use the genetic algorithm against 3 static strategies. Bottom: Evolution of the mean of the value of attacker genes for different strategies in the genetic case. The transparent areas indicate the standard deviation. Left: Attacker vs. Random defender. Middle: Attacker vs. Preferential defender. Right: Attacker vs. Balanced Replenishment.

5 Results

5.1 Scenarios Results

Static Defenders. Figure 2a shows that the dynamic attacker quickly approaches the fitness of the single best attacker strategy against a static random defender. The genes evolve accordingly (Figure 2c) , prioritizing high weights for the betweenness strategy and much lower weights for the other two strategies. It can also be observed that the standard deviation in the genes decreases over time, indicating that the individuals in the population converge. Playing against the other two static defender strategies show similar results (Figures 2e and 2f). The worst static defense strategy is preferential attachment which can be derived from the fact that the attacker fitness is highest in that case (middle in Figure 2b). The best possible static defense strategy is balanced replenishment as indicated by the low attacker fitness (Figure 2c). In all cases, the betweenness attack strategy is selected by the attacker's GA.

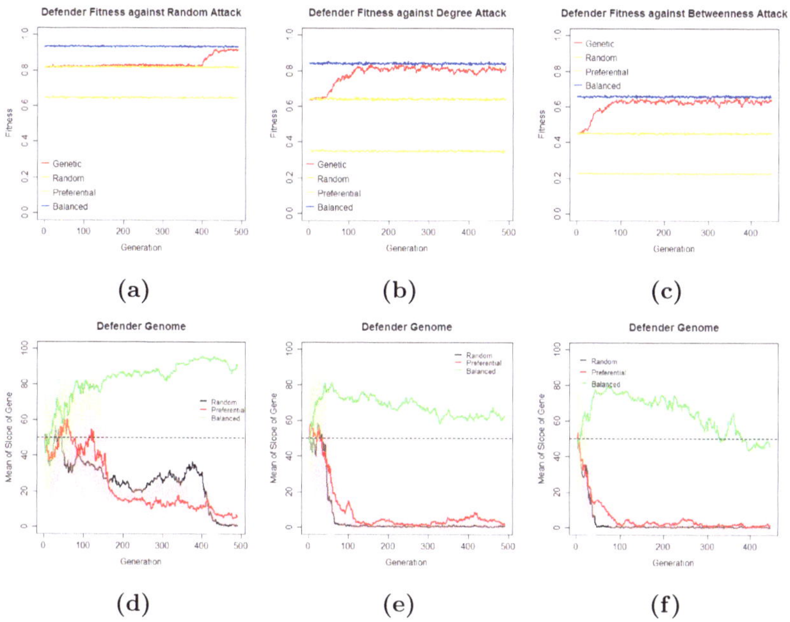

Fig. 3 Results of simulation runs: Defenders applying the genetic algorithm against 3 static attack strategies. Top: Mean of fitness of defender. Bottom: Evolution of value of mean of defender genes. The transparent areas indicate the standard deviation. Left: Random attack vs. Defender. Middle: Degree attack vs. Defender. Right: Betweenness attack vs. Defender.

Static Attackers. Also the defender has a preferred strategy, independent of the static attacker strategy. It is balanced replenishment. However, the GA takes more time to find the dominating strategy in comparison to the attacker's GA in some cases. Defending against a random attacker (Figure 3a) shows that the defender's fitness approaches the fitness of the best possible solution only after 400 generations - even though the balanced replenishment strategy is selected earlier as can be observed by the graph in Figure 3d. However, as long as the random strategy has a rather high weight, the fitness of the defender is not significantly increased. Only after ruling out the random defense, the fitness increases rapidly. That indicates that even a small amount of mixing of strategies may cause a rather bad performance of the defender. This is not the case for the second and third comparison in Figures 3b,3c, 3e, 3f - if the attacker applies the degree attack and betweenness strategy respectively, the defender evolves rapidly in using the balanced replenishment strategy only. The fitness, accordingly, increases quickly in both cases. The defender can deal best with the random attack strategy, as indicated by the comparativley high overall fitness in Figure 3a, while the best strategy for

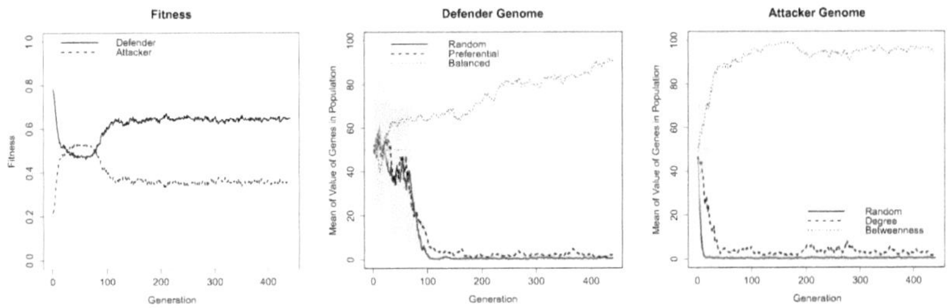

(a) (b) (c)

Fig. 4 Results of simulation runs. Left: evolution of the mean of fitness in the defender and attacker population in the co-evolution case. Middle: Evolution of the mean of value of defender genes for different strategies. The transparent areas indicate the standard deviation. Right: Evolution of the mean of value of attacker genes for different strategies. The transparent areas indicate the standard deviation.

the attacker seems to be betweenness attacks, as also confirmed by the results in the previous section.

Co-Evolution. In the case of co-evolution, i.e. both, defenders and attackers employ a genetic algorithm to select their strategy, attackers evolve quicker towards the more efficient strategy, causing a decline in the fitness of the defender (see Figure 4). However, after about 50 generations, there is a turn-around and the defender starts selecting the best defense strategy, causing an increase in the defender's fitness. After defenders and attackers have evolved into applying the balanced replenishment and betweenness attack strategies respectively, the fitness function stabilizes and no further major fluctuations are observed – an equilibrium is reached. This co-evolutionary process was tested for 25 different instances (while the cases described in the previous section was only tested for 1 instance) and the variance in the overall observed outcome of the gene weights and the fitness of defender and attacker was very low. The pattern shown in Figure 4 for one instance could, in a similar way, be observed in all instances of the problem.

5.2 Sensitivity Analysis

The sensitivity analysis assesses the effect of different defender budgets, i.e. the number of edges that are rewired after an attack, on the overall outcome. A high defender budget plus an efficient defense strategy (i.e. balanced replenishment) almost completely reduce the possibility of the attacker to increase her fitness (see Table 1, row Attacker GA vs. Balanced Replenishment and budget of 9). On the other hand, a low budget decreases the fitness improvements over time for the defender (see Table 1, budget of 3). This indicates that a meaningful game can only be played if the available budgets are in a certain, rather limited interval - too high of a budget for one of the two sides will make any response strategy inefficient. In the co-evolution case, the defender shows a lower fitness at the end of the evolution process than in the beginning if the budget is smaller or equal to 5 edges, while it is the other way round for a budget above that level.

Table 1 Fitness of attackers and defenders with varying budgets. FAS and FAE indicate the average fitness of the attacker at the start and the end of the simulation (i.e. generation 1 and generation 500), respectively. FDS and FDE indicate the average fitness of the defender at the start and at the end of the simulation, respectively.

Defender Budget	3		5		7		9	
Attacker GA vs.	FAS	FAE	FAS	FAE	FAS	FAE	FAS	FAE
Random Defense	0.38	0.63	0.22	0.52	0.12	0.37	0.08	0.18
Preferential Defense	0.48	0.76	0.40	0.76	0.36	0.64	0.32	0.62
Balanced Replenishment	0.37	0.54	0.10	0.34	0.01	0.02	0.01	0.01
Defender GA vs.	FDS	FDE	FDS	FDE	FDS	FDE	FDS	FDE
Random Attack	0.67	0.68	0.81	0.92	0.90	0.97	0.94	0.98
Degree Attack	0.49	0.54	0.63	0.81	0.81	0.98	0.90	0.98
Betweenness Attack	0.36	0.42	0.45	0.63	0.61	0.95	0.82	0.97
Co-Evolution	FDS	FDE	FDS	FDE	FDS	FDE	FDS	FDE
GA vs. GA	0.62	0.38	0.78	0.66	0.88	0.95	0.92	0.98

6 Related Work

Several researchers have assessed the robustness of networks in case of attacks on nodes or edges. Here we look more in detail to studies where the concepts of evolution of a network, in terms of its topology, is tied to the behavior of an attacker of the network. In a seminal paper by Albert *et al.* [2], the authors demonstrate that scale-free networks are vulnerable to targeted attacks of nodes of high degree, while fairly robust to random attacks. Holme *et al.* [8] consider attacks on edges as opposed to nodes, and suggest edge centrality as an effective target of an attacker.

As already mentioned in Section 1, the work of Nagaraja and Anderson [15] is relevant to our paper since it considers an evolutionary game theory approach that takes place on a network. In a way similar to our interpretation of the evolutionary game, their game is organized in rounds and each round consists of an attack followed by a recovery. The attack consists of targeting a number of nodes to be removed, depending on the attacker budget. However, the recovery is different than the one we propose in this paper, and consists in two stages, namely replenishment and adaptation. The first stage deals with inserting new nodes into the network and establishing new connections based on the defender's budget, while the second deals with rewiring existing links. The objective for the attacker is to split the network in separate components. The authors also consider betweenness as a type of attack and the effects are more disrupting against all types of defense. Our approach is more flexible giving the possibility to the attacker and defender to adapt or change their strategies (i.e., type of attack/defense) during the game, while in [15] the strategies are chosen and kept fixed through the game. Our model allows to identify the strategies for attackers and defenders that provide the maximum fitness out of a potentially broad set of strategies. In [15] the test performed takes into account scale free networks as initial topologies, whereas our approach starts with an initial topology that is already optimized by the defender under the assumption that the defender initially generates the network. One aspect that we prove through the evolution of the genome is the superiority in attack of the balanced replenishment strategy that is highlighted also in [15]. Nagaraja and Anderson's work is not without limitations, however. The cost of implementing an edge is essentially zero since the network is allowed to rewire with an arbitrary amount of newly added edges.

Kim and Anderson [11] expand upon the work of Nagaraja and Anderson. Kim and Anderson give each attacker and defender a fixed budget, or cost to add nodes and edges after an attack, and analyze the effect of attacks on a variety of different network topologies. They find a strategy of connecting low centrality nodes is the best defense strategy. However, as the edge to node ratio increases, the network becomes more robust, and even adding edges randomly is effective against targeted attacks. They find that there is a threshold value for the proportion of edges to nodes at which point the effectiveness of attacks decreases drastically.

The work of Domingo-Ferrer and Gonzalez-Nicolas [5] is based on the ideas and findings of previous work by Nagaraja and Anderson [15] and Kim and Anderson [11] and adds further properties to the networks and the experiment set. In the paper the authors analyze the evolution of the *order* and average path length of scale-free networks (weighted and unweighted) under attack and defense. The only strategy of attack considers betweenness centrality as the measure to identify the most critical node; whereas defense is achieved following two types of strategies: delegation and node replenishment. The results show basically that an important factor is the visibility that an attacker has of the network, while there is basically no difference in the disruption behavior of weighted and unweighted networks. Our approach is more flexible considering the possibilities of different strategies of attack and defense and networks that are not fixed a priori, but built by the defender that is usually the organization that has to defend from the attacks.

7 Conclusions and Future Work

We have shown that our approach to model interactions between attackers and defenders can be successfully modeled using genetic algorithms. Our results confirm what has been found in previous papers which compared various static strategies. In addition, our work shows that strategies for link placement can also be applied to generate networks from scratch, as we do in generating the networks, achieving already an initial strength against some types of attacks (in contrast to other papers, which only used them to rewire networks after they have been attacked)[2]. Obviously, the success of a defense and attack depends on the available resources. The choice of the strategy matters primarily when the defender's resources are slightly larger than the attacker's resources. In any other case, the results of the game are going to be biased towards the side with the resource advantage. If the defender resources are slightly higher than the attacker's and if the defender's goal is to maintain or increase the LCC and the attacker aims for the opposite, there are clear winning strategies among the ones tested in this study: the balanced replenishment and betweenness attack strategy, respectively, can be considered to be the most efficient ones, independent of which strategy is applied by the opponent. An equilibrium situation arises if the two opponents apply these strategies, although the defender appears to evolve slower than the attacker.

This result may be applied to social networks, computer networks, or any other kind of network. From an empirical perspective, it would be interesting if similar strategies are observed in real networks (i.e. where they have evolved 'naturally'). From a normative point of view, the results of this paper and related work can be used to design strategies to defend against attacks or to target attacks against certain nodes in networks.

Future work will include the development and testing of new defender and attacker strategies - currently, only three strategies are included. A larger number of strategies may make the game dynamics more complex than the current version, which allows for a stable equilibrium in the co-evolution case. Additionally, the current fitness function emphasizes connectedness of the network, but does not assess the efficiency of the network in providing transportation or communication services. Different fitness functions which may include a combination of the largest connected component with some measure of efficiency as, for example, the diameter or effective diameter of the network, therefore might be considered interesting options for future research.

[2] However, this difference is somehow minor if we consider that many attack-defense rounds applying the same defense strategies will cause the network topology to resemble a network that was built from scratch using the very same defense strategy.

Acknowledgements. This paper is the outcome of a group project started at the Complex Systems Summer School of the Santa Fe Institute in June, 2013. We are very grateful to the organizers of the summer school, and to all the lecturers and presenters. To all of our fellow summer school students: thanks for the good times – and the greenhouse sessions. Thanks to Mauricio Cantor and Bruno Pace for very valuable discussion on the subject of the paper. Very special thanks go to Tom Carter for all of his efforts and in particular for his enlightening performance of the "power-law blues". The simulations for this paper were partly executed on the Vienna Scientific Cluster (VSC) – thanks a lot to the VSC team for their support during the whole process.

References

1. Albert, R., Albert, I., Nakarado, G.: Structural vulnerability of the north american power grid. Physical Review E 69 (2004)
2. Albert, R., Jeong, H., Barabási, A.L.: Error and attack tolerance of complex networks. Nature 406(6794), 378–382 (2000)
3. Boss, M., Elsinger, H., Summer, M., Thurner, S.: The network topology of the interbank market. Quantitative Finance 4, 677–684 (2004)
4. Colizza, V., Barrat, A., Barthélemy, M., Vespignani, A.: Predictability and epidemic pathways in global outbreaks of infectious diseases: the sars case study. BMC Med. 5, 34 (2007)
5. Domingo-Ferrer, J., Gonzlez-Nicols, R.: Decapitation of networks with and without weights and direction: The economics of iterated attack an d defense. Computer Networks 55(1), 119–130 (2011)
6. Goldberg, D.E., Deb, K.: A comparative analysis of selection schemes used in genetic algorithms. Urbana 51, 61801–62996
7. Guimerà, R., Mossa, S., Turtschi, A., Amaral, L.A.N.: The worldwide air transportation network: Anomalous centrality, community structure, and cities' global roles. PNAS 102(22), 7794–7799 (2005)
8. Holme, P., Kim, B.J., Yoon, C.N., Han, S.K.: Attack vulnerability of complex networks. Physical Review E 65(5), 056109 (2002)
9. Iyer, S., Killingback, T., Sundaram, B., Wang, Z.: Attack robustness and centrality of complex networks. PloS One 8(4), e59613 (2013)
10. Jeong, H., Tombor, B., Albert, R., Oltvai, Z.N., Barabási, A.L.: The large-scale organization of metabolic networks. Nature 407(6804), 651–654 (2000)
11. Kim, H., Anderson, R.: An experimental evaluation of robustness of networks. IEEE Systems Journal 7(2), 179–188 (2013)
12. Krebs, V.E.: Mapping networks of terrorist cells. Connections 24(3), 43–52 (2002)
13. Latora, V., Marchiori, M.: Is the boston subway a small-world network? Physica A: Statistical Mechanics and its Applications 314(1-4), 109–113 (2002)
14. Nagaraja, S.: Topology of covert conflict. In: Christianson, B., Crispo, B., Malcolm, J.A., Roe, M. (eds.) Security Protocols 2005. LNCS, vol. 4631, pp. 329–332. Springer, Heidelberg (2007)

15. Nagaraja, S., Anderson, R.: The topology of covert conflict. Technical Report UCAM-CL-TR-637, University of Cambridge, Computer Laboratory (July 2005)
16. Newman, M.: Networks: an introduction. Oxford University Press (2009)
17. Pagani, G.A., Aiello, M.: The power grid as a complex network: A survey. Physica A: Statistical Mechanics and its Applications 392(11), 2688–2700 (2013)
18. Travers, J., Milgram, S.: An experimental study of the small world problem. Sociometry 32(4), 425–443 (1969)

Community Detection in Bipartite Networks Using Random Walks

Taher Alzahrani, Kathy J. Horadam, and Serdar Boztas

Abstract. Community detection plays a crucial role in many complex networks, including the increasingly important class of bipartite networks. Modularity-based community detection algorithms for bipartite networks are hampered by their well-known resolution limit. Unfortunately, the high-performing random walk based algorithm Infomap, which does not have the same constraint, cannot be applied to bipartite networks.To overcome this we integrate the projection method for bipartite networks based on common neighbors similarity into Infomap, to acquire a weighted one mode network that can be clustered by the random walks technique. We also compare results obtained from this process with results in the literature. We illustrate the proposed method on four real bipartite networks, showing that the random walks technique is more effective than the modularity technique in finding communities from bipartite networks as well.

Keywords: bipartite graph, community detection, random walks.

1 Introduction

One mode, or unipartite, networks are the typical framework for complex networks. Many techniques have been constructed to analyze them. However, many complex networks can best be described as bipartite [1]. A bipartite network is a network in which there are two different types of nodes, and the edges between nodes may occur only if nodes belong to different types. In the last few years, there has been increasing motivation to analyse bipartite networks as a separate network category, and in particular to investigate their community structure. For unipartite networks,

Taher Alzahrani · Kathy J. Horadam · Serdar Boztas
School of Mathematical and Geospatial Sciences,
RMIT University
Melbourne 3001, Australia
e-mail: {taher.alzahrani,kathy.horadam,serdar.boztas}@rmit.edu.au

P. Contucci et al. (eds.), *Complex Networks V*,
Studies in Computational Intelligence 549,
DOI: 10.1007/978-3-319-05401-8_15,

two approaches to community detection have been very popular, one based on modelling the community structure and one based on extracting it from flow calculations on the network. The best algorithms to cluster very large networks using each approach [2], compared using the LFR benchmark datasets [3], are now referred to as the Louvain algorithm [4] and the Infomap algorithm [5]. Unfortunately it is impossible to reformulate Infomap on a bipartite network, since then there is no stationary distribution for the probability of the walker to be at a given node on the bipartite graph. In other words, if you start in one class (set), then you will always be in that class after an even number of steps, so the probability of being at a particular vertex is zero at odd time steps. In the language of Markov chains, the random walk on a bipartite graph is periodic. The focus of this paper is on community detection in the weighted one mode networks which are projected from unweighted bipartite networks. This is a natural focus, as usually one set of nodes in a bipartite network, denoted the *primary set* P is of more interest for a particular purpose than the other, the *secondary set* S. The rôles of the two node sets can be switched for different applications. Our contribution is to apply a random walks based algorithm to the unipartite network projected on the primary set of the bipartite network. We also compare our results with the results in the literature that used bipartite modularity based algorithms. We investigate the communities found by Infomap in one case in detail, to demonstrate that the small communities found (below the resolution limit of modularity-based algorithms) represent real information.

2 Previous Work

Identifying communities, also called modules or clusters, in a network allows us to explore its hierarchical structure. This leads to better understanding of the major functions of the network, and more efficient spread of ideas, goods or services in the network.

2.1 Unipartite Networks

Girvan and Newman [6] initiated recent work on defining and evaluating communities, introducing the fast greedy technique which relies on a quality function called modularity. Modularity is a scalar measure of the quality of modules extracted from a network. The partition which maximises it is regarded as the best. The complexity of the algorithm is $O(n^3)$, where n is the number of nodes. Since the limited nodes size for previous algorithm is 10^3 many efforts have been devoted to upgrade the computational time of modularity optimization. For instance, the Radicchi et al [7] algorithm is in spirit the Girvan-Newman method but it iteratively removes the edges with highest clustering coefficient instead of edges with highest betweenness. The stated complexity of this algorithm is $O(n^2)$. Another algorithm that takes modularity optimization as its main quality function is that of Guimera and Amaral [8]. The fast modularity optimization algorithm by Blondel et al [4] has the best results compared with the previous algorithms. It is described as a multi level method in community detection. The complexity of the Louvain algorithm is linear in the number of edges in the network, that is $O(m)$. However, modularity-based algorithms have a known drawback: a resolution limit in detecting communities [9]; that is, communities with internal edge numbers $\leq O(\sqrt{m})$ cannot always be reliably detected. All these methods, which rely mainly on modularity, describe and reveal communities in networks according to how the networks are built or by modelling their structure. However, a different method using random walks, known as Infomap proposed by Rosvall and Bergstrom [5], identifies communities according to information flow in the networks. The quality function used, called the map equation, is based on minimum description length (MDL) [10]. This function measures the average length $L(M)$ in bits per step of a random walk on a network with the modular partition M, as follows:

$$L(M) = q_{\curvearrowright} H(Q) + \sum_{i} p^{i}_{\circlearrowright} H(P^{i}) \tag{1}$$

where $q_{\curvearrowright}$ is the probability that the random walk moves between modules, $H(Q)$ is the entropy of module names, $p^{i}_{\circlearrowright}$ the probability of movement within module i and $H(P^{i})$ is the entropy of the movements within module i. The complexity of the Infomap algorithm is also $O(m)$. However, Infomap does not apply to bipartite networks because a stationary distribution cannot be determined in general.

2.2 *Bipartite Networks*

Most authors follow Newman and Girvan's modularity method [6] to determine communities in bipartite networks. Michel et al [11] used unipartite Girvan-Newman modularity [12] as their standard model to derive a bipartite modularity model by building an unweighted "biadjacancy matrix" of a bipartite graph. Guimera [13] produces a modularity measurement for bipartite networks, denoting the sets of this network as actors and teams. The emphasis on finding modules here was in the actor set (P) after projection, with projection based on joint participation in teams. In [14], Barber developed a modularity matrix for bipartite networks, inspired by Newman's modularity matrix [15]. The previous approaches to finding modularity in bipartite networks were extended from Newman modularity [6]. A fast technique for unipartite networks, the Label Propagation Algorithm (LPA) [16], uses the local network structure as a guide for finding communities very efficiently (almost linearly in m). Barber and Clark [17] introduced a version of LPA, denoted LPAb, for bipartite networks. The speed of LPAb (complexity near linear in total number of edges m) makes it comparable with the fastest bipartite modularity optimization algorithm [18]. However, Liu and Murata [18] introduce a new version of LPAb, denoted LPAb+, which they claim as the most reliable algorithm with the highest bipartite modularity.

3 Method

The Infomap algorithm utilizes the information flow on a network in order to achieve its clustering. This information flow is approximated in practice by means of a random walk along the network, and iterating until a steady state distribution emerges, as it must, under the assumption that the network in question is strongly connected and aperiodic. Unfortunately for us, in general there is no stationary distribution of a random walk on bipartite networks that can be found from power iterations as discussed in Section 1. Thus Infomap cannot be directly used for our problem. Instead, we apply a projection method based on common neighbors similarity for our bipartite networks. The motivation is to be able to obtain a stationary distribution for the walk on the nodes in the unipartite network obtained by projection. This is achieved by integrating the projection process into the Infomap and Louvain algorithms. This allows us to compute the complexity time for the whole operation starting from converting bipartite networks to weighted unipartite networks followed by clustering them by the two algorithms. The reason the projection method is also applied to the Louvain algorithm is to be able to compare the performance of Infomap with that of Louvain, for the bipartite network case. The lack of existing benchmark bipartite networks motivated our work. Moreover, there is no evaluated community detection method in the literature that examines bipartite networks from a random walks perspective. We have programmed our projection algorithm in C++ for compatibility with the implementations we have of the Infomap and Louvain algorithms. We start by reading the bipartite network as a pair of nodes, the first from P and the second from S. The labels on the nodes in this dataset do not have to be numbers, they can be post codes, book serials, bank card numbers, names of social networks or even names of people. Then, we find the common neighbors between nodes i and j in P according to the following adjacency matrix A_{ij}:

- $A_{ij} = 1$, if nodes i and j have a common neighbor
- $A_{ij} = 1$, if node i has a neighbor which has no other neighbors in P (resulting in self loop, i=j)
- $A_{ij} = 0$, Otherwise

The weight of the edge between i and j in the projected unipartite network is the number of common neighbors of node i and node j.We also use special techniques in C++ that improved the efficiency of the projection method. Starting by using a C++ container called Mapvector which requests a key and a value, we choose the key to be the common neighbors and the value to be a vector of nodes $\{v_1, v_2, .., v_n\}$ where n is total number of nodes. Then, we create pairs in a one mode network and store the result in container called "Multiset".

4 Results and Discussion

Both algorithms were tested in four real world bipartite networks: the Southern women network, Newman's scientific collaboration network, a historical Australian

Table 1 Network sizes, where P and S are the number of primary set nodes and secondary set nodes respectively and m is the total number of edges

Network	P	S	m
Southern women	14	18	89
Scientific collaboration	16726	22016	58595
Australian government contracts	11924	1655	70019
NSW Crimes	155	22	9611

government contracts network and an Australian crime network. Their primary and secondary set sizes and total number of edges are given in Table 1.

Our results are summarised in Table 3. Since the Southern women network has been studied widely in the literature we investigate it in some detail first.

4.1 Southern Women Network

The "Southern women" network collected by Divas et al [19] has become a benchmark for testing community detection algorithms on bipartite networks. This network has 18 women (who form the primary set P) who attended 14 different events (the secondary set S). An edge exists between two women for each event they attend together. Most studies conducted before 2003 identify two (sometimes overlapping) communities of women while one identifies three communities [20]. In many studies, members within each community are further partitioned into core or peripheral members. More recent studies using bimodularity find more communities (3 and 4). Consequently, at least two communities are expected. Our implementation of the projection in Infomap produces 4 communities as shown in Figure 1. In Table 2, we list the community numbers found in the Southern women dataset by the more recent bipartite network algorithms described in Sections 2.2 and 3. We compare our results for the Southern women network with results in the literature, in more detail. Using Infomap, we have community A consisting of Evelyn and Theresa (women 1 and 3, respectively), community B consisting of Katherine and Nora (women 12 and 14, respectively), and two others $C = \{8, 9, 16, 17, 18\}$ and $D = \{2, 4, 5, 6, 7, 10, 11, 13, 15\}$, as shown in Figure 1. Our groups A and B consist of women frequently identified as core members of each of the two communities found in earlier studies. By contrast, Barber's two smaller communities consist of women who tended to be identified as peripheral members of each of the two communities found in earlier studies [20]. Barber also tested the success of his partition into four communities, found using the maximum bipartite modularity (as described in Section 2.2), as a partition in the corresponding *unweighted* projection network, and found it to have negative modularity [14]. As this is worse than considering the women as a single community, it further supports our use of the *weighted* projection network. Guimera et al [13] found only two communities of women (red and blue) whether modularity on the unweighted projection, the weighted projection or bipartite modularity was used. They found the communities were inaccurate with unweighted projection, but identical and in agreement with supervised results in [20] for the other two methods. The total number of edges in the Southern women network after weighted projection is 139 edges. Our community A (Evelyn and Theresa) has 7 internal edges and lies inside the red group, while our community B (Katherine and Nora) has 5 internal edges and lies inside the blue group. These two "core" communities are not detected by the modularity based algorithm, probably because their edge numbers fall below the resolution limit of modularity [9] which in this case is 12 (since $11 < \sqrt{139} < 12$). By comparison the 2 communities found by our Louvain algorithm have 45 and 33 internal edges. This demonstrates that the resolution limit for modularity applies to Louvain but is passed by Infomap, in this benchmark bipartite case.

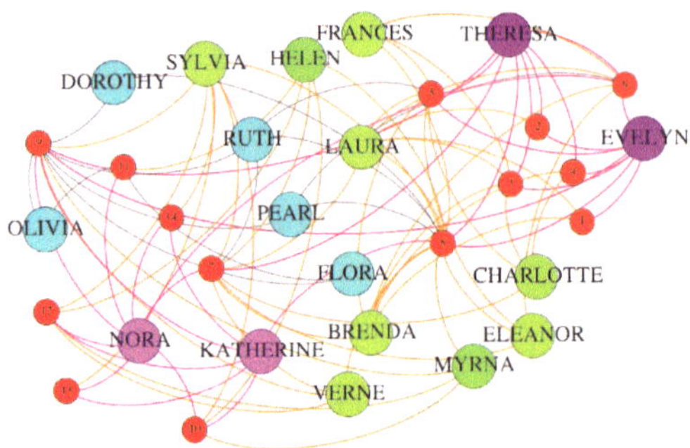

Fig. 1 The four communities of women found in the Southern women dataset. Red nodes represent S, the events the women attended, and the four other colors represent four communities within P, with nodes labelled by first name.

Table 2 Numbers of communities of women detected by different algorithms in the Southern women network

Algorithm	Quality function	Network applied to	Modules in P
Guimera [13]	modularity	weighted projection	2
Michel [11]	bimodularity	bipartite	3
Barber [14]	bimodularity	bipartite	4
LPAb(+) [18]	bimodularity	bipartite	4
This paper	map equation	weighted projection	4

4.2 *The Scientific Collaboration Network*

The scientific collaboration network in Newman [21], contains a bipartite network. It lists the relationships between publications and the scientists who are authors of these papers. There are 16726 scientists who wrote 22016 papers in this network. The number of edges between scientists and papers is 58595. The primary set in our projection is the scientists while the secondary set is the papers. Therefore, we are

Table 3 Community numbers obtained from our experiments, where L is the code length and Q is the modularity

	Infomap		Louvain	
Network	Comm.	L	Comm.	Q
Australian government contracts network	1114	8.340	836	0.530
Scientific collaboration network	2131	6.164	1266	0.877
Southern women network	4	3.992	2	0.352
Crime network	2	7.276	1	0.0

interested in detecting the communities of authors and determining who are more likely to collaborate together. The value of characterizing scientists in communities and describing the ties between distinct communities from different disciplines is important knowledge because it can help scientists collaborate. Our method utilizes a modified Infomap algorithm to characterize the scientific collaboration network into 2131 communities. The same method using Louvain algorithm finds fewer communities: 1266. We attribute these different results to the resolution limit of modularity optimization in Louvain algorithm. Scientists in one community have more in common than in another and they are likely to collaborate together and this increases the strength of the community.

4.3 The Historical Australian Government Contracts Network

The historical Australian government contract data has been published in 2012 [22]; it contains great detail about agencies and companies that undertake projects in Australia. We construct a bipartite network from this dataset. The network in this case has the ABN (Australian Business Number) a unique identifier number for agencies and companies, as the primary set. The postcode areas which these agencies have projects in them form the second set. The bipartite network has 11924 nodes as agencies and/or companies, and 1655 different postcode areas. The number of edges in this dataset is 70019, which is number of projects from 1999 until 2012 [22]. The weighted one mode projected network relates agencies that have common projects in the specific postcode area. The results produced from our method implemented in Infomap illustrate 1114 communities are found which contained agencies working on projects in the same postcode area. However, there were different results from Louvain where only 836 communities have been identified. The investigation of the historical Australian government contract data could lead to more collaboration between agencies/companies if they have projects in the same postcode area.

4.4 Crime Network

This monthly data is collected from January 1995 to December 2009, and shows the crimes and offences committed in New South Wales (NSW) in Australia [23].

Moreover, it provides the location of the crimes, thus we are interested in the location and the crime itself, which form our bipartite network. The primary set in this data is the location of the crime, whereas the offence is the secondary set. The intention in finding communities in the crime dataset is to identify and illustrate where similar crimes have occurred. There are 155 locations and 22 types of offences committed. The number of crimes committed during almost 14 years is 9611, which is the number of edges in this network. Results from applying our integrated Infomap algorithm show that there are two communities (one with 73 locations and the other with 82) where similar crimes are being committed. However, only one community that is the entire state of NSW is found when applying the Louvain algorithm, so this algorithm provides no useful information.

5 Conclusion

In this paper, we have integrated the projection method for bipartite networks into Infomap, to acquire a weighted one mode network that can be clustered by the random walks technique. The results from this process reflect valuable information compared with bipartite network based algorithms in the literature. Experiments on four real world bipartite networks show that a random walk based algorithm is more functional in detecting the communities in the primary set of a bipartite network than a modularity based algorithm. For future work, we intend to modify our approach to project the two sets P and S of a bipartite network in parallel, cluster them under random walks algorithms and then merge the whole into a clustered bipartite network. Moreover, weighted bipartite networks, overlapping communities, measuring quality of the communities found and difference of the quality between (projection + Infomap) and a method which would compute bipartite communities and then project the will be taken into consideration.

Acknowledgement. The first author would like to thank the Ministry of Finance of Saudi Arabia for supporting his research. The work of the second and third authors was partly supported by Department of Defence of Australia Agreement 4500743680.

References

1. Nishikawa, T., Motter, A.E., Lai, Y.C., Hoppensteadt, F.C.: Heterogeneity in oscillator networks: Are smaller worlds easier to synchronize? Physical Review Letters 91, 014101 (2003)
2. Lancichinetti, A., Fortunato, S.: Community detection algorithms: A comparative analysis. Physical Review E 80, 056117 (2009)
3. Lancichinetti, A., Fortunato, S., Radicchi, F.: Benchmark graphs for testing community detection algorithms. Physical Review E 78, 046110 (2008)
4. Blondel, V.D., Guillaume, J.L., Lambiotte, R., Lefebvre, E.: Fast unfolding of communities in large networks. Journal of Statistical Mechanics: Theory and Experiment 2008, P10008 (2008)

5. Rosvall, M., Bergstrom, C.T.: Maps of random walks on complex networks reveal community structure. Proceedings of the National Academy of Sciences 105, 1118–1123 (2008)
6. Newman, M.E., Girvan, M.: Finding and evaluating community structure in networks. Physical Review E 69, 026113 (2004)
7. Radicchi, F., Castellano, C., Cecconi, F., Loreto, V., Parisi, D.: Defining and identifying communities in networks. Proceedings of the National Academy of Sciences of the United States of America 101, 2658–2663 (2004)
8. Guimera, R., Sales-Pardo, M., Amaral, L.S.A.N.: Modularity from fluctuations in random graphs and complex networks. Physical Review E 70, 025101 (2004)
9. Fortunato, S., Barthelemy, M.: Resolution limit in community detection. Proceedings of the National Academy of Sciences 104, 36–41 (2007)
10. Grunwald, P.D., Myung, I.J., Pitt, M.A.: Advances in minimum description length: Theory and applications. MIT press (2005)
11. Crampes, M., Plantie, M.: A Unified Community Detection, Visualization and Analysis method. arXiv preprint arXiv:1301.7006 (2013)
12. Girvan, M., Newman, M.E.: Community structure in social and biological networks. Proceedings of the National Academy of Sciences 99, 7821–7826 (2002)
13. Guimera, R., Sales-Pardo, M., Amaral, L.S.A.N.: Module identification in bipartite and directed networks. Physical Review E 76, 036102 (2007)
14. Barber, M.J.: Modularity and community detection in bipartite networks. Physical Review E 76, 066102 (2007)
15. Newman, M.E.: Finding community structure in networks using the eigenvectors of matrices. Physical Review E 74, 036104 (2006)
16. Raghavan, U.N., Albert, R.K., Kumara, S.: Near linear time algorithm to detect community structures in large-scale networks. Physical Review E 76, 036106 (2007)
17. Barber, M.J., Clark, J.W.: Detecting network communities by propagating labels under constraints. Physical Review E 80, 026129 (2009)
18. Liu, X., Murata, T.: An Efficient Algorithm for Optimizing Bipartite Modularity in Bipartite Networks. JACIII 14, 408–415 (2010)
19. Davis, A., Gardner, B.B., Gardner, M.R.: Deep south: A Social Anthropological Study of Caste and Class University of Chicago Press Chicago (1941)
20. Freeman, L.C.: Finding social groups: A meta-analysis of the southern women data. In: Dynamic Social Network Modeling and Analysis, pp. 39–97. National Academies Press (2003)
21. Newman, M.E.: The structure of scientific collaboration networks. Proceedings of the National Academy of Sciences 98, 404–409 (2001)
22. Department of Finance and Deregulation. Dataset. Historical Australian Government Contract Data (February 27, 2013), `http://data.gov.au/dataset/historical-australian-government-contract-data/`
23. NSW Bureau of Crime Statistics and Research. Dataset. NSW Crime data (December 2008), `http://data.gov.au/dataset/nsw-crime-data/`

A Coevolutionary Model of Strategic Network Formation

Ibrahim Al-Shyoukh and Jeff S. Shamma

Abstract. In foundational models of network formation, the mechanisms for link formation are based solely on network topology. For example, preferential attachment uses degree distributions, whereas a strategic connections model uses internode distances. These dynamics implicitly presume that such benefits and costs are instantaneous functions of the network topology. A more detailed model would include that benefits and costs are themselves derived through a dynamic process, which, in the absence of time-scale separation, necessitates a coevolutionary analysis. This paper introduces a new coevolutionary model of strategic network formation. In this model, network formation evolves along with the flow of benefits from one node to another. We examine the emergent equilibria of this combined dynamics of network formation and benefit flow. We show that the class of strict equilibria is stable (or robust to small perturbations in the benefits flows).

1 Introduction

Networks involving benefit exchanges between the different nodes are ubiquitous. Examples include information exchange in social networks, goods exchange in economic markets, and scientific collaboration networks. The abundance and importance of such networks have manifested a growing area of research that looks into the theory of network formation and the relevance of emerging structures. A number of different models for the network formation in multiple disciplines have been proposed that encompass a range of ideas [1–9]. A common feature of these models is that there is no interdependence or feedback between the network formation dynamics, and the dynamics on

Ibrahim Al-Shyoukh · Jeff S. Shamma
School of Electrical & Computer Engineering
Georgia Institute of Technology
Atlanta, GA 30308 USA

P. Contucci et al. (eds.), *Complex Networks V*,
Studies in Computational Intelligence 549,
DOI: 10.1007/978-3-319-05401-8_16,

the network. Recent work has begun to investigate models with endogenously formed (i.e., coevolutionary) network topologies in a wide class of systems including opinion dynamics [10–19].

The work in this paper concerns the study of strategic coevolutionary network formation. In foundational models of network formation, the mechanisms for link formation are based solely on network topology. For example, preferential attachment [20] uses degree distributions, whereas a strategic connections model [21] uses internode distances. These dynamics implicitly presume that such benefits and costs are instantaneous functions of the network topology. A more detailed model would include that benefits and costs are themselves derived through a dynamic process, which, in the absence of time-scale separation, necessitates a coevolutionary analysis.

Here we present a model that captures the dynamic flow of benefits in a network. The model is inspired by and builds upon the strategic network formation model of Bala & Goyal [21]. In the model, and upon link establishment, benefits flow from one node to another over time. The amount of benefit and speed of flow are distance dependent. As the distance between nodes increase, the total attainable benefits becomes smaller and it takes longer for the benefits to be attained. Another feature of the model is that when links are severed, then benefits are not immediately lost. Rather, they are dissipated over time.

By allowing time to propagate, then a node can realize the full benefits from an established link, and nodes can seek to maximize such asymptotically realized benefits. However, this analysis presumes a separation of time scales. Instead, we consider the case when nodes are myopic decision makers that seek to maximize the one time step flow of benefits. We examine the conditions for equilibria of this model and the stability of such equilibria. We also show that this model admits equilibria that can only be realized at a higher cost in the case of immediate benefit availability. This formulation gives rise to a richer set of network topologies without additional cost constraints. Section 2 introduces some preliminaries, and Section 3 presents the model and relevant analysis.

2 Preliminaries

Let us recall the strategic network formation model of Bala & Goyal [21]. The model represents the flow of benefits in a network of $N \geq 3$ nodes. Consider for example the network shown in Fig. 1. A directed edge $i \leftarrow j$ indicates flow of benefits from j to i, e.g, node 8 is an immediate (one-hop) beneficiary from node 5, and an indirect (two-hop) beneficiary from nodes 2, 6, and 7. Nodes dynamically form and sever links based on the rewards of benefit flow and costs of link formation and maintenance.

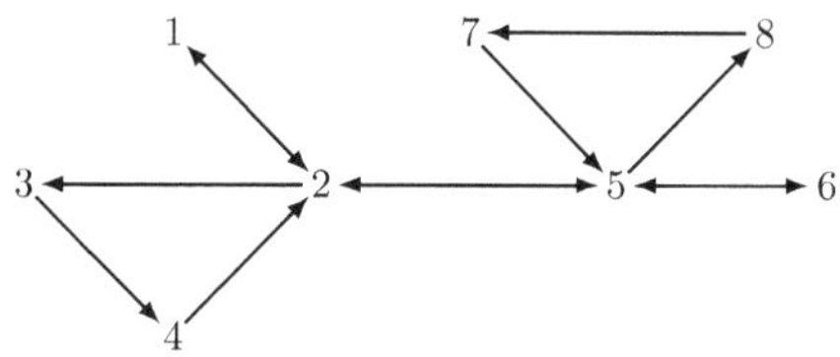

Fig. 1 An example of a directed network of information flow. The arrow direction indicates the direction of flow.

2.1 The Static Game

First, we will consider the static network formation game. Let $\mathcal{N} = \{1, 2, \dots, N\}$ be the set of all nodes of the network. Given that a node can connect to $N-1$ nodes, a node's strategy can be represented by the binary valued vector $g_i = (g_{i,1}, \dots, g_{i,i-1}, g_{i,i+1}, \dots, g_{i,N})$,, where $g_{i,j} = 1$ whenever node i has a link with node j, $g_{i,j} = 0$ otherwise. A network g can be represented by the joint strategies of all nodes as $g = (g_1, g_2, \dots, g_N)$. We shall use g_{-i} to refer to the network constructed from g by excluding node i's links, i.e., $g_{-i} = (g_1, \dots, g_{i-1}, g_{i+1}, \dots, g_N)$. A path from node j to node i is denoted by $\overline{ij}$. Let $|\overline{ij}|$ denote the length of path $\overline{ij}$. Define $d_{ij}(g_i, g_{-i}) = \min_{\overline{ij} \in g} |\overline{ij}|$ as the length of the shortest path from j to i. For compactness, in the remainder of this paper, we will write d_{ij} instead of $d_{ij}(g_i, g_{-i})$ whenever the arguments are clear.

Immediate Benefit Availability. Whenever node i establishes a connection with node j, benefits become accessible to i. In the existing models of strategic network formation, the benefits are fully transferred from node j and its neighbors to i immediately upon link establishment. The amount of benefits transferred can be distance dependent. Thus, the value of benefits from a direct connection can generally be assumed to be $\delta \in (0, 1]$. Whereas if j is an indirect connection of i, then the value of benefits is $\delta^{d_{ij}}$. Additionally, let c denote the cost of establishing a connection with another node. The cost is only incurred by the node establishing/maintaining the link. In the directed flow network, the benefits will only flow to the node establishing the connection.

For a given network g, let $\mathcal{N}_i^+(g) = \{k \in \{1, \dots, N\} : \overline{ik} \in g\}$ denote the set of all nodes that have a path to node i. This set defines all the neighbors of i, direct or indirect. As such, benefits can flow from these nodes to i. We shall define $\mu_i(g_i) = \sum_k g_{i,k}$ as the number of links, or direct neighbors, of node i. The utility of a given strategy can be defined as the net value of the benefits available through the connections established by the strategy minus the cost of establishing these connections.

$$u_i(g_i, g_{-i}) = \sum_{j \in \mathcal{N}_i^+(g)} \delta^{d_{ij}(g_i, g_{-i})} - c\mu_i(g_i). \tag{1}$$

A best response strategy of node i to g_{-i}, hereafter denoted by $\mathrm{BR}(g_{-i})$, is a strategy g_i such that

$$\mathrm{BR}(g_{-i}) \in \arg\max_{g_i \in \mathcal{G}_i} u_i(g_i, g_{-i}), \tag{2}$$

where $\mathcal{G}_i$ is the set of all possible pure strategies of node i[1]. Hence for any best response g_i,

$$u(g_i, g_{-i}) \geq u(g_i', g_{-i}) \qquad \forall g_i' \in \mathcal{G}_i.$$

Definition 1. A network g is said to be a *Nash network* if $g_i = \mathrm{BR}(g_{-i})$, $\forall i \in \mathcal{N}$.

2.2 Repeated Myopic Play

Consider the case when the network formation game described above is played repeatedly at time steps $t = 1, 2, \ldots$. At the beginning of every time step[2], every node plays the same strategy it used in the last time step with probability p_i. That is, the nodes' strategies exhibit inertia from one time step to another. With probability $1 - p_i$, the nodes update their strategies based on myopic best response to the observed network structure from the previous time. In the case that the best response strategy is not unique, the node randomizes its decision over the set of best response strategies. As a result, a node playing a best response to the same network observed in the previous time step might switch strategies.

Let g^{t-1} denote the network at time $t-1$, then the dynamics of network formation for agent i are

$$g_i^t = \mathrm{BR}(g_{-i}^{t-1}). \tag{3}$$

As an example, consider the 3-node networks shown in Fig. 2. Starting with the network in (a), the networks in (b) and (c) can be constructed by having node 1 switch its connection from 2 to 3, or by adding a connection to node 3 respectively. Therefore, $g_1^{(a)} = (1,0)$, $g_1^{(b)} = (0,1)$, $g_1^{(c)} = (1,1)$ and $g_{-1} = (g_2, g_3) = ((1,1),(0,1))$. For node 1, the utility for the different strategies are, $u_1(g_1^{(a)}, g_{-1}) = \delta + \delta^2 - c$, $u_1(g_1^{(b)}, g_{-1}) = \delta + \delta^2 - c$, and $u_1(g_1^{(c)}, g_{-1}) = 2\delta - 2c$. For node 2, if $c \leq \delta$, then $u_2((1,1), g_{-2}) \geq u_2(g_2', g_{-2})$ for any other strategy $g_2' \in \mathcal{G}_2$ of node 2. Because of the symmetry between nodes 1 and 3, then the network in (a) is a Nash network if $c < \delta$ and

[1] Here we are restricting our attention to the set of pure strategies

[2] Except at $t = 1$.

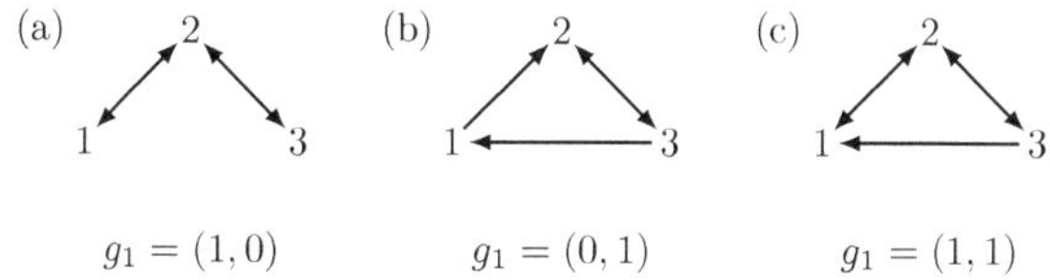

Fig. 2 A 3-node network showing the three strategies for node 1 with nodes 2, and 3 using the strategies $g_2 = (1, 1)$ and $g_3 = (0, 1)$

$$\begin{aligned} u_1(g_1^{(a)}, g_{-1}) &\geq u_1(g_1^{(c)}, g_{-1}) \\ \delta + \delta^2 - c &\geq 2\delta - 2c \\ c &\geq \delta - \delta^2. \end{aligned} \tag{4}$$

Notice that node 1 would be indifferent between the strategies $g_1 = (1, 0)$ and $g_1 = (0, 1)$. Hence, networks (a) and (b) would be Nash networks, and node 1 would switch between these two configurations provided the other two nodes do not change their strategies.

If nodes 1 and 3 are allowed to change strategies simultaneously based on a best response to the previous network, then it is conceivable that both nodes would switch strategies where they switch to connections from nodes 3 and 1 instead of the existing connections to node 2. Hence, the network becomes $g = ((0, 1), (1, 1), (1, 0))$. As such the utility of this network for either node becomes $u_1 = u_2 = \delta - c$, which is less than the current utility of $u_1 = u_2 = \delta + \delta^2 - c$. Therefore, in the event that players are allowed to switch strategies simultaneously, the network in (a) is not stable.

3 Coevolutionary Model

3.1 Dynamic Flow of Benefits

This work is concerned with the case of dynamic flow of the benefits. In this case, the benefits flow over time from one node to another. If the timescale for flow is fast compared to the network formation dynamics, then there is a separation of time scales, and this situation would closely resemble the above mentioned case of immediate benefit availability. However, if the time scales for benefit flow and network formation are comparable, then this presents a coevolutionary process through which benefit flows and network formation occur simultaneously and the emergent behavior can be different. We consider the case where the benefits obtained are derived through a dynamic process. Upon establishing a link, a node will realize a portion of the direct benefit of the connected node, and with time, the benefits are asymptotically realized. This model represents delay in the flow of benefits from a node another. The same applies to benefits from non-direct connections, and the delay is distance dependent, i.e., the further away two nodes are from each other, then the slower is the flow of benefits from one to another. The distance dependence is very relevant to a number of systems including physical transfer of goods, and information or knowledge transfer.

Additionally, when a path between two nodes is severed, then the benefits available from a node to another are not lost immediately, but are forgotten over time. Here, the rate is also distance dependent. Formally, we define the benefit flow model to be

$$b_{ij}^t = f(b_{ij}^{t-1}, g_i, g_{-i}) = \begin{cases} \alpha_{d_{ij}} b_{ij}^{t-1} + (1-\alpha_{d_{ij}})\delta^{d_{ij}}, & \delta^{d_{ij}} \geq b_{ij}^{t-1} \quad (5a) \\ \beta_{d_{ij}} b_{ij}^{t-1} + (1-\beta_{d_{ij}})\delta^{d_{ij}}, & \delta^{d_{ij}} < b_{ij}^{t-1} \quad (5b) \end{cases}$$

such that $\beta_i, \alpha_i \in [0,1]$, $\alpha_1 \leq \alpha_2 \leq \ldots$ and $\beta_1 \geq \beta_2 \geq \ldots$.

Here, b_{ij}^t is the benefit available to node i from node j at time t. Let B denote the matrix whose elements are b_{ij}, and b_i be the ith row of the matrix B[3].

Examining Eq. (5a) closely, notice that the benefits are increasing since $\delta^{d_{ij}} \geq b_{ij}$. That is, when the attainable benefit is higher that the current flow of benefits from a given node, then the benefits will increase. The rates for increase $\alpha_{d_{ij}}$ are distance dependent and hence the subscript. Note that the higher the value of $\alpha_{d_{ij}}$, then the slower that benefits flow. As such, the benefits flow slower as the distances between nodes increase.

When the attainable benefit from a given node, $\delta^{d_{ij}}$ is less that the current flow of benefits, then the benefits will decrease according to Eq. (5b). The rates of decrease are distance dependent. The lower the value of $\beta_{d_{ij}}$, the higher the decrease in benefits. When there is no path between two nodes i and j, then d_{ij} is infinite and $\delta^{d_{ij}} = 0$, and benefits will decrease at a rate of β_∞. In the case when $\alpha_{d_{ij}} = 1$, then no benefits will flow. Similarly, when $\beta_{d_{ij}} = 1$, benefits will not decrease. Furthermore, when $\alpha_{d_{ij}} = 0$, $\beta_{d_{ij}} = 0$, then the dynamics in (5) become equivalent to the instantaneous benefit availability model.

Compared to the instantaneous model of benefit availability, for a fixed network, the attainable benefits of both models are the same. However, when the network is fixed, the dynamic flow model reaches that benefit in the limit, as the distances between nodes d_{ij} do not change for a fixed network $\lim_{t\to\infty} b_{ij}^t = \delta^{d_{ij}}$.

For a given network $g = (g_i, g_{-i})$, and given the benefits b_i, the utility for node i can be given by

$$u_i(b_i, g_i) = \sum_j b_{ij} - c\mu(g_i).$$

In the instantaneous benefit flow model, the feedback law to select strategies assumed an instantaneous realization of the full benefits from other nodes. In the dynamic benefit flow case, a similar model can be obtained by allowing

[3] b_{ii} will be assumed to be equal to 1.

the dynamics to propagate to infinite time and and then selecting a new strategy based on the limit of the average utility over time. This model, however, introduces a separation of time scales where the dynamics of benefit flow have no effect on the outcome of repeated play of the strategic network formation game.

Alternatively, the strategy of a node can dynamically depend on the available benefits at a given time. Here, a node can be selecting a strategy at a given point in time such that it maximizes a utility dependent cost function, for example the total discounted utility

$$J = \sum_{t} \rho^t u(b_i^t, g_i).$$

The complexity introduced by the dynamic interdependence of benefits on the strategies of other nodes, renders the computation of strategies a challenging task. Alternatively, we will consider the case when nodes are myopic decision makers, whose interest is to maximize the projected utility based on the benefit flow in the next time step, and assuming the strategies of other nodes remain unchanged from the currently observed topology.

Hence, for a given strategy g_i, strategies of other nodes g_{-i}^{t-1}, and benefits vector at time $t-1$, the utility is

$$u_i(g_i, g_{-i}^{t-1}, b_i^{t-1}) = \sum_{j} f(b_{ij}^{t-1}, g_i, g_{-i}^{t-1}) - c\mu(g_i). \tag{6}$$

As such, at time steps $t = 1, 2, \ldots$, a randomly selected node plays a best response to the currently observed benefit flow and network topology,

$$g_i^t = \mathrm{BR}(g_{-i}^{t-1}, b_i^{t-1}) \in \arg\max_{g_i \in \mathcal{G}_i} u_i(g_i, g_{-i}^{t-1}, b_i^{t-1}). \tag{7}$$

To that end, at time $t-1$ a node i evaluates, for every possible strategy, the benefits available using (5). With the selected strategy, the benefit flow dynamics are propagated one time step and a new node is selected randomly to update its strategy.

3.2 *Equilibria of the Coupled Dynamics*

We shall consider the limiting behavior of the interconnection of the dynamic benefit flow model (5) and myopic best response network formation (7).

Definition 2. The pair (B^*, g^*) is an equilibrium of the coupled dynamics in (5) and (7) if $\forall i \in \mathcal{N}$, $g_i^* = \mathrm{BR}(g_{-i}^*, b_i^*)$ and $\forall j$ $b_{ij}^* = f(b_{ij}^*, g_i^*, g_{-i}^*)$.

Here the equilibrium involves both network topology g^* and a steady-state benefit flow B^*. One class of equilibria that can emerge is when the topology of the network remains unchanged, i.e., $g^t = g^*$, $\forall t \geq t_0$ for some network topology g^*. As a consequence, the shortest distances between nodes remain unchanged, i.e, $d_{ij}(g_i^t, g_{-i}^t) = d_{ij}(g_i^*, g_{-i}^*)$, $\forall i, j$ and $\forall t \geq t_0$. Therefore, the benefits for each node will correspond to $b_{ij}^* = \delta^{d_{ij}(g_i^*, g_{-i}^*)}$.

Definition 3. The pair (B^*, g^*) is a strict equilibrium of the coupled dynamics in (5) and (7) if and only if $u(g_i^*, g_{-i}^*, b_i^*) - u(g_i', g_{-i}^*, b_i^*) > 0$, $\forall g_i' \in \mathcal{G}_i \backslash g_i^*$, $\forall i \in \mathcal{N}$.

Now let $d_{ij}^* = d_{ij}(g_i^*, g_{-i}^*)$ and $d_{ij}' = d_{ij}(g_i', g_{-i}^*)$ for some strategy $g_i' \in \mathcal{G}_i \backslash g_i^*$. Also define $\mathcal{S}_1 = \{j : d_{ij}' \leq d_{ij}^*\}$ and $\mathcal{S}_2 = \{j : d_{ij}' > d_{ij}^*\}$, these are the sets of nodes whose distance to node i given strategy g_i' are respectively smaller than and greater than their distances given the equilibrium strategy g_i^*. In retrospect, the sets would correspond to those nodes whose benefit dynamics will be updated using Equations (5a) and (5b) respectively.

Proposition 1. *An equilibrium (B^*, g^*) is strict if $\forall i \in \mathcal{N}$, and $\forall g_i' \in \mathcal{G}_i \backslash g_i^*$*

$$\sum_{j \in \mathcal{S}_1} (1 - \alpha_{d_{ij}'})(\delta^{d_{ij}^*} - \delta^{d_{ij}'}) + \sum_{j \in \mathcal{S}_2} (1 - \beta_{d_{ij}'})(\delta^{d_{ij}^*} - \delta^{d_{ij}'}) + c(\mu(g_i') - \mu(g_i^*)) > 0. \tag{8}$$

Proof. For any equilibrium such that $g_i^t = g_i^{t-1}$, $\forall t \geq t_0$, we know that $b_{ij}^* = \delta^{d_{ij}(g_i^*, g_{-i}^*)} = \delta^{d_{ij}^*}$. For a strict equilibrium we have

$$\begin{aligned}
u(g_i^*, g_{-i}^*, b_i^*) - u(g_i', g_{-i}^*, b_i^*) &> 0 \\
\sum_j f(b_{ij}^*, g_i^*, g_{-i}^*) - c\mu(g_i^*) - \sum_j f(b_{ij}^*, g_i', g_{-i}^*) - c\mu(g_i') &> 0 \\
\sum_{j \in \mathcal{S}_1 \cup \mathcal{S}_2} \delta^{d_{ij}^*} - \sum_{j \in \mathcal{S}_1} \alpha_{d_{ij}'} \delta^{d_{ij}^*} + (1 - \alpha_{d_{ij}'}) \delta^{d_{ij}'} & \\
- \sum_{j \in \mathcal{S}_2} \beta_{d_{ij}'} \delta^{d_{ij}^*} + (1 - \beta_{d_{ij}'}) \delta^{d_{ij}'} + c(\mu(g_i') - \mu(g_i^*)) &> 0 \\
\sum_{j \in \mathcal{S}_1} (1 - \alpha_{d_{ij}'})(\delta^{d_{ij}^*} - \delta^{d_{ij}'}) + \sum_{j \in \mathcal{S}_2} (1 - \beta_{d_{ij}'})(\delta^{d_{ij}^*} - \delta^{d_{ij}'}) + c(\mu(g_i') - \mu(g_i^*)) &> 0
\end{aligned}$$

□

Notice that for $j \in \mathcal{S}_1$, $\delta^{d_{ij}(g_i^*, g_{-i}^*)} \leq \delta^{d_{ij}(g_i', g_{-i}^*)}$, and for $j \in \mathcal{S}_2$, $\delta^{d_{ij}(g_i^*, g_{-i}^*)} > \delta^{d_{ij}(g_i', g_{-i}^*)}$. Therefore, the first term on the left in (8) is nonpositive and the second term is positive.

An equilibrium of the combined dynamics of network formation and benefit flow involves both a network topology and benefit flow matrix. The equilibria here concern the limiting behavior of the coupled dynamics and not just a best response network like a Nash network of the static game. One of the questions to consider is whether the coevolutionary dynamics can induce some network topologies to become equilibria while they are not equilibria of the non-coevolutionary network formation (static game). In the following we show through an example that given a common set of parameters, such topologies exist.

Proposition 2. *For $N = 3$, if $(1-\beta_\infty)\delta \geq c \geq (1-\alpha_1)(\delta-\delta^2)$, and $\beta_2 < \alpha_1$, then the pair (B^*, g^*) given by*

$$g^* = ((1,0),(1,1),(0,1)), \quad B_g^* = \begin{bmatrix} 1 & \delta & \delta^2 \\ \delta & 1 & \delta \\ \delta^2 & \delta & 1 \end{bmatrix},$$

is an equilibrium of the coupled dynamics in (5) and (7), and g^* is not a Nash equilibrium of the static game when $\alpha_1 > 0$.

Proof. Consider the node utilities of the network in Fig. 2(a) and the associated benefits matrix B^*. By symmetry of nodes 1 and 3, we will only consider the utilities of nodes 1 and 2. For node 1, comparing strategies (1,0) and (0,1), using (8), we have

$$\begin{aligned} (1-\alpha_{d_{13}})(\delta^{d^*_{13}} - \delta^{d'_{13}}) + (1-\beta_{d'_{12}})(\delta^{d^*_{12}} - \delta^{d'_{12}}) &> 0 \\ (1-\alpha_1)(\delta^2-\delta) - (1-\beta_2)(\delta^2-\delta) &> 0 \\ \alpha_1 &> \beta_2. \end{aligned}$$

Moreover, comparing strategies (1,0) and (1,1) using (8) we have

$$\begin{aligned} (1-\alpha_{d_{13}})(\delta^{d^*_{13}} - \delta^{d'_{13}}) + c &> 0 \\ (1-\alpha_1)(\delta-\delta^2) &< c. \end{aligned}$$

For node 2, comparing strategies (1,1) and (1,0) or equivalently (0,1) we have

$$\begin{aligned} (1-\beta_{d'_{23}})(\delta^{d^*_{23}} - \delta^{d'_{23}}) - c &> 0 \\ (1-\beta_\infty)\delta &> c. \end{aligned}$$

Notice that when $\alpha_1 = 0$, then g^* is an equilibrium if $c \geq \delta - \delta^2$ which retrieves the conditions for the static game shown before in (4). □

The above shows that the coupled coevolutionary dynamics can create equilibria that can only be possible at higher costs of link formation in the non-coevolutionary case. Additionally, the equilibrium also highlights the requirement that nodes need to forget benefits of distant or severed nodes faster than receiving benefits.

Characterizing equilibria for all N is a difficult problem that is yet to be tackled. Instead, we present some equilibria of some small networks to highlight some of the typical topologies of such equilibria. For $c < \delta - \delta^2$, the equilibrium in the model of Bala & Goyal [21] is the complete network. For the coevolutionary model, the equilibria are quite diverse and examples of these equilibria for 4- and 5-node are presented in Fig. 3. A sample run converging to a non Nash-network of the static game is shown in Fig. 4.

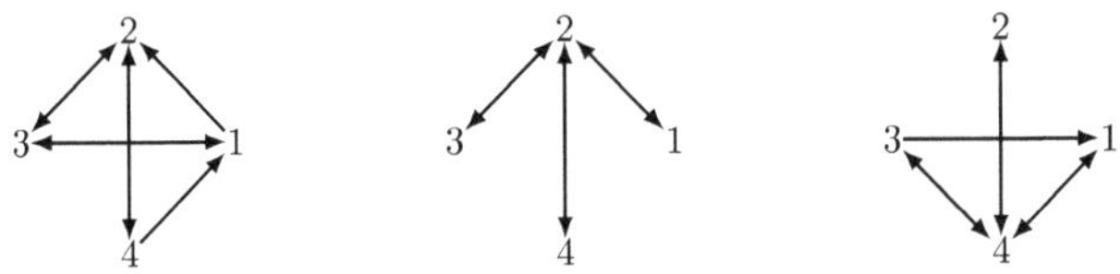

$$\begin{bmatrix} 1 & \delta^2 & \delta & \delta \\ \delta & 1 & \delta & \delta \\ \delta & \delta & 1 & \delta^2 \\ \delta^2 & \delta & \delta^2 & \delta^2 \end{bmatrix} \quad \begin{bmatrix} 1 & \delta & \delta^2 & \delta^2 \\ \delta & 1 & \delta & \delta \\ \delta^2 & \delta & 1 & \delta^2 \\ \delta^2 & \delta & \delta^2 & 1 \end{bmatrix} \quad \begin{bmatrix} 1 & \delta^2 & \delta & \delta \\ \delta^2 & 1 & \delta^2 & \delta \\ \delta^2 & \delta^2 & 1 & \delta \\ \delta & \delta & \delta & 1 \end{bmatrix}$$

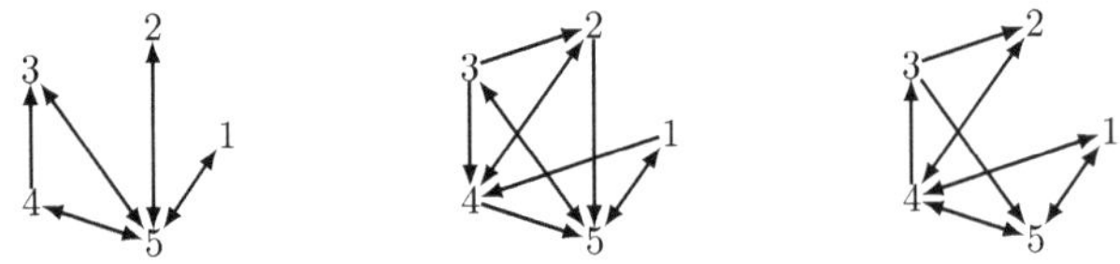

$$\begin{bmatrix} 1 & \delta^2 & \delta^2 & \delta^2 & \delta \\ \delta^2 & 1 & \delta^2 & \delta^2 & \delta \\ \delta^2 & \delta^2 & 1 & \delta & \delta \\ \delta^2 & \delta^2 & \delta^2 & 1 & \delta \\ \delta & \delta & \delta & \delta & 1 \end{bmatrix} \quad \begin{bmatrix} 1 & \delta^2 & \delta^2 & \delta^2 & \delta \\ \delta^3 & 1 & \delta & \delta & \delta^2 \\ \delta^2 & \delta^3 & 1 & \delta^2 & \delta \\ \delta & \delta & \delta & 1 & \delta^2 \\ \delta & \delta & \delta & \delta & 1 \end{bmatrix} \quad \begin{bmatrix} 1 & \delta^2 & \delta^2 & \delta & \delta \\ \delta^2 & 1 & \delta & \delta & \delta^2 \\ \delta^2 & \delta^2 & 1 & \delta^2 & \delta^2 \\ \delta & \delta & \delta^2 & 1 & \delta^2 \\ \delta & \delta & \delta & \delta & 1 \end{bmatrix}$$

Fig. 3 Examples of equilibria of a 4- and 5-node network, when $\delta = 0.9$, $c = 0.05$, $\alpha_1 = 0.6$, $\alpha_2 = 0.7$, $\alpha_3 = 0.8$, $\alpha_1 = 0.9$, $\beta_1 = 0.4$, $\beta_2 = 0.3$, $\beta_3 = 0.2$, $\beta_4 = 0.1$, $\beta_\infty = 0.01$

3.3 Equilibrium Stability

In this section, we will examine the behavior of the coupled dynamics when the network topology is at that of an equilibrium whereas the benefits available are close to the equilibrium values.

Proposition 3. *Let (g^*, B^*), where $g^* = (g_i^*, g_{-i}^*)$, $B^* = [b_{ij}^*]$, be a strict equilibrium such that $\forall i$, $u_i(g_i^*, g_{-i}^*, b_i^*) - u_i(g_i, g_{-i}^*, b_i^*) \geq \gamma$, $\forall g_i \in \mathcal{G}_i \backslash g_i^*$, for some $\gamma > 0$. Also, let $g_{t_0} = g^*$ and $b_{ij}^{t_0} = b_{ij}^* \pm \epsilon$ $\forall i, j$, for some time t_0. If ϵ is sufficiently small, then $g_t = g^*$, $\forall t \geq t_0$, and $\lim_{t \to \infty} b_{ij}^t = b_{ij}^*$, $\forall i, j$.*

Proof. We shall examine the utility of a strategy $g_i^t = g_i'$ compared to $g_i^t = g_i^*$. First define,

$$\mathcal{I}_1' = \{j : \delta^{d_{ij}(g_i', g_{-i}^*)} \geq b_{ij}^{t_0}\}, \quad \mathcal{I}_1 = \{j : \delta^{d_{ij}(g_i^*, g_{-i}^*)} \geq b_{ij}^{t_0}\},$$
$$\mathcal{I}_2' = \{j : \delta^{d_{ij}(g_i', g_{-i}^*)} < b_{ij}^{t_0}\}, \quad \mathcal{I}_2 = \{j : \delta^{d_{ij}(g_i^*, g_{-i}^*)} < b_{ij}^{t_0}\}.$$

Then,

$$\begin{aligned}
u_i(g_i', g_{-i}^*, b_i^{t_0}) =& \sum_j f(b_{ij}^{t_0}, g_i', g_{-i}^*) - c\mu(g_i') \\
=& \sum_{j\in\mathcal{I}_1'} \alpha_{d_{ij}(g_i',g_{-i}^*)} b_{ij}^{t_0} + (1-\alpha_{d_{ij}(g_i',g_{-i}^*)})\delta^{d_{ij}(g_i',g_{-i}^*)} \\
&+ \sum_{j\in\mathcal{I}_2'} \beta_{d_{ij}(g_i',g_{-i}^*)} b_{ij}^{t_0} + (1-\beta_{d_{ij}(g_i',g_{-i}^*)})\delta^{d_{ij}(g_i',g_{-i}^*)} - c\mu(g_i') \\
=& \sum_{j\in\mathcal{I}_1'} \alpha_{d_{ij}(g_i',g_{-i}^*)} b_{ij}^{*} + (1-\alpha_{d_{ij}(g_i',g_{-i}^*)})\delta^{d_{ij}(g_i',g_{-i}^*)} \\
&+ \sum_{j\in\mathcal{I}_2'} \beta_{d_{ij}(g_i',g_{-i}^*)} b_{ij}^{*} + (1-\beta_{d_{ij}(g_i',g_{-i}^*)})\delta^{d_{ij}(g_i',g_{-i}^*)} \\
&- c\mu(g_i') \pm \sum_{j\in\mathcal{I}_1'} \alpha_{d_{ij}(g_i',g_{-i}^*)}\epsilon \pm \sum_{j\in\mathcal{I}_2'} \beta_{d_{ij}(g_i',g_{-i}^*)}\epsilon \\
=& u_i(g_i', g_{-i}^*, b_i^*) \pm \sum_{j\in\mathcal{I}_1'} \alpha_{d_{ij}(g_i',g_{-i}^*)}\epsilon \pm \sum_{j\in\mathcal{I}_2'} \beta_{d_{ij}(g_i',g_{-i}^*)}\epsilon.
\end{aligned}$$

Similarly, we can write

$$u_i(g_i^*, g_{-i}^*, b_i^{t_0}) = u_i(g_i^*, g_{-i}^*, b_i^*) \pm \sum_{j\in\mathcal{I}_1} \alpha_{d_{ij}(g_i^*,g_{-i}^*)}\epsilon \pm \sum_{j\in\mathcal{I}_2} \beta_{d_{ij}(g_i^*,g_{-i}^*)}\epsilon.$$

Therefore,

$$\begin{aligned}
u_i(g_i^*, g_{-i}^*, b_i^{t_0}) - u_i(g_i', g_{-i}^*, b_i^{t_0}) =& u_i(g_i^*, g_{-i}^*, b_i^*) - u_i(g_i', g_{-i}^*, b_i^*) \\
&\pm \sum_{j\in\mathcal{I}_1} \alpha_{d_{ij}(g_i^*,g_{-i}^*)}\epsilon \pm \sum_{j\in\mathcal{I}_2} \beta_{d_{ij}(g_i^*,g_{-i}^*)}\epsilon \\
&\mp \sum_{j\in\mathcal{I}_1'} \alpha_{d_{ij}(g_i',g_{-i}^*)}\epsilon \mp \sum_{j\in\mathcal{I}_2'} \beta_{d_{ij}(g_i',g_{-i}^*)}\epsilon \\
\geq& \gamma - \epsilon(\sum_{\mathcal{I}_1} \alpha_{d_{ij}(g_i^*,g_{-i}^*)} + \sum_{\mathcal{I}_2} \beta_{d_{ij}(g_i^*,g_{-i}^*)}) \\
&- \epsilon(\sum_{\mathcal{I}_1'} \alpha_{d_{ij}(g_i',g_{-i}^*)} + \sum_{\mathcal{I}_2'} \beta_{d_{ij}(g_i',g_{-i}^*)}).
\end{aligned}$$

Since $\alpha_k, \beta_k \in [0,1]$, then for small ϵ we have

$$u_i(g_i^*, g_{-i}^*, b_i^{t_0}) - u_i(g_i', g_{-i}^*, b_i^{t_0}) > 0.$$

This implies that g_i^* is a best response and that $g_i^{t_0+1} = g_i^*$. Since the topology remains unchanged, then $\delta^{d_{ij}}, \forall i, j$ remain unchanged. Furthermore, the stable dynamics in (5) results in $|b_{ij}^{t_0+1} - b_{ij}^*| \leq \epsilon' < \epsilon$. Using the same arguments recursively, the results follow. □

Here we have shown that local stability is guaranteed, for small deviations in the benefit flows from their equilibrium values, for strict equilibria. Strict equilibria are equilibria such that their utilities are strictly greater than the utilities of other strategies given a unilateral deviation of strategy.

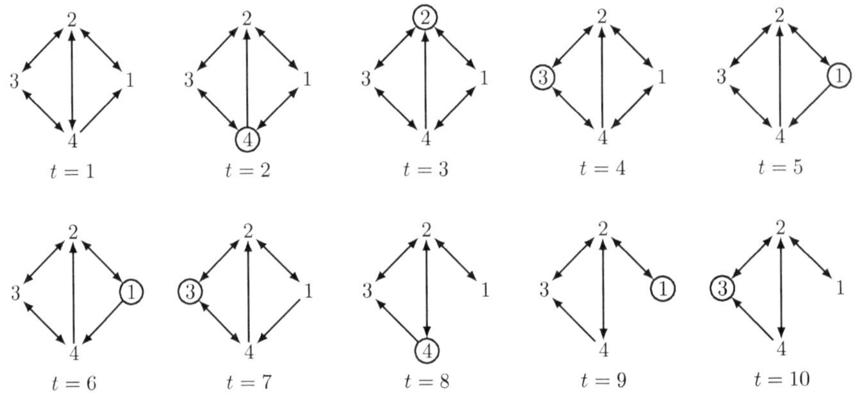

Fig. 4 A sample run of the algorithm converging to a non-Nash network of the static game. Initially $b_{12} = 5.835e^{-1}$, $b_{13} = 8.893e^{-1}$, $b_{14} = 1.893e^{-2}$, $b_{21} = 2.966e^{-4}$, $b_{23} = 1.934e^{-2}$, $b_{24} = 1.299e^{-1}$, $b_{31} = 6.847e^{-1}$, $b_{32} = 5.124e^{-2}$, $b_{34} = 9.707e^{-1}$, $b_{41} = 1.099e^{-3}$, $b_{42} = 6.443e^{-1}$, $b_{43} = 1.711e^{-1}$. A circle around the node number denotes the node updating its network.

4 Conclusions

In this work, we presented a coevolutionary model of network formation based on dynamic flow of benefits between nodes. We showed that the combined dynamics can induce network topologies to be equilibria of the dynamics, whereas these topologies are not Nash networks of the static game. These equilibria can emerge at a lower cost than the non-coevolutionary case. We also showed the stability of a class of equilibria of the combined network formation and benefit flow dynamics. The model can be extended to cases where each edge has a weight that corresponds to the strength of the connection. However, this setup can manifest different behaviors and will be the subject of further studies.

Acknowledgement. Supported by AFOSR project #FA9550-09-1-0420 and ONR project #N00014-09-1-0751.

References

1. Jackson, M.O.: A survey of network formation models: stability and efficiency. In: Group Formation in Economics: Networks, Clubs and Coalitions, pp. 11–57. Cambridge University Press, Cambridge (2005)
2. Tardos, E., Wexler, T.: Network formation games and the potential function method. In: Algorithmic Game Theory, pp. 487–516 (2007)
3. Bloch, F., Jackson, M.O.: The formation of networks with transfers among players. Journal of Economic Theory 133(1), 83–110 (2007)
4. Jackson, M.O.: Social and economic networks. Princeton University Press (2010)
5. Arcaute, E., Johari, R., Mannor, S.: Network formation: Bilateral contracting and myopic dynamics. IEEE Transactions on Automatic Control 54(8), 1765–1778 (2009)
6. Christakis, N.A., Fowler, J.H., Imbens, G.W., Kalyanaraman, K.: An empirical model for strategic network formation. Technical report, National Bureau of Economic Research (2010)
7. Cowan, R.: Network models of innovation and knowledge diffusion. In: Clusters, Networks and Innovation, pp. 29–53 (2005)
8. Young, H.P.: The dynamics of social innovation. Proceedings of the National Academy of Sciences 108(suppl. 4), 21285–21291 (2011)
9. Young, H.P.: Individual strategy and social structure: An evolutionary theory of institutions. Princeton University Press (2001)
10. Ehrhardt, G., Marsili, M., Vega-Redondo, F.: Diffusion and growth in an evolving network. International Journal of Game Theory 34(3), 383–397 (2006)
11. Kozma, B., Barrat, A.: Consensus formation on adaptive networks. Physical Review E 77(1), 016102 (2008)
12. Burda, Z., Krzywicki, A., Martin, O.: Adaptive networks of trading agents. Physical Review E 78(4), 046106 (2008)
13. Blondel, V.D., Hendrickx, J.M., Tsitsiklis, J.N.: On Krause's multi-agent consensus model with state-dependent connectivity. IEEE Transactions on Automatic Control 54(11), 2586–2597 (2009)
14. Skyrms, B., Pemantle, R.: A dynamic model of social network formation. In: Adaptive Networks, pp. 231–251. Springer (2009)
15. Marceau, V., Noël, P.A., Hébert-Dufresne, L., Allard, A., Dubé, L.J.: Adaptive networks: Coevolution of disease and topology. Physical Review E 82(3), 036116 (2010)
16. Acemoglu, D., Mostagir, M., Ozdaglar, A.: State-dependent opinion dynamics. Working paper (2012)
17. Fox, M.J., Piliouras, G., Shamma, J.S.: Medium and long-run properties of linguistic community evolution. In: 9th International Conference on the Evolution of Language (Evolang IX) (2012)
18. Mirtabatabaei, A., Bullo, F.: Opinion dynamics in heterogeneous networks: convergence conjectures and theorems. SIAM Journal on Control and Optimization 50(5), 2763–2785 (2012)
19. Kianercy, A., Galstyan, A.: Coevolutionary networks of reinforcement-learning agents. Physical Review E 88(1), 012815 (2013)

20. Barabási, A.L., Albert, R.: Emergence of scaling in random networks. Science 286(5439), 509–512 (1999)
21. Bala, V., Goyal, S.: A noncooperative model of network formation. Econometrica 68(5), 1181–1229 (2000)

One-Max Constant-Probability Models for Complex Networks

Mark Korenblit, Vadim Talis, and Ilya Levin

Abstract. This paper presents a number of the tree-like networks that grow according to the following newly studied principles: i) each new vertex can be connected to at most one existing vertex; ii) any connection event is realized with the same probability p; iii) the probability Π that a new vertex will be connected to vertex i depends not directly on its degree d_i but on the place of d_i in the sorted list of vertex degrees. The paper proposes a number of models for such networks, which are called *one-max constant-probability models*. In the frame of these models, structure and behavior of the corresponding tree-like networks are studied both analytically, and by using computer simulations.

1 Introduction

According to the well-known Barabási-Albert model [1], scale-free networks are characterized by two main mechanisms: continuous growth and preferential attachment. That is, a) the networks expand continuously by addition of new vertices, and b) there is a higher probability that a new vertex will be linked to a vertex already having many connections (high-degree vertex). Most vertices have only a few connections while there are a few highly connected hubs. Vertices of a scale-free network are the elements of any system and its edges represent the interaction between them.

Mark Korenblit
Holon Institute of Technology, Israel
e-mail: korenblit@hit.ac.il

Vadim Talis
Jerusalem College of Technology, Israel
e-mail: talisv@yahoo.com

Ilya Levin
Tel-Aviv University, Israel
e-mail: ilial@post.tau.ac.il

P. Contucci et al. (eds.), *Complex Networks V*,
Studies in Computational Intelligence 549,
DOI: 10.1007/978-3-319-05401-8_17,

The Barabási-Albert random graph model is described as follows:

Starting with a small number m_0 of vertices, at every time step we add a new vertex with $m \leq m_0$ edges that link the new vertex to m different vertices already present in the system. To incorporate preferential attachment, we assume that the probability Π that a new vertex will be connected to vertex i depends on the degree d_i of that vertex.

The mechanism of preferential attachment is assumed to be linear in the model, i.e., $\Pi(d_i)$ is proportional to d_i [1]. However, as noted in the same work, in general relationship between $\Pi(d_i)$ and d_i could have an arbitrary form and, therefore, different types of preferential attachment may be considered.

It is of interest to consider a special case when in every step a new vertex is connected to only one of the old vertices ($m = 1$). In this case the resulting graph is a tree known as a *nonuniform random recursive tree*. The probability of linking to its vertex depends on its degree. The structure and properties of such trees are investigated in [2], [5], [6], and many other works. When the probability of linking to a vertex is proportional to its degree, this gives a *random plane-oriented recursive tree*.

Nonuniform random recursive trees have a number of applications. They may serve for modeling pyramidal structures based on the principle "success breeds success". In a pyramid scheme where each entrant competes with those already participating, the experience gained in successful recruiting enhances the prospects for further success as captured by the growth rule of these trees [6]. The example of simulation of stock markets with these trees is given in [4].

In our paper we introduce a number of new network models based on nonuniform random recursive trees, so called *one-max constant-probability models*. These models are characterized by the following features: i) each new vertex may be connected to at most one old vertex, i.e., in every time step at most one new edge appears in the network; ii) any connection event is realized with the same probability p due to external factors; iii) the probability Π that a new vertex will be connected to vertex i depends not directly on its degree d_i but on the place of d_i in the sorted list of vertex degrees.

The proposed network model is rather realistic because in real life the choice of an object may be determined not by an absolute characteristic of the object but by a relative status of this object among other objects. The status itself depends, in its turn, on the objects' characteristics. Besides, this model explicitly defines the order of priorities in the search of appropriate connection and, therefore, it allows not just to analyze the topology of networks, but also to examine the network dynamics step-by-step.

2 Constant-Probability Search Model

The first model (we call it *Constant-Probability Search Model* or *CPSM*) is based on a regular linear search of a vertex with a maximum degree realized by consecutive comparisons of a current maximum degree with a degree's value of a current checked vertex. If this value is greater than a current maximum, the maximum is updated. For vertices with equal degrees, an earlier arrived vertex is preferable. However, in contrast to the standard search, every comparison is performed not always but with probability p. A new vertex is connected to a vertex v with a found maximum degree which, correspondingly, is equal to a true maximum degree with probability p. The degree of vertex v is incremented by 1 and the new vertex's degree is assigned to 1 if it has been connected to any vertex.

Therefore, the vertex with the 1-st largest degree will be chosen for connection by a new vertex with probability p, the vertex with the 2-nd largest degree – with probability $(1-p)p, \ldots$, the vertex with the i-th largest degree – with probability $(1-p)^{i-1}p$ (for equal degrees, the degree of a vertex checked earlier is quasi larger). For n existing vertices, the probability that the new vertex will connect to no vertex is equal to $(1-p)^n$.

Proposition 1. *Given an n-vertex network which starts with a single vertex and is based on CPSM, the lower bound of the expected number M_n of the maximum degree in the network is equal to $p(n-1)$.*

Below, one can see that Proposition 1 holds not only for CPSM but also for all other one-max constant-probability models.

It is clear that the higher is p, the larger is degree of the first vertex in the network and the rather this degree is maximum. That is, older vertices increase their connectivity at the expense of the younger ones and a "rich-get-richer" phenomenon [1] is detected for high p.

Diagrams of two 100-vertex networks simulated for different values of p are presented in Fig. 1. Three the largest degrees in a network are indicated (degree of a vertex arrived in time step t is denoted by d_t).

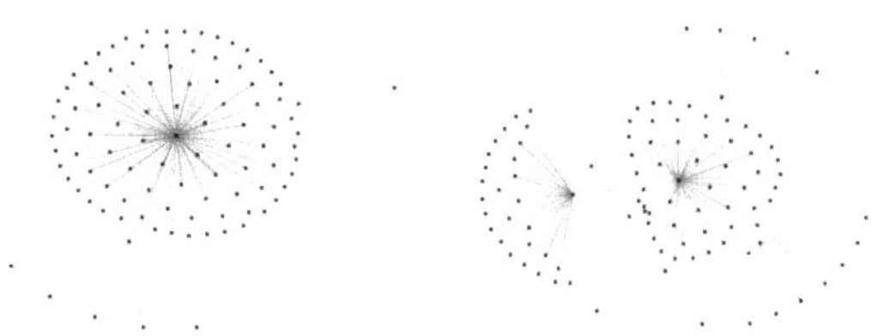

(a) p=0.95, d_1=93, d_2=6, d_3=2 (b) p=0.5, d_2=51, d_5=28, d_8=7

Fig. 1 100-vertex networks based on CPSM

3 Constant-Probability Ordered Model

The second model, so called *Constant-Probability Ordered Model* (*CPOM*) is similar to CPSM. However, in contrast to CPSM, the list of existing vertices is kept sorted in decreasing order of their degrees so that the vertex with a maximum degree is in the top of the list. The list is scanned from the top and a new vertex is connected to the first vertex v which "is allowed to be connected by the probability p". The degree of vertex v is incremented by 1 and this vertex is moved toward the top of the list to find a proper new place for it. The new vertex's degree is assigned to 1 and this vertex is inserted into the list above vertices with degrees 0 (*isolated vertices*) if it has been connected to any vertex.

The running time of the search of appropriate connection in an n-vertex CPSM network is $O(n)$ for any p since always all existing vertices of the network have to be checked. At the same time, CPOM gives $O(n)$ running time in the average case only, while in the best case its running time is $O(1)$. Besides, CPOM exhibits a real network that has a mechanism which keeps most referred sites in the top of the list and makes them, correspondingly, more reachable than others.

Despite the different algorithms used by CPSM and CPOM, both models provide identical network topologies and diagrams illustrated in Fig. 1 are appropriate to CPOM as well.

CPOM (as CPSM) is characterized by the following phenomenon that becomes apparent for low p. Some vertices which come first may remain isolated since while a network is not large, a new vertex may rather connect to no existing vertices and find oneself at the bottom of the list. Next later vertices will find more vertices in the network and the probability of their connecting to one of existing vertices will be higher. At that, they will be linked with a higher probability to vertices with larger degrees and their degrees after connection will be 1. Therefore, as the size of the network increases, the chance of vertices with zero degrees "to be found" by new vertices decreases.

Fig. 2 illustrates the above phenomenon for $p = 0.1$. A network after 100 time steps (Fig. 2 (a)) and the same network after 1000 time steps (Fig. 2 (b)) have the same 6 isolated vertices with order numbers 1, 5, 11, 15, 23, 27.

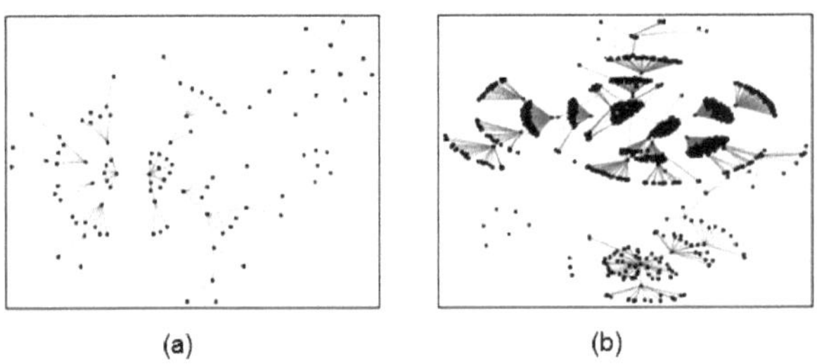

Fig. 2 The phenomenon of first isolated vertices for CPOM

Proposition 2. *Given an n-vertex network based on CPSM or CPOM, the expected number I_n of isolated vertices in the network is defined recursively as follows:* $I_1 = 1$; $I_{n+1} = I_n + 2(1-p)^n - (1-p)^{n-I_n}$.

The result is well reasoned. For $p = 0$, $I_{n+1} = I_n + 1$ (the number of isolated vertices increases in every time step). For $p = 1$, $I_{n+1} = I_n$ (all new vertices are connected to the first one and the number of isolated vertices does not increase at all). For large n, I_{n+1} tends to I_n (probabilities of appearance of new isolated vertices and of connecting new vertices to old isolated vertices decrease).

Corresponding computational results for p from 0 to 1 are presented in Fig. 3. One can see that for $p < 0.5$, the higher is p, the smaller is n for which I_n reaches saturation and the smaller is I_n in saturation itself. For $p > 0.5$, the expected number of isolated vertices is less than 1.

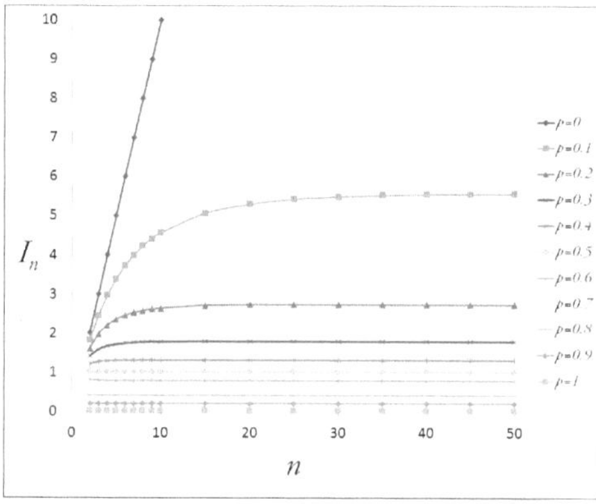

Fig. 3 Expected numbers of isolated vertices in CPOM and CPSM networks

4 Constant-Probability Ordered Non-0 Model (CPOM-N0)

In order to neutralize the negative effect described in the previous section, when some vertices which come first may remain isolated, we slightly modify CPOM. A new vertex connected to one of existing vertices is not inserted above isolated vertices and remains at the bottom of the list. Thus old vertices with zero degrees will not be at the bottom and the list will be sorted only concerning degrees exceeding 1. Such a model is appropriate to be called *Constant-Probability Ordered Non-0 Model* (*CPOM-N0*).

The example of this model's behavior for $p = 0.1$ is shown in Fig. 4. In Fig. 4 (a) one can see a network after 100 time steps. This network has 3 isolated vertices: 5, 12, and 17. The same network after 300 time steps is presented in Fig. 4 (b). It has the only isolated vertex 5. At last, after 1200 time steps, there are no isolated vertices in this network (Fig. 4 (c)).

CPOM-N0 is evidence that the additional advantage of CPOM in contrast with CPSM is its flexibility. The list of existing vertices in CPOM is actually the priority list. While in CPSM a vertex's degree directly determines the vertex' priority, in CPOM the vertex's place in the list is this criterion. One can define this place not only as a function of a degree but as a function of additional parameters as well.

There are also other differences in behavior of CPOM and CPOM-N0. Isolated vertices not only disappear in networks based on CPOM-N0 for large n. For the same small n, the expected number of vertices with zero degree in a CPOM-N0

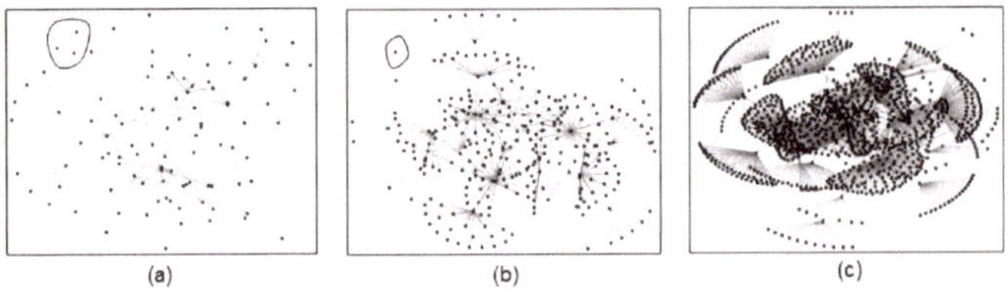

Fig. 4 A network based on CPOM-N0 ($p = 0.1$)

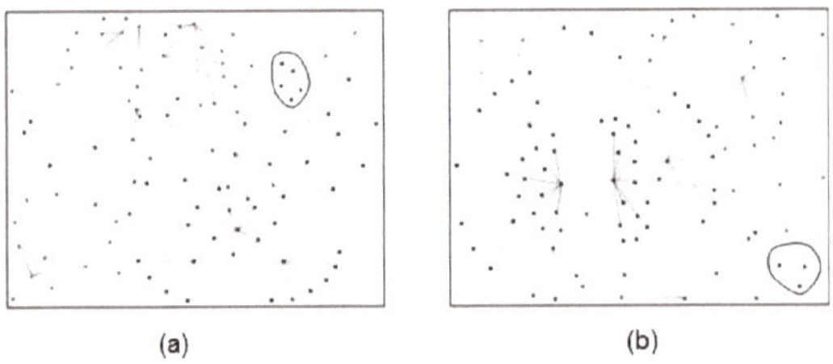

Fig. 5 100-vertex networks

network is less than in a CPOM network. On the other hand, the expected number of *connected components* (collections of connected vertices which have no connections to one another) consisting of more than one vertex in a CPOM-N0 network is greater than in a CPOM network of the same size. The explanation of this phenomenon is the following. An isolated vertex of a CPOM network may rather remain isolated in the next time steps than in a CPOM-N0 network in which this vertex has a higher probability to become a start vertex of a new autonomous part of the network. In any case, both networks are characterized by the same expected number of connected components including isolated vertices that is equal to the number of vertices which were isolated some time, i.e., to the number of appearances of isolated vertices.

Two corresponding examples are illustrated in Fig. 5. In Fig. 5 (a) one can see the CPOM network after 100 time steps. This network has 11 connected components, 5 of which are isolated vertices (5, 12, 14, 30, 57). The CPOM-N0 network after 100 time steps presented in Fig. 5 (b) has also 11 connected components and only 3 of them are isolated vertices (25, 33, 40). With increase of the network in Fig. 5 (b), new vertices will connect to these 3 vertices sooner or later, while the probability of connecting new vertices to 5 isolated vertices in Fig. 5 (a) will decrease in every time step. Herewith, both networks will consist of 11 connected components, and the probability of appearance of new connected components will decrease with increase of the networks.

The expected numbers of connected components including isolated vertices are equal in networks of the same size based on all one-max constant-probability models. This fact allows to formulate and to prove the following proposition:

Proposition 3. *Given an n-vertex network based on a one-max constant-probability model, the expected number C_n of connected components in the network is defined recursively as follows:* $C_1 = 1$; $C_{n+1} = C_n + (1-p)^n$.

Corollary 1. *Given a network discussed in Proposition 3, the expected number C_n of connected components in the network is expressed explicitly as follows:* $C_n = 1 + (1-p)\frac{1-(1-p)^{n-1}}{p}$. *With increase of n, C_n tends to* $\frac{1}{p}$.

5 Constant-Probability Ordered Directed Model

Previous models assume that connecting a new vertex to an old one leads to increase of a number of connections both of the old and the new vertices. However, not always a subject that initiates a connection is considered as acquiring this connection. At the same time, a referred object is regarded as a possessor of this connection in any case. Thus while most networks (from social to biological ones) are undirected, there are systems that should be simulated by directed networks. For example, Web pages are connected by directed links [3], [7], software modules are taken as vertices of a directed graph with links according to their interaction [3].

We slightly modify CPOM and introduce a *Constant-Probability Ordered Directed Model* (*CPODM*). An edge corresponding to a new connection leaves the new vertex and enters the old one. The list of vertices is sorted by their in-degrees. It is clear that the in-degree of a new vertex is 0 even if it has been connected to any existing vertex and, therefore, a new vertex is always in the bottom of the list.

Out-degree of any vertex in a network based on CPODM is 1 (if the vertex has been connected to any vertex when arriving) or 0 (if the vertex has been connected to no vertex when arriving). As follows from the model's description, the list of vertices does not distinguish between vertices with zero and non-zero out-degrees. For two vertices with zero in-degrees, the older vertex will be nearer to the top. Thus old *isolated vertices* (with zero in-degrees and out-degrees) will not be at the bottom of the list.

One can see that CPODM is similar to CPOM-N0. Although CPOM-N0 describes an undirected network, it distinguishes in special cases between a vertex that is connected to another one and a vertex to which another vertex is connected. In fact, both CPOM-N0 and CPODM identically process new vertices. For this reason, the same characteristic features inherent in both models. Like in CPOM-N0 networks, isolated vertices disappear in networks based on CPODM for large n. For small n, expected numbers of isolated vertices and of connected components consisting of more than one vertex for CPODM are the same as for CPOM-N0.

6 Conclusion

In this paper we proposed a number of new models of tree-like networks and studied genesis and evolution of these networks' topology. Some remarkable network effects were observed. We provided the interpretation of the network behavior on the base of analysis of simulation results.

Specifically, we have discovered the phenomenon of the existence of isolated vertices when subjects that were at the origins of a complex network creation may ultimately find oneself out of the network. We have interpreted the cause of this phenomenon and have shown how it can be prevented. The absence of isolated vertices in a large network, in turn, does not prevent it from splitting on unlinked autonomous parts (connected components) whose number tends to $\frac{1}{p}$ with increase of the network.

References

1. Barabási, A.-L., Albert, R.: Emergence of scaling in random networks. Science 286, 509–512 (1999)
2. Devroye, L., Fawzi, O., Fraiman, N.: The height of scaled attachment random recursive trees. In: DMTCS Proc. AM, AofA 2010, pp. 129–142 (2010)
3. Hein, O., Schwind, M., König, W.: Scale-free networks – the impact of fat tailed degree distribution on diffusion and communication processes. Wirtschaftsinformatik 48(4), 267–275 (2006)
4. Hoffmann, A.O.I., Jager, W., Von Eije, J.H.: Social simulation of stock markets: taking it to the next level. Journal of Artificial Societies and Social Simulation 10(2) (2007), http://jasss.soc.surrey.ac.uk/10/2/7.html
5. Katona, Z.: Levels of a scale-free tree. Random Structures and Algorithms 29(2), 194–207 (2006)
6. Mahmoud, H.M., Smythe, R.T., Szymański, J.: On the structure of random plane-oriented recursive trees and their branches. Random Structures and Algorithms 4(2), 151–176 (1993)
7. Raigorodski, A.M.: Models of random graphs and their use. Trudy MFTI 2(4), 130–140 (2010) (in Russian)

Efficient Routing in Data Center with Underlying Cayley Graph*

Miguel Camelo, Dimitri Papadimitriou, Lluís Fàbrega, and Pere Vilà

Abstract. Nowadays data centers are becoming huge facilities with hundreds of thousands of nodes, connected through a network. The design of such interconnection networks involves finding graph models that have good topological properties and that allow the use of efficient routing algorithms. Cayley Graphs, a kind of graphs that represents an algebraic group, meet these properties and therefore have been proposed as a model for these networks. In this paper we present a routing algorithm based on Shortlex Automatic Structure, which can be used on any interconnection network with an underlying Cayley Graph (of some finite group). We show that our proposal computes the shortest path between any two vertices with low time and space complexity in comparison with traditional routing algorithms.

1 Introduction

The growing demand for cloud computing services is leading to an increasing deployment of large-scale Data Center (DC) as its underlying infrastructure [1]. A fundamental component of such DCs is the interconnection network that provides communication among the different components of the physical infrastructure. The design of such networks has the goal of finding graph models that i) have good properties topological (e.g. high connectivity, small degree, etc.) to ensure good

Miguel Camelo · Lluís Fàbrega · Pere Vilà
IIiA, Universitat de Girona, Spain
e-mail: {miguel.camelo,lluis.fabrega,pere.vila}@udg.edu

Dimitri Papadimitriou
Alcatel-Lucent Bell, Belgium
e-mail: dimitri.papadimitriou@alcatel-lucent.com

* This research work is partially funded by the European Commission (EULER project, FP7 258307), the Spanish Government (RoGER project, TEC 2012-32336), the Generalitat of Catalonia (CSI project, SGR-1202), and the SUR of the Generalitat of Catalonia and the European Social Fund (PhD grant FI-DGR 2011).

P. Contucci et al. (eds.), *Complex Networks V*,
Studies in Computational Intelligence 549,
DOI: 10.1007/978-3-319-05401-8_18, © Springer International Publishing Switzerland 2014

performance in terms of throughput, delay, robustness, etc., and that ii) allow to have routing algorithms with both low time complexity (time to take routing decision) and low space complexity (memory resources to build the routing table) with respect to the number of nodes of the network [2]. Cayley Graphs (CGs) [3], a kind of graphs that represents an algebraic group, meet these 2 properties and therefore have been proposed as a model of DCs interconnection networks [4, 5, 6].

Concerning topological properties, CGs have high symmetry, hierarchical structure, recursive construction, high connectivity and fault tolerance, among others [2]. The definition of the CG implies that the vertices are elements of some group but it does not imply any specific group. This flexibility allows to get a graph that meets the desired requirements on diameter, vertex degree, number of nodes, etc [7]. Moreover, it has been demonstrated that CGs can also be used as models of deterministic small world networks [8]. With respect to routing algorithms, the traditional Dijkstra or Bellman-Ford routing algorithms can be used in any kind of graph but requiring large amount of memory and/or with slow convergence time for large graphs [4]. Unlike them, there are routing algorithms for specific type of graphs that take advantage of their particular topological characteristics, reducing their time and space complexity. This is the case of the routing algorithms for network topologies based on hypercubes [9], butterflies [10] and star graphs [11], among others, which actually are CGs of some specific groups.

A routing algorithm for two specific classes of Cayley-based networks, the star and pancake graphs, is presented in [13]. Since these graphs have a representation by permutations, they propose a routing algorithm based on permutation sort. However, this approach does not ensure a shortest path routing. K. Tang and B. Arden prove in [14] that all finite CGs can be represented by generalized chordal rings (GCR) and then propose an iterative routing algorithm based on table look-up. The space complexity of such algorithm is $O(n^2)$ and its time complexity O(D), where n and D are the size and diameter of the network, respectively. Wang and Tang [15] propose a topology-based routing for Xmesh with CGs as the underlying topology. They prove that the average path lengths between nodes is smaller and the averaged power consumed is less than the original Xmesh. They use a CG from the Borel Subgroup, which is also known as Borel Cayley Graph (BCG), as underlying topology. Their routing algorithm computes off-line a shortest path routing table from the node *Id* to all other nodes, and then they use this table to create the routing table for the rest of nodes based on the vertex transitivity property of CG. This algorithm is is bounded by $O(\log_4 n)$ and its space complexity is given by $O(n^2D)$.

A distributed and fault-tolerant routing algorithm for BCG is presented in [16]. This two-phase algorithm uses two types of routing tables according to link failures: (1) a Static Routing Table (SRT) (computed using [14]) and (2) a Dynamic Routing Table (DRT). The first phase performs routing through the shortest path according to the Static Routing Table. If there is a link failure making the shortest path unavailable, DRT is updated and other shortest paths will be used. In the case that all shortest paths are disconnected, the phase 2 exploits the path length information in the SRT to search additional routes besides the shortest paths. Finally, authors in [5] present a routing algorithm on a special class of CGs used as underlying graph for a wireless data center. A two-level routing algorithm is proposed to send messages between 1) servers in the same rack and 2) servers in different racks. This algorithm is a geographical routing that exploits the uniform structure of the underlying topology. The identification of each server is defined by composition of three values: the coordinates of the rack, the story that contains the server within the rack and the index of the server in the story. In addition, each server uses three routing tables to forward package from source to destination using a shortest path route.

Note that the works presented above could be grouped into the one or more of the following routing algorithm categories: a) those ones that are designed to specific CGs, b) general purpose ones with high space/time complexity and c) low complexity ones that do not ensure shortest paths. In contrast with them, in this paper we present a low space and time complexity routing algorithm for any interconnection network where its underlying graph is a CG of some finite group. The input of the algorithm can be either the group presentation $G = \langle S|R \rangle$, where S and R are the generators and relators of G /citewmagnus, or the permutation (or matrix) representation of the group. The proposed algorithm is based on the fact that finite groups are Automatics and have an Shortlex Automatic Structure (SAS) [12]. These structures solve the shortest path problem in CGs of finite groups by solving the equivalent Minimum Word Problem, which is NP-Hard [18], in quadratic time with respect to the length of the equivalent path written as a sequence of group generators.

The paper is organized as follows. In Section 2 we present the theoretical background about group theory, geometrical group theory and automatic structures. In Section 3 we describe our proposed shortest path algorithm, its time and space complexity and an example of the application of our routing algorithm to a 3-cube graph. Conclusions and future work are presented in Section 4.

2 Preliminaries

In this section, we establish terminology, notation, and background material about group theory and Automatic Groups. For more definitions and results on combinatorial group theory we refer the reader to [3], and for groups and graphs to [20].

Let G be a finite group. The identity of the group G is denoted by Id and the group operation is the multiplication. Let $S = \{s_1, \ldots, s_n\}$ be a set of elements in a group G. We say that S generates G if every element of G can be expressed as a product of elements from S and their inverses. A group G is finitely generated if it has a finite generating set. A word is a sequence $w = (s_1 s_2 \ldots)$, where $s_i \in S \cup S^{-1}$, for all i. We say that w is freely reduced if it does not contain any sub-word $s_i s_i^{-1}$. We say that a group is a free group with basis S, represented by $F(S)$, if S is a set of generators for the group and no freely reduced word $w \in F(S)$ represents the identity.

Definition 1. *Let G be a group with generating set S. The Cayley Graph $\Gamma(G,S)$ of G with respect to S is the graph with vertex set $V(\Gamma(G,S)) = \{g \mid g \in G\}$ and edge set $E(\Gamma(G,S)) = \{(g, gs) \mid s \in S, g \in G\}$.*

The group G acts on $\Gamma(G,S)$ by multiplication on the left: the element $g \in G$ defines a map $\phi_g : h \to gh$ that maps a vertex $h \in \Gamma(G,S)$ to the vertex gh, while it brings adjacent vertices to adjacent vertices, preserving edges. The graph $\Gamma(G,S)$ is directed but it also can be considered undirected if we take an inverse-closed generating set, i.e. $S = S^{-1}$. If $\Gamma(G,S)$ has not auto-loops, then $Id \notin S$. If $\Gamma(G,S)$ has no loops and no multiple edges, then we say that $\Gamma(G,S)$ is reduced, and then we say that it is simple if in addition is undirected. Finally, for any finite presentation of a group in terms of generators and relators, there exists an associated Cayley Graph, i.e. the geometry and structure of the $\Gamma(G,S)$ is directly related with a group presentation and specifically with its generator set [3].

A metric on the CG is defined by assigning unit length to each edge and defining the distance between two points to be the minimum length of paths joining them. In this case, the action by left multiplication of G on $\Gamma(G,S)$ is by isometries. Finally, the algebraic structure of the group, which is encoded into its group presentation, permits to define the word length and metric on such group:

Definition 2. *Let $\pi : F(S) \to G$ be a group homomorphism and let $\pi(w)$ be the element of G represented by w under π. The length of g, identified by $l_s(g)$, is the length of the shortest word in the free group $F(S)$ representing g, i.e. $l_s(g) = min\{l_s(w) \mid w \in F(S), \pi(w) = g\}$.*

Definition 3. *Let G be a group with generating set S. The corresponding word metric (i.e. distance function) d_s is the metric on G satisfying $d(Id,s) = d(Id,s^{-1}) = 1$ for all $s \in S$, and $d(g,h) = min\{l_s(w) \mid w \in F(S), \pi(w) = g^{-1}h\}$, for all $g,h \in G$.*

Note that the word metric on a group G is a way to measure the length of the shortest path between any two elements of G on $\Gamma(G,S)$. This metric measures how efficient the difference $g^{-1}h$ can be expressed as a word in the generating set for G. Thus, it is possible to visualize the geometry of a group G by looking at its CG, because the word metric of the group corresponds to the graph metric induced on $\Gamma(G,S)$.

Additionally to the algebraic and geometric structure of the group, there exists a third point of view to work in an efficient way on groups: they can be seen as languages. Let G be a group, A an alphabet and A^* the set of strings (or words) on the alphabet A. By interpreting concatenation as an associative multiplication on G, we define a monoid homomorphism $\pi : A^* \to G$. If w is a string over A, we say that $\pi(w)$ is the element of G represented by w. If the homomorphism is surjective, i.e. $\pi(A)$ generates G as a group, then A is the set of group generators for G. We also define a bijective map $\phi : A \to S$, where S is the set of generators of G, to indicate that each element of the set A represents an element of S. In the rest of the paper, we will use the set S to reference both generators of the group and the alphabet A.

Given any word w, there is an associated edge path in $\Gamma(G,S)$. The path starts at the identity vertex and then traverses edges of $\Gamma(G,S)$ as dictated by w. Conversely, every finite edge path in $\Gamma(G,S)$ describes a word in terms of the generators and their inverses: reading off the labels of edges being traversed, and adding an inverse if they are traveling in the opposite direction of the orientation of the edge. Given this relationship between languages and groups, D. Epstein et. al. in [12] present a complete work about algebraic groups treated by finite state automaton.

Definition 4. *Let G be a group. An automatic structure on G consists of a set S of generators of G, a finite state automaton W over S, and a finite state automaton M_s over (S,S), for $s \in S \cup Id$, satisfying the following conditions: 1) the map π : $L(W) \to G$ is surjective and 2) for $s \in S \cup Id$, we have $(w_1, w_2) \in L(M_s)$ if and only if $\pi(w_1)s = \pi(w_2)$ and both w_1 and w_2 are elements of L(W).*

In the definition, W is called the *word acceptor*, M_{Id} the *equality recognizer*, and each M_s, for $s \in S$, a *multiplier automaton* for the automatic structure. An automatic group is one that admits an automatic structure. Note that M_{Id} recognizes equality in G between words in $L(W)$. From a given automatic structure, it is always possible to use M_{Id} to construct another one, such that W accepts a unique word mapping onto each element of G; choosing the lexicographically least among the shortest words that map onto each element as the normal form representative of that element. This W is a word-acceptor with uniqueness.

Definition 5. *Let $\leq_S$ be some total order on the alphabet S, the automatic structure is called shortlex if L(W) consists of the shortlex least representatives of each element $g \in G$; therefore the map $\pi_1 : L(W) \to G$ is bijective and all paths in $\Gamma(G,S)$ according to the words of L(W) are shortest ones. In other words, $L(W) = \{w \in S^* \mid w \leq_S v, \forall w, v \in S^*, w =_G v\}$*

Thus, given a group G with generator set S, a word $w \in S^*$ is called a geodesic if it has minimal length among all strings representing the same element as w. Since the language of all geodesic strings maps finite-to-one onto G, SAS is an automatic structure for G that contains a geodesic representative for each element of G. The package KBMAG [17] implements a procedure for computing SAS for groups with finite presentation.

3 A Greedy Routing Algorithm Using Automatic Structures

Let g and h be two vertices in $\Gamma(G,S)$ represented by the words (or labels) w_g and w_h in the set $S \cup S^{-1}$. If $w_h = w_g s_1 s_2 ... s_t$ with $s_i \in S \cup S^{-1}$, $1 \leq i \leq t$, then $s_1 s_2 \dots s_t$ defines a path from vertex g to h with edges labeled by $s_1 s_2 ... s_t$ in $\Gamma(G,S)$. This is equivalent to finding a path from $g^{-1}h = s_1 s_2 \dots s_t$ to the vertex Id. Notice that given g and h, solving the shortest path problem between any pair of vertices in the CG turns into finding a word $w = s_1 s_2 \dots . s_t$ with minimum length such that $w_g^{-1} w_h = w$. This problem is called the Minimum Word Problem [18], and the SAS is an efficient tool to solve it:.

Theorem 1. *(Theorem 2.3.10, [12]) Let G be an group and $(S, L(W))$ an automatic structure for G. For any word w over S, we can find a word in L(W) representing the same element of G as w, in time proportional to the square of the length of w.*

Note that if the word acceptor has uniqueness, then this structure can solve the Minimum Word Problem by reducing words to their (shortlex) normal forms. Because of finite groups have Automatic Structures (in fact they have a SAS [12]), it is possible to find a SAS for a given group presentation G and to use it as a shortest path computation mechanism using only local information. We assume that an incoming message, whether originating at the vertex or in transit from another vertex, contains the shortlex word that represents the destination vertex.

Our routing algorithm consists of two procedures, the vertex labeling and the message forwarding. The vertex labeling procedure is the following:

- Given a group presentation $G = \langle S|R \rangle$ of the underlying $\Gamma(G,S)$, compute the SAS = $(S,L(W))$ of G. If $\Gamma(G,S)$ is given by either its matrix or permutation representation, then construct the group presentation by using the fundamental cycles of $\Gamma(G,S)$ (e.g. by using [19]).
- Select a random vertex of $\Gamma(G,S)$ and construct a spanning tree $T(\Gamma)$ rooted on it by using the Breadth First Search (BFS) algorithm. In the same process, label all vertices with an integer from 1 (the root vertex) to $|G| = n$, according their order of discovery.
- Use $(S,L(W))$ to enumerate the elements of the group G according to its shortlex ordering (corresponding to a BFS through S^*) and re-label each vertex in the spanning tree with its corresponding shortlex word enumerated before. Note that there exists natural one-to-one mapping between vertices and the elements of the group represented by their shortlex words following their BFS ordering.
- Create a table with d rows in each vertex $g \in T(\Gamma)$, where d is the vertex degree, and keep the label of each vertex at distance 1 from itself.

The message forwarding procedure uses the word metric on G to perform a greedy routing strategy. Given two vertices $g,h \in \Gamma(G,S)$ with labels w_g and w_h, the procedure to find the shortest path between them is the following:

- When a message arrives to g, compare the label of g with the destination label and verify whether they are equal or not.
- If the labels are equal, the destination is reached. Otherwise, send the message to the neighbor p_i of g, where $i \in [1,\ldots,d]$, such that $l_s(w_{p_i}^{-1} w_h)$ is minimum. If there exists more than one neighbor p_i with equal minimum length on $l_s(w_{p_i}^{-1} w_h)$, then the message is sent to the neighbor with shortlex w_{p_i}.

The space complexity of our algorithm is bounded by $O(dn)$ because each vertex keeps a list of its d neighbors. On the other hand, the time to make a routing decision is bounded by $O(D^2)$, where D is the diameter of $\Gamma(G,S)$. Note that any two vertices $g,h \in \Gamma(G,S)$ with labels w_g and w_h will have $l_s(w_g) \leq D$ and $l_s(w_h) \leq D$. In fact, any resulting word from $w_g^{-1} w_h$ has $l_s(w_g^{-1} w_h) \leq 2D$. Since $(S,L(W))$ can reduce any word w of length $l_s(w)$ in a time proportional to $O(l_s(w)^2)$ (see Theorem 1), any word $v = w_g w_h$ will have a length $l_s(v) \leq 2D$, and then it can be reduced in $O(D^2)$ to a shortlex equivalent word.

The following is an example of the application of our routing algorithm to a 3-cube graph modeling a 8-node interconnection network (see Figure 1a). This graph is isomorphic to the $\Gamma(G,S)$ of the elementary Abelian group of order 2^3 with group presentation equal to $G = \langle S|R \rangle$, where $S = \{a,b,c\}$ and $R = \{a^2,b^2,c^2,aba^{-1}b^{-1},aca^{-1}c^{-1},bcb^{-1}c^{-1}\}$. We start with the labeling procedure.

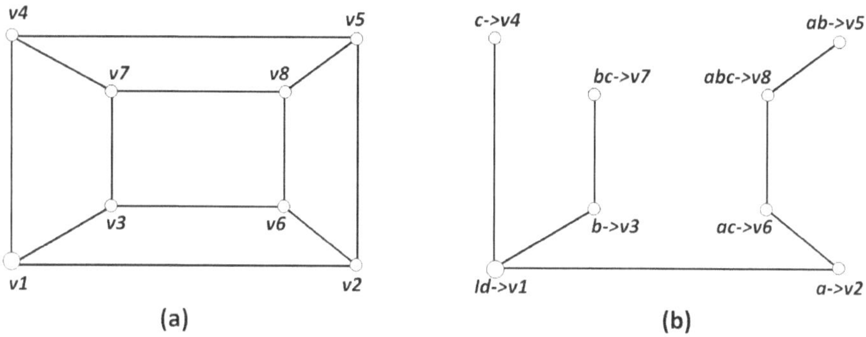

Fig. 1 a) The 3-Cube graph and b) the resulting BFS tree with shortlex labels on vertices

Given a group presentation for $\Gamma(G,S)$, the SAS is computed. Then a random vertex in $\Gamma(G,S)$ (left-bottom corner) is selected and a BFS spanning tree rooted on that vertex is constructed. The resulting tree $T(\Gamma)$ has the following vertices (is discovery order): $\{v_1,v_2,v_3,v_4,v_5,v_6,v_7,v_8\}$. At the same time, we use the SAS to enumerate the elements of the group G according to the shortlex ordering: $L(W)=\{Id,a,b,c,ab,ac,bc,abc\}$. Next, the map $\pi : L(W) \to G$ is performed, i.e. the word $w_i \in L(W)$ labels the vertex $v_i \in T(\Gamma)$, for $1 \leq i \leq |G|$ (See Figure 1b). Finally, each vertex creates a table with the labels of its neighbors at distance 1.

Assume now that vertex v_4 sends a message to vertex v_8 (labeled as *abd*). v_4 uses the labels of its neighbors v_1 and v_7 to compute the length of the shortlex words that represent the following multiplications: $(Id)^{-1}(abc)$ and $(bc)^{-1}(abc)$. As a result of the reduction process using $(S,L(W))$, $(Id)^{-1}(abc)=abc$ and $(bc)^{-1}(abc)=a$. Therefore v_4 sends the message to v_7. Vertex v_7 does the same process with the labels of its neighbors v_6 and v_8. The reduced words that represent $(b)^{-1}(abc)$ and $(abc)^{-1}(abc)$ are ac and Id, respectively. Since Id is the empty word, i.e. the word of length 0, v_7 sends the message to v_8, the final destination. Note that although the labeling process is based on a rooted spanning tree, the algorithm has found the shortest path between v_4 and v_8 in the whole graph and not only in that tree.

4 Conclusions and Future Work

We have proposed a routing algorithm based on SAS for route computation in DC interconnection networks with underlying CG. The input of the algorithm is either a group presentation $G=\langle S|R\rangle$ or the matrix/permutation representation of G. Our routing algorithm is a shortest path one and each node in the network only needs information about its neighbors to take its routing decision. This decision is taken in time proportional to the square of the diameter of $\Gamma(G,S)$. Our proposal uses the fact that finite groups have SAS and these structures can efficiently solve the

Minimum Word Problem in G, which is equivalent to find the shortest path between any pair of vertices of $\Gamma(G,S)$. Moreover, since any finite group is isomorphic to some group of permutations, our algorithm can work on any interconnection network with underlying CG. Finally, although the topologies for DCs are usually very static, it would be important to consider as a future work the network dynamics and to propose fault tolerance mechanism in our algorithm.

References

1. Cisco System Inc.: Cisco Cloud Computing - Data Center Strategy, Architecture, and Solutions. Point of View White Paper for U.S. Public Sector, 1st edn. (2009)
2. Heydemann, M.-C., Ducourthial, B.: Graph Symmetry. NATO ASI Series, pp. 167–224. Springer Netherlands (1997) ISBN 978-90-481-4885-1
3. Meier, J.: Groups, Graphs and Trees: An Introduction to the Geometry of Infinite Groups. Cambridge University Press (2008) ISBN 978-0521719773
4. Schibell, S.T., Stafford, R.M.: Processor Interconnection Networks from Cayley Graph. Discrete Applied Mathematics 40(3), 333–357 (1992) ISSN 0166-218X
5. Shin, J., Sirer, E., Weatherspoon, H., Kirovski, D.: On the feasibility of completely wireless datacenters. IEEE/ACM Transaction on Networking 21(5), 1666–1679 (2013)
6. Xiao, W., Liang, H., Parhami, B.: A Class Of Data-Center Network Models Offering Symmetry, Scalability, and Reliability. Parallel Processing Letters (2012)
7. Loz, E., Siran, J.: New record graphs in the degree-diameter problem. Australasian Journal of Combinatorics 41, 63–80 (2008)
8. Xiao, W., Parhami, B.: Cayley graphs as models of deterministics small-world networks. Information Processing Letters 97(3), 115–117 (2006)
9. Stamoulis, G.D., Tsitsiklis, J.N.: Efficient routing Scheme for Multiple Broadcasts in Hypercubes. IEEE Trans. on Parallel and Distributed Systems 4(7), 725–739 (1993)
10. Stamoulis, G.D., Tsitsiklis, J.N.: The Efficiency of Greedy Routing in Hypercubes and Butterflies. IEEE Transaction on Communication 42(11), 3051–3061 (1994)
11. Kiasari, A.E., Sarbazi-Azad, H.: Analytic performance comparison of hypercubes and star graphs with implementation constraints. Journal of Computer and System Sciences (2007)
12. Epstein, D., Cannon, J., Holt, D., Levy, S., Paterson, M., Thurson, W.: Word Processing in Groups. Jones and Bartlett Publishers (1992) ISBN 0-86720-244-0
13. Akers, S., Krishnamurthy, B.: A group-theoretic model for symmetric interconnection networks. IEEE Transactions on Computers 38(4), 555–566 (1989)
14. Tang, K., Arden, B.: Vertex-transitivity and routing for Cayley graphs in GCR representations. In: ACM Symposium on Applied Computing, SAC, pp. 1180–1187 (1992)
15. Wang, L., Tang, K.: Topology-Based Routing for Xmesh in Wireless Sensor Networks. In: Powell, S., Shim, J.P. (eds.) Wireless Technology. LNEE, vol. 44, pp. 229–239. Springer, Heidelberg (2009)
16. Ryu, J., Noel, E., Tang, K.: Fault-tolerant Routing on Borel Cayley Graph. In: Next Generation Networking Symposium, IEEE ICC 2012, pp. 2872–2877 (2012)
17. Holt, D.: The Warwick Automatic Groups Software. In: Geometrical and Computational Perspectives on Infinite Groups. Amer. Math. Soc. DIMACS Series, vol. 25, pp. 69–82 (1995)

18. Even, S., Goldreich, O.: The minimal-length generating sequence problem is NP-Hard. Journal of Algorithms 2, 311–313 (1981)
19. Cannon, J.: Construction of defining relators for finite groups. Discrete Math. 5, 105–129 (1973)
20. Magnus, W., Karrass, A., Solitar, D.: Combinatorial Group Theory - Presentation of Groups in Terms of Generators and Relations, 2nd revised edn. Dover Publications (2004)

Distributed Generation of Billion-node Social Graphs with Overlapping Community Structure

Kyrylo Chykhradze, Anton Korshunov, Nazar Buzun, Roman Pastukhov, Nikolay Kuzyurin, Denis Turdakov, and Hangkyu Kim

Abstract. In the field of social community detection, it is commonly accepted to utilize graphs with reference community structure for accuracy evaluation. The method for generating large random social graphs with realistic community structure is introduced in the paper. The resulting graphs have several of recently discovered properties of social community structure which run counter to conventional wisdom: dense community overlaps, superlinear growth of number of edges inside a community with its size, and power law distribution of user-community memberships. Further, the method is by-design distributable and showed near-linear scalability in Amazon EC2 cloud using Apache Spark implementation.

Keywords: random graph, social network, community detection, benchmark network, graph generation, LFR benchmark, Affiliation Graph Model, SNAP, distributed algorithms, Amazon EC2, Apache Spark.

1 Introduction

Community structure is a natural property of human networks, including online social networks where users tend to unite either explicitly (by means of

Kyrylo Chykhradze · Anton Korshunov · Nazar Buzun · Roman Pastukhov · Nikolay Kuzyurin · Denis Turdakov
Institute for System Programming of the Russian Academy of Sciences
Moscow, Russia
e-mail: {chykhradze,korshunov,nazar,pastukhov,nnkuz,turdakov}@ispras.ru

Hangkyu Kim
Data Intellegence Lab, Software Research Center,
Samsung Electronics Co., Ltd.
Suwon, South Korea
e-mail: hangkyu.kim@samsung.com

P. Contucci et al. (eds.), *Complex Networks V*,
Studies in Computational Intelligence 549,
DOI: 10.1007/978-3-319-05401-8_19,

grouping functionality of network software) or implicitly (by establishing ties based on shared affiliation, role, activity, social circle, interest, function, or some other property). Social data scientists widely employ intuitive notions of separability, density, and cohesiveness of social groups for discovering and evaluating implicit communities from social networks [4, 10, 12].

Recent advances in studying modular structure of social networks [13, 14] helped to reveal several fundamental properties that appear to be common in human interaction networks: dense community overlaps, superlinear growth of number of edges inside a community with its size, power law distribution of user-community memberships and communities size, etc. This suggests the need for revisiting accuracy evaluation techniques for community detection methods and adequacy of the methods themselves.

Despite the availability of community detection benchmarks based on real networks, it is desirable to learn some fundamental properties from it and develop a tool for producing synthetic benchmarks with similar properties and different characteristics. For a reliable and comprehensive evaluation, a community detection method must be tested on benchmark networks of variable size and other parameters as they may have significant impact on the results.

The main contributions of the paper could be summarized as follows:

- we introduce a novel approach to benchmark networks generation for community detection methods based on *Community-Affiliation Graph Model (AGM)*, where memberships of users to communities are modeled with a bipartite graph and links among people stem from shared community affiliations [13];
- we introduce *CKB* - a method for distributed generation of large benchmark networks with realistic properties of social graph and social community structure;
- we make our method particularly suitable for benchmarking community detection algorithms by providing the set of parameters for tuning the most important structural properties: number of nodes, mean node degree, edge probability inside a community, power law exponents for distributions of community size and node-community memberships, etc;
- we introduce simple and efficient distributed algorithms for building user-community affiliation network and linking nodes inside communities;
- we develop and evaluate a distributed implementation of the proposed method capable of producing billion-node random social graphs with reference community structure;
- we make our implementation accessible to the research community by providing it as a web service with possibility to download the generated graphs[1].

The rest of the paper is organized as follows. Section 2 contains problem description and section 3 describes the details of CKB. In section 4 accuracy

[1] http://ckb.at.ispras.ru/

and performance of the method are evaluated. We conclude in section 5 with possible future directions.

2 Problem

Let's consider a graph $G = (V, E)$, where $|V| = N_1$ and $|E| = m$. A community C_i with $|C_i| = n_{c_i}$ is defined as an induced subgraph. The number of communities is N_2, all communities together constitute a *cover* of a graph. The number of entries of j-th node into different communities (*node-community memberships*) is m_j.

The *internal* (d_j^{int}) and *external* (d_j^{ext}) degree of vertex $j \in C_i$ are defined as the number of edges connecting j to other vertices in C_i or to the rest of the graph respectively. So the *total* degree of vertex j is $d_j = d_j^{int} + d_j^{ext}$. Number of edges inside C_i is $d_{c_i} = \frac{1}{2} \sum_{\forall j \in C_i} d_j^{int}$.

The task is to generate G with the following properties:

1. power law degree distribution: $p(d) \sim d^{-\beta}$ [3];
2. giant connected component presence: $\exists\ G^* \subset G : |V^*| \sim N_1, \forall\ i, j \in V^*\ \exists\ w_1, w_2, ..., w_k \in E^*\ \exists\ t_l \in V$: $w_0 = (i, t_0), w_1 = (t_0, t_1), ..., w_D = (t_{D-1}, j)$ [11];
3. small effective diameter: $\forall\ i, j \in V^* \exists\ w_1, w_2, ..., w_k \in E^*\ \exists\ t_l \in V$:$w_0 = (i, t_0), w_1 = (t_0, t_1), ..., w_{D_{i,j}} = (t_{D_{i,j}-1}, j)$ such that $\mathbb{P}\left((1-\epsilon)\frac{\ln N_1}{\ln \ln N_1} \le D_{i,j} \le (1+\epsilon)\frac{\ln N_1}{\ln \ln N_1}\right) \to 1$ for $N_1 \to \infty$ [1];
4. users could have zero degrees and memberships: $d_j \ge 0$, $m_j \ge 0$;
5. communities are overlapping: $\exists\ C_i, C_j : C_i \cap C_j \ne \emptyset$ [6];
6. each community C_i is connected with high probability: $\mathbb{P}\left(C_i \text{ is connected}\right) \ge 1 - \frac{1}{n_{C_i}}$;
7. intra-community density is larger than the average link density of whole graph G: $\frac{d_{C_i}}{n_{C_i}(n_{C_i}-1)} > \frac{m}{N_1(N_1-1)}$ [7];
8. number of edges inside the community is greater than number of edges linking vertices of the community with the rest of the graph: $d_{C_i} > \sum_{\forall j \in C_i} d_j^{ext}$;
9. number of edges in the community increases superlinearly with the community size: $d_{C_i} \propto n_{C_i}^{1+\gamma}$, where $\gamma \in (0, 1)$ [14];
10. user-community memberships have power-law distribution: $p(m_i) \sim m_i^{-\beta_1}$ [14];
11. size of communities has power-law distribution: $p(n_{C_i}) \sim n_{C_i}^{-\beta_2}$ [5];
12. overlaps of communities are more densely connected than the non-overlapping parts: $\forall\ C_i \forall\ C_j (i \ne j), C_{i \cap j} = C_i \cap C_j \Rightarrow \frac{d_{C_{i\cap j}}}{n_{C_{i\cap j}}(n_{C_{i\cap j}}-1)} > \frac{d_{C_i}}{n_{C_i}(n_{C_i}-1)}$ [13];
13. low-degree nodes tend to be part of very few communities, while high-degree nodes tend to be members of multiple groups: $d_i \sim m_i$ [8].

3 Method

The main steps of CKB graph generator are:

1. Degree sequences for users and communities are generated on the assumption of input parameters;
2. Users are assigned to communities using modified configuration model [9];
3. Edges inside each community are generated using configuration model.

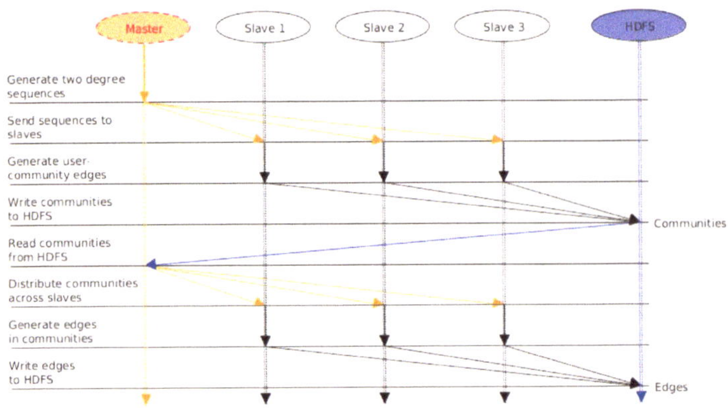

Fig. 1 General workflow

General workflow is shown in Figure 1. *Master node* is the central node of computational cluster and by *slave nodes* we mean the rest of cluster nodes. HDFS files are distributed across local file systems of slave nodes. On the master node, the non-distributed part of computations is carried. During the distributed computations, the master node assigns tasks to slave nodes and aggregates the results. First step of the generation process is done once on the master node while second step is distributed across slave nodes. During the third step edge generation inside each community is performed in a distributed way, so that each slave node generates some part of edges which are then merged.

3.1 Users-Communities Bigraph Generation

Bipartite graph (or **bigraph**) is a graph whose vertices can be divided into two disjoint sets U and V and such that every edge connects a vertex in U to one in V. In our case V ($|V| = N_1$) is a set of nodes and U ($|U| = N_2$) is a set of communities. User-community affiliations are modeled as bigraph edges.

1. Number of users (nodes) N_1 is a parameter. At first on the master node number of communities N_2 is computed from the equation:

$$M_0 = N_1 \cdot \mathbb{E}[m] = N_2 \cdot \mathbb{E}[x], \tag{1}$$

Table 1 Parameters of CKB

Parameter	Meaning	Default value
N_1	number of nodes	–
d_{mean}	mean node degree	–
x_{min}	minimum user-community memberships	1
m_{min}	minimum community size	2
x_{max}	maximum user-community memberships	10,000
m_{max}	maximum community size	10,000
$\beta_1 > 1$	power law exponent of user-community membership distribution	2.5
$\beta_2 > 1$	power law exponent of community size distribution	2.5
$\alpha > 0$	affects edge probability inside communities	4
$0 < \gamma < 1$	affects edge probability inside communities	0.5
ϵ	controls the number of edges in ϵ-community	$2N_1^{-1}$

where $\mathbb{E}[m]$ and $\mathbb{E}[x]$ are the expectation of node memberships and community sizes respectively. The number of generated edges is defined as

$$M = (1 + \mathbb{E}[\mathbb{P}^{mult}_{c_i,j}])M_0, \tag{2}$$

where $\mathbb{P}^{mult}_{c_i,j}$ is multiple edge probability (section 3.3) which helps to reduce the bias introduced by deleting multiple edges.
k-th moment of the random variable distributed by power law with exponent (β_1 for membership distribution and β_2 for community size distribution are parameters) is

$$\mathbb{E}[x^k] = \int_{x_{min}}^{x_{max}} x^k p(x)dx = \int_{x_{min}}^{x_{max}} x^k \frac{1-\beta}{x_{max}^{1-\beta} - x_{min}^{1-\beta}} x^{-\beta} dx, \tag{3}$$

since $p(x) = \frac{1-\beta}{x_{max}^{1-\beta} - x_{min}^{1-\beta}} x^{-\beta}$. So,

$$\mathbb{E}[x^k] = \frac{(1-\beta)(x_{max}^{k+1-\beta} - x_{min}^{k+1-\beta})}{(x_{max}^{1-\beta} - x_{min}^{1-\beta})(k+1-\beta)} \tag{4}$$

Note that for $k - \beta + 1 = 0$ the expectation equals to

$$\mathbb{E}[x^k] = \frac{1-\beta}{x_{max}^{1-\beta} - x_{min}^{1-\beta}} \ln\left(\frac{x_{max}}{x_{min}}\right)$$

2. Identical power law degree sequences are generated on each slave node.
3. Each vertex is associated with degree (d_i^1 for i-th user-node and d_j^2 for j-th community-node) from degree sequence that was generated at previous step.

4. Numbers $D_1^1 = d_1^1$, $D_2^1 = D_1^1 + d_2^1$, ... , $D_{k+1}^1 = D_k^1 + d_{k+1}^1$, ..., $D_{N_1}^1 = D_{N_1-1}^1 + d_{N_1}^1$ and $D_1^2 = d_1^2$, $D_2^2 = D_1^2 + d_2^2$, ... , $D_{k+1}^2 = D_k^2 + d_{k+1}^2$, ... ,$D_{N_2}^2 = D_{N_2-1}^2 + d_{N_2}^2$ are computed.
5. For the sequence of natural numbers

$$[M] = \{1, 2, 3, \dots, \lfloor \frac{M}{s} \rfloor\},$$

where $\lfloor x \rfloor = max\{n \in \mathbb{Z} | n \leq z\}$ and s is the number of slave nodes, compute in a loop on each slave node:
for $t = 1$ to $\lfloor \frac{M}{s} \rfloor$ do:
   a. choose random natural numbers p and q from $[M]$ with uniform distribution;
   b. find the interval $[D_i^1, D_{i+1}^1]$ to what the number p belongs;
   c. find the interval $[D_j^2, D_{j+1}^2]$ to what the number q belongs;
   d. if $i \neq j$ add to the bigraph an edge (i, j).
6. Merge all generated edges and remove multiple edges.

Complexity of this stage is $O(M \log(N_1 N_2))$.

3.2 Intra-community Edges Generation

At this stage edges between nodes are generated in conformity with their belonging to communities. We sample number of edges in community C_j from Binomial distribution (considering the number of multiple edges):

$$M_{c_j} = \frac{1}{s}(1 + \mathbb{P}_{c_k}^{mult})\text{Bin}(x_{c_k}, p_{c_k}), \tag{5}$$

where x_{c_k} is community size, p_{c_k} is edge probability in the community c_k, s is the number of slave nodes, and $\mathbb{P}_{c_k}^{mult}$ is multiple edge probability (section 3.3) which helps to reduce the bias introduced by deleting multiple edges.

Then, on each slave node M_{c_j} edges are generated using configuration model. Finally, all generated edges are merged and multiple edges are removed. Self-loops are filtered during the generation process.

For each pair of nodes i and j in the community c_k the probability of edge (i, j) is defined as

$$p_{c_k} = \frac{\alpha}{x_{c_k}^{\gamma}}, \tag{6}$$

where α and γ are parameters ($0 < \gamma < 1$, $\alpha > 0$) [14].

The total probability of an edge between i and j in overall graph is

$$p(i, j) = 1 - \prod_{c_k \in C_{ij}} (1 - p_{c_k}), \tag{7}$$

where C_{ij} is a set of communities that i and j share [14].

Therefore, overlaps of communities are more densely connected than the non-overlapping parts. Also low-membership nodes will have low-degree. And vice versa, if membership of node increases then its degree is growing too.

ϵ-Community. To allow for edges between nodes that don't share any common communities, we add an additional ϵ-community [13] which connects any pair of nodes with a small probability ϵ. This step is also necessary to ensure the existence of zero membership nodes with non-zero degrees in the resulting graph. In other words, some part of users with low degrees are not members of any community.

The number of edges generated on each slave node is

$$M_\epsilon = \frac{1}{s}\frac{N_1(N_1-1)}{2}\epsilon, \tag{8}$$

where ϵ is a parameter.

Complexity of this stage is $O(K_m)$, where $K_m = \sum_{c_j} M_{c_j}$.

3.3 Multiple Edges

Knowing multiple edge probability for each step of generation helps to reduce the bias introduced by deleting multiple edges produced by configuration models. So for **users-communities bigraph generation** the probability that in bipartite graph an edge appears two or more times will be

$$\mathbb{P}^{mult}_{c_i,j} \approx \left(\frac{x_{c_i} m_j}{M}\right)^2 \tag{9}$$

For **intra-community edges generation** the probability of multiple edge is

$$\mathbb{P}^{mult}_{c_k} \approx \frac{\alpha^2}{4x^{2\gamma}_{c_k}} \tag{10}$$

3.4 Mean Degree

Since mean degree is an important feature for graph analysis and community detection algorithm testing, we obtained the dependence between input parameters α and γ and mean degree. Calculation of the mean degree allows to prove that density of edges inside the community is increased than in overall graph. Due to limited space we provide only the final equations. So the mean degree is:

$$d_{mean} \approx \frac{(S_1 - S_2 + S_3)\binom{N_1}{2}}{N_1}, \tag{11}$$

where N_1 is the number of nodes in the whole graph and S_1, S_2 and S_3 are

$$S_r = \alpha^r \mathbb{E}\Big[\sum_{c_1 < \ldots < c_r} \prod_{k=\{1,\ldots,r\}} \frac{1}{x_{c_k}^{\gamma}} \prod_{t=\{i,j\}} \Big(\frac{x_{c_k} m_t}{M}\Big)\Big] \tag{12}$$

Each moment $\mathbb{E}[x^y]$ can be computed from (4). Now after solving the cubic equation in variable α (and fixed γ) we can compute probability p_{c_i}. But the (11) is valid only with some constrains, that are not provided due to limited space.

3.5 Connectedness of Community

Using known results on evolution of random graphs in Erdos-Renyi model [2] we can claim the following:

Theorem 1. *The community C_i is connected with high probability for*

$$\alpha > \ln(x_{c_i}) x_{c_i}^{\gamma - 1} \tag{13}$$

4 Evaluation

We implemented the proposed method in Scala using Apache Spark[2] - a framework for efficient computations in distributed environment.

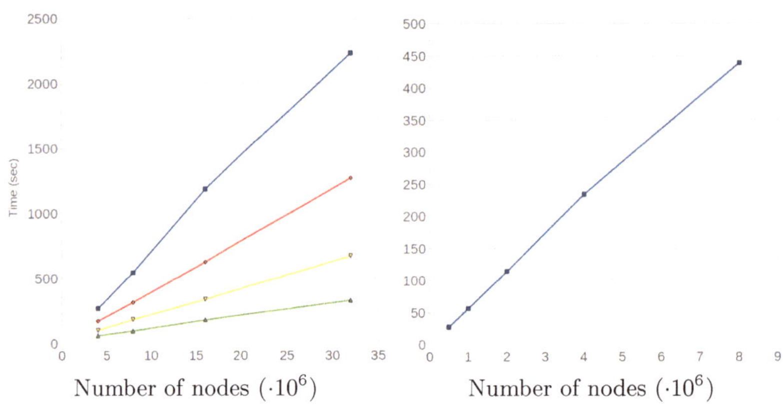

Fig. 2 Scalability evaluation results. **Left: Amazon EC2 clusters of *m1.large* instances**, blue line - 2 slave nodes, red line - 4 slave nodes, yellow line - 8 slave nodes, green line - 16 slave nodes. **Right: single machine**.

[2] `http://spark.incubator.apache.org/`

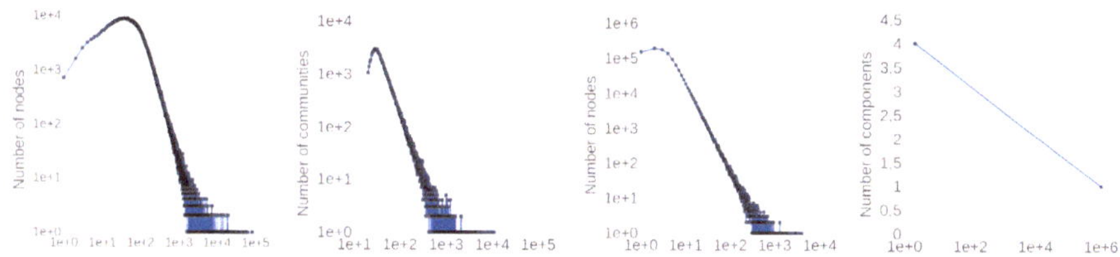

Degree Community size Membership Connected component

Fig. 3 Community size, user-community memberships, degree distribution and connected component distribution for $N_1 = 10^6$, $\beta_1 = \beta_2 = 2.5$

Table 2 Comparison of CKB against SNAP networks and LFR

	Orkut	**LiveJournal**	**YouTube**	**CKB**		**LFR**
Number of nodes	3M	4M	1.1M	3M	97.5K	100K
Mean degree	76.2	17.3	5.3	109.9	68.8	66.7
Community size power law exponent $\beta_{comm_{size}}$	2.12	2.14	2.36	2.19	2.57	2.54
Membership power law exponent β_{memb}	1.59	2.22	2.83	2.28	2.62	–
Degree distribution power law exponent β_{graph}	1.58	2.15	2.53	2.22	2.54	2.56
Community size distribution median	16	2	3	5	49	40
Membership distribution median	14	2	1	7	1	1
Average clustering coefficient	0.169	0.353	0.172	0.039	0.055	0.226
Effective diameter d_{eff}	4.8	6.4	6.5	4.38	3.88	3.98
Generation time (sec)	–	–	–	160	11	863

The results of running time evaluation on Amazon EC2 clusters and single machine are shown in Figure 2. Near-linear scalability on the number of nodes in the generated graph allows to produce synthetic networks of huge size in reasonable time: one billion nodes graph generation took < 2 hours on Amazon EC2 cluster with 100 *m1.large* instances.

Table 2 summarizes the most important statistics of LiveJournal, ORKUT and YouTube datasets from *Stanford Large Network Dataset Collection*[3] and compares them with CKB and LFR benchmarks [5]. Comparing the tables suggests that CKB graphs have very similar structural properties to real networks. The only difference is low average clustering coefficient of the generated networks. However, achieving more realistic clustering coefficient requires some changes in the edge generation process and is a subject of future work.

[3] `http://snap.stanford.edu/data/index.html`

5 Conclusion

A method for distributed generation of large benchmark networks with realistic properties of social graph and social community structure has been introduced and evaluated. Possible directions for future work include:

- distributed computation of Normalized Mutual Information or other measures for comparing covers of communities;
- testing different community detection algorithms;
- allow to control clustering coefficient and degree correlation of nodes.

References

1. Albert, R., Jeong, H., Barabasi, A.-L.: Diameter of the world wide web. Nature 401, 130–131 (1999)
2. Erdos, P., Renyi, A.: On the evolution of random graphs. Bull. Inst. Int. Statist. Tokyo 38, 343–347 (1961)
3. Faloutsos, M., Faloutsos, P., Faloutsos, C.: On power-law relationships of the internet topology. In: SIGCOMM, pp. 251–262 (1999)
4. Fortunato, S.: Community detection in graphs. Physics Reports 486(3), 75–174 (2010)
5. Lancichinetti, A., Fortunato, S.: Benchmarks for testing community detection algorithms on directed and weighted graphs with overlapping communities. Phys. Rev. 80(1) (2009)
6. Lancichinetti, A., Fortunato, S., Kertsz, J.: Detecting the overlapping and hierarchical community structure in complex networks. New J. Phys. (2009)
7. Leskovec, J., Lang, K.J., Dasgupta, A., Mahoney, M.: Statistical properties of community structure in large social and information networks
8. Mislove, A., Marcon, M., Gummadi, K., Druschel, P., Bhattacharjee, B.: Measurement and analysis of online social networks. In: IMC (2007)
9. Molloy, M., Reed, B.: A critical point for random graphs with a given degree sequence. Random Structures and Algorithms 6, 161–180 (1995)
10. Plantié, M., Crampes, M.: Survey on social community detection. In: Social Media Retrieval, pp. 65–85. Springer (2013)
11. Spencer, J.: The giant component: The golden anniversary. Not. Am. Math. Soc. 401, 130–131 (1999)
12. Xie, J., Kelley, S., Szymanski, B.K.: Overlapping community detection in networks: the state of the art and comparative study. arXiv preprint arXiv:1110.5813 (2011)
13. Yang, J., Leskovec, J.: Community-afliation graph model for overlapping network community detection. In: IEEE 12th International Conference on Data Mining (2012)
14. Yang, J., Leskovec, J.: Structure and overlaps of communities in networks. In: ACM SIGKDD Conference on Knowledge Discovery and Data Mining (2012)

Using the Entropy of the DFT of the Laplacian Eigenvalues to Assess Networks

Danilo R.B. de Araújo, Carmelo J.A. Bastos-Filho, and Joaquim F. Martins-Filho

Abstract. There are several metrics that are very useful to analyze and to design networks. These metrics, including the spectral-based ones, can be used to retrieve topological properties from the network. We observed that if one applies the Discrete Fourier Transform (*DFT*) over the eigenvalues of the Laplacian matrix, it is possible to observe different patterns in the *DFT* depending on some properties of the analyzed networks. In this paper, we propose a new metrics based on the entropy of the *DFT* samples, that can be used to identify the type of network. We evaluated this metrics in networks generated by four different procedures (k-Regular, Erdos-Renyi, Watts-Strogatz and Barabasi-Albert) and in well-known datasets of real networks. The results indicate that one can use the proposed metrics to identify the generational model of the network.

1 Introduction

Important advances in Network Science are highly correlated to the recent proposition of generative procedures to create graphs with similar topological properties of real world networks. In 1960, Erdos and Renyi presented important studies regarding random graphs. Watts and Strogatz proposed the first procedure for generating small world networks (SW) in 1998. In 1999, Barabasi and Albert presented a model based on preferential attachment to generate scale-free networks (SF). Variations of the preferential attachment mechanism were presented by Dorogovtsev and Mendes in 2002. Since these models can generate networks with specific characteristics, many models to generate networks have been proposed in the literature. A review on generative procedures can be found in [1].

Danilo R.B. de Araújo · Joaquim F. Martins-Filho
Federal University of Pernambuco, Recife, 50740-550, Pernambuco, Brazil
e-mail: jfmf@ufpe.br

Carmelo J.A. Bastos-Filho
University of Pernambuco, Recife, 50720-001, Pernambuco, Brazil
e-mail: carmelofilho@poli.br

P. Contucci et al. (eds.), *Complex Networks V*,
Studies in Computational Intelligence 549,
DOI: 10.1007/978-3-319-05401-8_20,

In most of cases, real networks do not present a random topological structure, such as the networks that can be generated by the Erdos-Renyi (ER) model. In general, real world networks present characteristics that are similar to regular networks, small-world networks, scale-free networks, or a combination of them. These topological features can be used to classify networks. It is quite interesting for real world applications, since these characteristics can present high correlation with some desired behavior of some real world networks. For example, scale-free networks generated by using the Barabasi-Albert (BA) model are related to real networks that present high resilience to random node failures, but these type of network is very vulnerable to targeted attacks (*i.e.* the network can be seriously damaged when high connected hubs are attacked). On the other hand, random networks are robust to targeted attacks, but they are more vulnerable to random failures [2].

Planning real networks can be associated to choose a network topology that promote specific patterns of dynamic behaviour. In general, these patterns are related to network requirements during the operational phase. Thus, it is key that we can assess whether topologies follow theoretical models. For example, metrics to assess robustness include: the algebraic or natural connectivity, the average path length (APL), the largest connected component (LCC). In general, the studies about network planning have adopted this approach to forecast if the network performance will obey the requirements [3]. In this context we emphasize the importance of using expressive metrics to assess networks. Currently, there is a need to use several metrics to capture specific properties of the network under design. The use of several metrics for this type of analysis can be computationally expensive and error prone. Furthermore, it would be desirable to have a metrics that can be used for a variety of graphs, including small, large, sparse and dense ones.

In this paper we propose to use the entropy of the DFT coefficients of the eigenvalues of the Laplacian Matrix to classify networks as: regular, random, SW or SF networks. The paper is organized as follows: Section 2 provides a brief review of topological analysis of complex networks; Section 3 explains the proposed metrics; Section 4 details the experimental setup used to obtain the results; Section 5 presents the results and Section 6 presents conclusions and suggestions for future work.

2 Characterization of Complex Networks

Networks with the same topological properties define a family of graphs. This section presents a brief review of the most used topological metrics and explains how these metrics have been used to characterize networks. Several surveys [4, 5] are available and can be used for further studies on the concepts briefly presented in this section.

In this work we consider a Complex Network as a graph $G = (\mathcal{N}, \mathcal{L})$, in which $\mathcal{N}$ and $\mathcal{L}$ denote the set of vertices and the set of edges, respectively. In this paper we just considered unweighted and undirected graphs. Besides, a graph cannot contain self-loops (connections beginning and ending at the same node). We can also define the amount of nodes and links in a network as $n = |\mathcal{N}|$ and $l = |\mathcal{L}|$.

The *link density* (d) of a network is defined as the ratio between the number of links that actually exists and the maximum number of links that could exist in the network. The *node degree* (k) describes the number of links or neighbor nodes of a given node. The largest value of k, among all the nodes of a graph, defines the *hub degree*, $k(hub)$. The *node degree distribution* defines the probability, $Pr(k)$, of a randomly selected node to have a certain degree k. The *node degree distribution* of a real network has been commonly used to indicate the canonical model of this network. The average number of links that are connected to the nodes is called the *average node degree*. The *entropy* of graph G, that is, $I(G)$, is a measure of the graph "randomness" and it is calculated over its *node degree distribution*. The *shortest path* (SP) describes the number of hops between a given pair of nodes. The *average path length APL* is the average of the SP between all source-destination pairs of nodes of the network. The *clustering coefficient* (c_i) is the ratio between the number of triangles that contains node i and the number of triangles that could possibly exist if all neighbors of i were interconnected [6]. The *clustering coefficient for the entire graph* (CC) is the average of the clustering coefficients of all the network nodes.

In the graph theory, a network can be analyzed by its *Adjacency matrix* (A) or the *Laplacian matrix* (L). The matrix A of an undirected graph with n nodes is a n x n matrix, in which the non-diagonal entries (i, j) are equal to "1" if the nodes i and j are adjacent (connected), or "0" otherwise. In A, the entries (i,i) are always equal to "0", since we are considering that a node can not be connected to itself. Since we are considering undirected graphs in this paper, A is a symmetric matrix for this case. The *Node Degree matrix* (D) is used together with A to build L as $L = D - A$, in which the non-diagonal entries (i, j) are either "-1" or "0", depending on whether nodes i and j are adjacent or not, respectively, and the diagonal entries (i,i) are equal to the degree of the nodes D_i. The study on the relationship between a graph and its eigenvalues (and eigenvectors) is referred in the literature as spectral graph theory. All eigenvalues are real for A, whereas all eigenvalues are real and nonnegative for L. The ordered set of n eigenvalues of A or L is called the spectrum of the matrix. If there are two graphs with similar sets of eigenvalues, this means that they probably present similar graph structures or graph isomorphism [7]. The *density function* of the eigenvalues $f_\lambda(t)$ is also used to recognize specific families of networks and it is suitable to analyze the eigenvalues $\{\lambda_m\}_{1 \leq m \leq n}$ in large graphs [8].

The most basic topological characterization of a graph G can be obtained in terms of the node degree distribution $Pr(k)$. Information regarding the node distribution of a undirected network can be obtained either by a plot of $Pr(k)$ or by the calculation of the moments of the distribution [1]. In order to adopt the degree distribution directly to characterize networks, one must calculate the distribution for a specific network and then select the distribution that better fits the data. In this case, if the distribution best fits a power law distribution, then the network is a typical SF; if the distribution best fits a Poisson distribution, the network should be random or SW (SW if the curve is taller and thinner than an equivalent random network). This approach is not suitable to be used automatically. The most known models of complex networks can be also analyzed according to the level of structuring of the topology. Models rely between two extremes cases: networks highly structured and predictable (such as a ring network), and completely random networks (such as networks provided by ER model). In this sense, one can classify SW networks and SF networks as intermediate cases between these two extremes. The SW networks are more structured than random, and SF networks are more random than structured networks [1]. However, this approach is very imprecise if used alone. In order to properly classify an arbitrary real world network, it is common to consider a combination of various properties derived from the network topology. Lewis [1] proposed the use of four topological properties (entropy, CC, APL and the hub degree) in order to understand whether a network presents properties more related to regular, random, SW or SF networks. If one considers dense networks and only these metrics for the analysis, one can observe that basic topological properties for each model begin to disappear and all the networks appear to have been generated by the same model. Besides, we know that each of these properties can be tuned to match a different value if the parameters of the generative procedures are modified. Araújo *et al.* identified characteristic points in the DFT of Laplacian spectrum and proposed two metrics [9], but these metrics are imprecise for dense networks.

3 Our Proposal

We propose a new metrics to analyze networks based on the entropy of the Fourier Transform coefficients over the Laplacian spectrum. L contains a suitable summary of the network topology, because it contains information about the node degrees and connected links simultaneously. Thus, a metrics derived from its spectrum should summarize properly the network topology. Besides, a metrics based on entropy can maintain the information about randomness of the network. The use of the entropy metrics directly over the Laplacian spectrum (without using the Fourier Transform) do not offer the same reasoning for different values of d. Thus, we have considered the entropy of the DFT coefficients over the Laplacian spectrum in order to classify graphs according to their topology and also to provide a measure of randomness. The proposed metrics can be calculated according to the Algorithm 1. Eq. (1) summarizes the metrics value.

Algorithm 1: The algorithm used to calculate $I(\hat{\mathscr{F}})$.

Let A the adjacency matrix of a graph G;
Calculate the degree matrix D;
Calculate the Laplacian matrix $L = D - A$;
Calculate the real eigenvalues of L and store it in E;
Calculate the Discrete Fourier Transform (DFT) over E and store the values in $\mathscr{F}$;
Normalize the $\mathscr{F}$ set in order to obtain values between 0 and 1 and store it in $\hat{\mathscr{F}}$;
Calculate the entropy of $\hat{\mathscr{F}}$ values using Eq. (1).

$$I(\hat{\mathscr{F}}) = -\sum_{i=1}^{|\hat{\mathscr{F}}|}(\hat{\mathscr{F}}_i \cdot \log_2 \hat{\mathscr{F}}_i). \tag{1}$$

4 Experimental Setup

We obtained all the results for this paper based on experiments by using a simulation platform for complex networks developed in the Java programming language.

We aim to show that the entropy of the *DTF* calculated over the eigenvalues of the *Laplacian matrix* presents different characteristics for different types of networks independently of the network density. In order to show this, we generated different networks using the ER, BA and WS generative procedures. We used density values d from 0.02 to 0.98 with a step value of 0.02. For each pair ($generative procedure$, q), we created 30 different networks with different sizes (100 and 1000 nodes). Our implementation of Random Graphs establishes a link between a pair (i, j) if a uniform random variable assumes a value below a probability value p. In order to generate SW networks we created a k-regular graph and change existing links (i, j) to a new one (k, l) considering a rewiring probability rp. We use the value of d to calculate a value to k. Finally, to generate a SF networks we used the preferential attachment process. The networks start with $n = 3$ nodes and each of the $(N-3)$ remaining nodes are attached to the network by adding Δm links to the existing nodes. We use d to determine the value of Δm. The probability for a new link to be attached to an existing node i is proportional to: $P(i) \propto \frac{k_i^{\tau}}{\sum_{j=1}^{n} k_j^{\tau}}$.

In our work one also used real networks from datasets of previous studies from different applications (biological, contact, communication, interaction and social networks). For each real network, we generated equivalent ER, BA e WS networks (with the same n and d). The value of $I(\hat{\mathscr{F}})$ was analyzed in order to find the best model to fit the real networks. After this, we compared other topological properties between the networks obtained by our process and the original ones. We considered the following datasets: Highland tribes [10]; Zachary karate club [11]; Hypertext 2009 [12]; Manufacturing emails [13]; Infectious [12]; Caenorhabditis elegans metabolic [14]; U. Rovira i Virgili [15].

We analyzed our results based on tables and scatter graphs. The statistical behavior of the values for the proposed metrics was analyzed by using the box-plot graphs and by using a hypothesis test from the 50th percentile (median).

5 Results

Figure 1 presents the curves of "$I(\hat{\mathscr{F}})$ *versus* density" for networks with 100 and 1000 nodes. One can observe that $I(\hat{\mathscr{F}}, d)$ presents the same meaning for small and large networks. BA networks present the greater value to $I(\hat{\mathscr{F}}, d)$. ER networks presents their $I(\hat{\mathscr{F}}, d)$ below the equivalent values of BA and WS networks presents their $I(\hat{\mathscr{F}}, d)$ below ER and BA. $I(\hat{\mathscr{F}}, d)$ for k-Regular networks decays near to zero quickly. One can associate this behavior to the lack of randomness of k-Regular networks and the capacity of our metrics to capture randomness. $I(\hat{\mathscr{F}},d)$ of BA networks present huge values if we increase the network size, but for the ER model the metrics remains constant for different n. If we consider two different values $I_A(\hat{\mathscr{F}},d_1)$ and $I_B(\hat{\mathscr{F}},d_1)$, related to the same d but different models, A and B, we can observe for any two pair of models that $I_A(\hat{\mathscr{F}},d_1)/|I_A(\hat{\mathscr{F}},d_1)-I_B(\hat{\mathscr{F}},d_1)|>0.10$ $(0<d<0.94)$. Thus, one can conclude that our metrics is very sensitive to the topological properties of each model and it can be properly used to classify sparse, dense, small and large networks. This is an advantage over FZC and HVC [9], because these metrics are imprecise for dense networks.

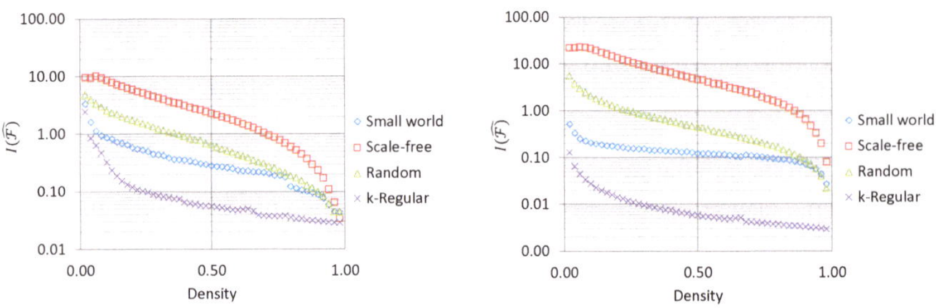

(a) Networks with $n=100$. (b) Networks with $n=1000$.

Fig. 1 $I(\hat{\mathscr{F}})$ *versus* density for k-regular, BA, random and WS networks

If we consider a set of networks with the same value to n, d and generative procedure, our metrics should present similar values, because the topological characteristics are the same. Figure 2 presents the box-plot charts in terms of $I(\hat{\mathscr{F}})$ for 30 different networks with $n=100$. Figure 2a summarizes the statistics for sparse networks that were generated from the BA, ER and WS algorithms. One can notice that the minimum value of BA networks is above the maximum value of ER networks and the minimum value of ER networks is above the maximum value of WS networks. Figure 2b summarizes the statistics for networks with $d=0.30$. One can notice the same behavior observed for sparse networks, but the height of the boxes was lowered. Figure 2c summarizes the statistics for very dense networks. As expected, the variation in the values for the metrics was reduced to denser networks.

Table 1 present the value of $I(\hat{\mathscr{F}})$ calculated for the real network and for each canonical model. In Table 1, we considered $rp=0.05$ for SW networks and $\tau=1$ for SF networks. We used the values of $I(\hat{\mathscr{F}})$ for each canonical network to drive a choice to the best model that fits the real network. For example, if $I(\hat{\mathscr{F}})=0.64$ for "Highland tribes" network, we conclude that a canonical small-world network should better represent this network than a canonical scale-free network. On the other hand, the "Infectious" network should be best represented by Scale-Free networks with "stretched exponential". For the sake of comparison, we have created 30 networks with a similar value for $I(\hat{\mathscr{F}})$ between the original networks and the generated ones. One can infer important topological metrics if this generative process is

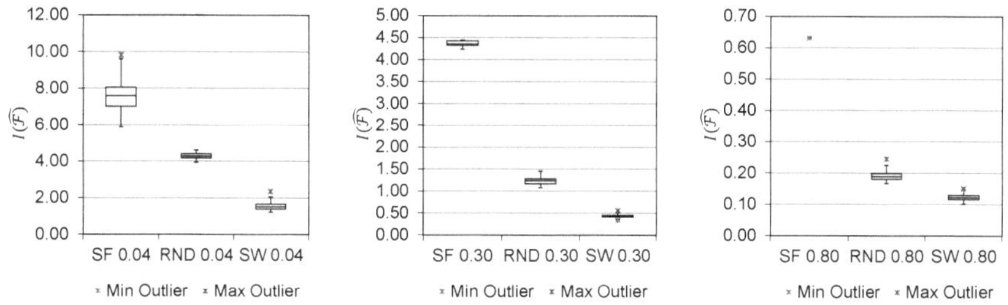

(a) Networks with $d = 0.04$. (b) Networks with $d = 0.30$. (c) Networks with $d = 0.80$.

Fig. 2 Box-plot for 30 different networks with $n = 100$ and $d = 0.04$, 0.30 and 0.80

Table 1 Real networks and the equivalent WS, ER and BA networks

Network	n	d	$I(\hat{\mathscr{F}})$ for each scenario Original	Small-world	Random	Scale-free	Model and parameter to best fit
Highland tribes	16	0.48	0.64	0.51	0.74	0.90	SW ($rp = 0.15$)
Zachary karate club	34	0.13	2.76	0.90	1.82	2.67	SF ($\tau = 1.29$)
Hypertext 2009	113	0.35	3.46	0.21	1.01	3.18	SF ($\tau = 1.42$)
Manufacturing emails	167	0.23	5.60	0.37	1.64	5.43	SF ($\tau = 1.07$)
Infectious	410	0.10	8.67	0.55	4.86	12.89	SF ($\tau = 0.77$)
C. elegans metabolic	453	0.02	20.69	0.76	6.35	14.58	SF ($\tau = 1.13$)
U. Rovira i Virgili	1,133	0.01	18.19	0.71	8.55	26.09	SF ($\tau = 0.79$)

taken. For example, the original "Highland tribes" present $I(G) = 2.58$, $APL = 1.54$ and $k(hub) = 10$. If one calculates the average value for each metrics for the SW approximated by $I(\hat{\mathscr{F}})$ ($rp = 0.15$), the obtained values are $I(G) = 2.33$, $APL = 1.54$ and $k(hub) = 10$. According to our results, it was possible to create networks with $I(G)$, APL and hub degree with average error near 0.10 among all the networks analyzed if one searches the correct generative algorithm by using $I(\hat{\mathscr{F}})$.

6 Conclusion

In this paper we proposed a new approach to analyze and to classify networks according to their topology. The proposed metrics consists in calculating the entropy of the *DFT* over the eigenvalues of the Laplacian matrix. We evaluated the ability of our method to classify networks with different sizes and densities. According to our results on real datasets, the proposed metrics can summarize well known metrics such as entropy of node degrees, APL and hub degree, i.e., even if the entropy of the DFT over the eigenvalues of L is used alone, one can capture information about these three metrics. Thus, we emphasize the expressivity of our metrics due to the possibility of replacing the combined analysis of several metrics by a simpler method in order to assess networks.

Further analysis aims to investigate the impact of using an approximate set of eigenvalues due to scenarios related to very large sparse matrices (the computation of the entire eigenvalues set is prohibitive). We also suggest investigations about the behaviour of the proposed metrics when it is applied on variations of the generative procedures we analyzed in this work.

Acknowledgements. The authors acknowledge the financial support from CNPq, UFPE and UPE.

References

1. Lewis, T.G.: Network Science - Theory and Applications. John Wiley & Sons (2009)
2. Albert, R., Jeong, H., Barabasi, A.L.: Error and attack tolerance of complex networks. Nature 406(6794), 378–382 (2000)
3. Sha, Z., Panchal, J.: Towards the design of complex evolving networks with high robustness and resilience. Procedia Computer Science 16, 522–531 (2013)
4. van Mieghem, P.: Graph Spectra for Complex Networks. Cambridge University Press (2011)
5. Cvetkovic, D.M., Gutman, I.: Selected topics on applications of graph spectra. Matematicki institut SANU, Beograd (2011)
6. Watts, D.J., Strogatz, S.H.: Collective dynamics of small-world networks. Nature 393, 440–442 (1998)
7. van Dam, E.R., Haemers, W.H.: Which graphs are determined by their spectrum? Linear Algebra and its Applications 373, 241–272 (2003)
8. Farkas, I.J., Derényi, I., Barabási, A.L., Vicsek, T.: Spectra of "real-world" graphs: Beyond the semicircle law. Phys. Rev. E 64, 026704 (2001)
9. Araújo, D.R.B., Bastos-Filho, C.A., Martins-Filho, J.F.: Towards Using DFT to Characterize Complex Networks. In: Proceedings of the XXXI Brazilian Symposium on Telecommunications, SBrT 2013, pp. 1–5 (2013)
10. Read, K.E.: Cultures of the central highlands. Southwestern J. of Anthropology 10(1), 1–43 (1954)
11. Zachary, W.W.: An information flow model for conflict and fission in small groups. Journal of Anthropological Research 33, 452–473 (1977)
12. Isella, L., Stehlé, J., Barrat, A., Cattuto, C., Pinton, J.F., Broeck, W.: What's in a crowd? analysis of face-to-face behavioral networks. J. of Theoretical Biology 271(1), 166–180 (2011)
13. Michalski, R., Palus, S., Kazienko, P.: Matching organizational structure and social network extracted from email communication. In: Abramowicz, W. (ed.) BIS 2011. LNBIP, vol. 87, pp. 197–206. Springer, Heidelberg (2011)
14. Jeong, H., Tombor, B., Albert, R., Oltvai, Z.N., Barabási, A.L.: The large-scale organization of metabolic networks. Nature 407, 651–654 (2000)
15. Guimera, R., Danon, L., Diaz-Guilera, A., Giralt, F., Arenas, A.: Self-similar community structure in a network of human interactions. Physical Review E 68, 065103 (2003)

EGIA – Evolutionary Optimisation of Gene Regulatory Networks, an Integrative Approach

Alina Sîrbu, Martin Crane, and Heather J. Ruskin

Abstract. Quantitative modelling of gene regulatory networks (GRNs) is still limited by data issues such as noise and the restricted length of available time series, creating an under-determination problem. However, large amounts of other types of biological data and knowledge are available, such as knockout experiments, annotations and so on, and it has been postulated that integration of these can improve model quality. However, integration has not been fully explored to date. Here, we present a novel integrative framework for different types of data that aims to enhance model inference. This is based on evolutionary computation and uses different types of knowledge to introduce a novel *customised initialisation and mutation operator* and *complex evaluation criteria*, used to distinguish between candidate models. Specifically, the algorithm uses information from (i) knockout experiments, (ii) annotations of transcription factors, (iii) binding site motifs (expressed as position weight matrices) and (iv) DNA sequence of gene promoters, to drive the algorithm towards more plausible network structures. Further, the evaluation basis is also extended to include structure information included in these additional data. This framework is applied to both synthetic and real gene expression data. Models obtained by data integration display both quantitative and qualitative improvement.

1 Introduction

Gene regulatory network reverse engineering is an important aim of Systems Biology [7], as models obtained can be used for analysis and simulation in contexts

Alina Sîrbu
Institute for Scientific Interchange Foundation, Turin, Italy
e-mail: alina.sirbu@isi.it

Alina Sîrbu · Martin Crane · Heather J. Ruskin
Center for Scientific Computing and Complex Systems Modelling
School of Computing, Dublin City University, Dublin, Ireland

P. Contucci et al. (eds.), *Complex Networks V*,
Studies in Computational Intelligence 549,
DOI: 10.1007/978-3-319-05401-8_21,

often difficult to realise in laboratory experiments. Approaches using mathematical modelling, ranging from qualitative to quantitative, have been applied to discovery of GRNs from gene expression data [10]. However, the size of GRNs and the nature of the data (high dimensional, noisy, insufficient for analysis of dynamics), limit robustness when mimicking natural behaviour. This is particularly true for *quantitative* models, which aim to simulate very detailed patterns of expression, increasing the number of parameters to be inferred. However, such models can provide extremely useful insight on the gene expression process, where improvement of reverse engineering techniques is an ongoing aim of Systems Biology [16].

Given the challenges posed by available gene expression data and poor model robustness to date, a new direction is integration of several data types, [16], and these reports have started to appear, mostly for coarse-grained analysis [8]. These integrate expression data with other types of measurements, such as binding affinities or protein interactions, to better discriminate between candidate models, but usually are limited, (i.e. use only one additional data type, besides time-course data). However, several such data-types are available, and the hypothesis is that combining all of these, can further increase modelling power. Recently, *Drosophila Melanogaster* datasets have been integrated, but again for qualitative analysis only [3]. Here, a novel inferential framework for *quantitative* models, based on Evolutionary Computation (EC), is presented (EGIA - Evolutionary optimisation of GRNs - an Integrative Approach). Although other methods are also possible, the EC approach has been selected as it provides increased flexibility, implicit parallelism and has proved to be a suitable search method for underdetermined problems, noisy data and large search spaces [1]. The hypothesis tested is that integration of diverse large-scale biological data improves qualitative and quantitative performance of models inferred.

The strength of the newly-introduced platform is the number of data types to be combined and the flexibility of integration. The novel customisation of different stages of the Evolutionary Algorithm permits more knowledgeable exploration of the search space and more informative evaluation criteria, based on the data available. This is crucial for improving the performance of the models inferred, both quantitatively and qualitatively. Furthermore, a general methodology for GRN inference from multiple data types is developed. This includes an *error structure analysis* to identify the stage of the algorithm at which each data type should be integrated.

2 Methods

2.1 Data

Both synthetic and real datasets are used to assess algorithm performance. Synthetic networks are from the DREAM4[12] competition. This is a research community competition where data from known GRNs are published and researchers have the task of reverse engineering the original networks. These networks are carefully generated so as to resemble real GRNs. The data used here, generated by networks of 10 and 100 genes respectively, contain both time-series measurements and knockout experiments. The set of known interactions are used for qualitative evaluation, and MSE for dual-knockout experiments for quantitative.

For real data, a sub-network of 27 genes involved in *Drosophila melanogaster* embryo development is analysed. A single-channel (SC) microarray dataset [21], is used for training, while a dual-channel (DC) dataset [11] is used for quantitative evaluation. Cross-platform normalisation (namely XPN, [17]) has been performed prior to model inference. For qualitative evaluation, 16 interactions from the Drosophila Interactions Database (DROID) [13], version 2010_10, are considered gold-standard. Additional data types are also integrated: (i) knockout experiments for 8 genes, which were used to compute log-ratios against wild-type experiments [11, 5, 20, 4, 6], (ii) pair-wise correlation between gene expression patterns, (iii) Gene Ontology (GO) [14] annotations, which assign the function of *transcriptional regulation* to 17 of these genes and (iv) binding site affinities for 11 transcription factors (computed using known cis-regulatory modules and position specific weight matrices - PSWMs [15, 2]).

Algorithm performance is evaluated both quantitatively and qualitatively. *Qualitative* evaluation analyses GRN *topology*, to assess whether known interactions between gene pairs are retrieved by the algorithm. This means that the known adjacency matrix of the network is compared to the one retrieved by our algorithm. Specifically, the AUROC (Area Under the ROC Curve) and AUPR (Area Under the Precision-Recall Curve) are computed, measures used also in the DREAM4 competition. Given that the algorithm is stochastic in nature, predictions of interactions have been performed by using multiple models obtained in different runs, and employing a voting procedure for possible interactions. In this way, an interaction that appears in more models is considered to be more plausible. The ranking of possible interactions is used for AUROC/AUPR computation. *Quantitative* evaluation assesses whether the inferred models are able to predict the real-valued expression levels seen in the data. This is performed by simulating a set of *test* data, *not used for model inference*, and by computing average MSE (Mean Squared Error) values over multiple runs.

2.2 Algorithm

EGIA seeks to exploit several types of data related to the process of gene expression, which contribute at different stages of the evolutionary algorithm. The framework is based on a previously introduced inferential algorithm, [9]. This algorithm has been shown to be among the most scalable and least sensitive to noise of several methods from the literature [18]. Based on this, we have chosen to extend it further for data integration, by introducing *novel mutation, initialisation and evaluation operators*.

2.2.1 The Basic Algorithm

In [9], a neural-genetic hybrid approach to GRN inference was introduced. This models the GRN as a single-layered Artificial neural Network (ANN), consisting of one neural unit per gene. Each unit i takes as input the expression values of the regulators of gene g_i (i.e. g_j) at time point t and computes the expression level for gene g_i at time $t+1$, using the input weights w_{ij} and the logistic function $S(x) = \frac{1}{1+e^{-x}}$ for activation:

$$g_i(t+1) = S\left(\sum_j w_{ij} g_j(t) + b_i - d_i g_i(t)\right) \tag{1}$$

where b_i accounts for external input, while d_i represents the degradation rate.

The basic algorithm divides optimisation into two phases: *structure* and *parameter* search. The first involves optimising network topology, i.e. the set of regulators for each gene. This is implemented as a Genetic Algorithm, where each individual encodes a candidate structure, as a subset of the possible regulators for the current gene. Each candidate structure is assigned a fitness value during the parameter search phase, which employs Gradient Descent to optimise the input weights for the neural unit for the current gene. The final error obtained is considered the fitness of the candidate structure. A divide-and-conquer approach is used to optimise parameters for each gene at a time, i.e. training small networks with one neural unit, independently of the other units.

2.2.2 Algorithmic Schema Extension

The basic algorithm [9] optimises parameters for each gene separately, in a divide-and-conquer manner. This approach reduces dimensionality of the system for each optimisation run. However, the model obtained by directly combining sub-models may not be able to correctly simulate the whole system, as separate optimisation disregards the feed-back from the full gene set. In consequence, we have added a second optimisation stage, which combines single-gene models and performs a fine-tuning of complete-network parameters, using the same structure and parameter optimisation.

One way of obtaining models that are robust to noise involves creating noisy replicates from the available data [22]. This simulates technical replicates, and results in multiple time series to be used during inference. Here, a larger set of time-series has been derived from available data through addition of random Gaussian noise. This has been performed during the parameter optimisation phase, for ANN training.

2.2.3 Custom Initialisation and Mutation

The basic algorithm achieves an initial population of candidate structures by randomly selecting possible transcription factors for a specific gene. Similarly, mutation is performed by replacing one of the regulators with a randomly chosen gene. However, many data types provide indications on which interactions between genes are most likely. For example, binding site affinities can indicate what transcription factors can bind to a specific gene promoter. This type of information is very valuable, and can be used to explore the search space in a more knowledgeable manner.

For this, we have developed a *customised initialisation and mutation* procedure, which uses likelihood assignment for gene regulation. This results, for each gene g, in a non-uniform probability mass function, which describes which of the genes in the network are more likely to be regulators of gene g. When performing mutation or initialisation, this function is used to select a candidate regulator for gene g. This is similar to *Wheel of Fortune (WOF)* selection [1], (also known as the *roulette wheel*), so will be addressed henceforth as WOF mutation and initialisation.

In order to build the probability mass function for each gene g, the strategy is to assign segments on the WOF to each gene in the network, if there is any indication in the data of a possible effect of that gene on the current gene g. This number of segments has to be defined by the user based on the reliability of the data used. In the following we provide the values used in our experiments, empirically determined through multiple applications of the algorithm. Of course, these values can be changed to produce a higher or lower effect on the resulting WOF. Several different types of data can be used for this, as follows.

Correlation Patterns. Althought dependences between genes can be non-linear, a good correspondence between linear gene expression correlation-based networks and GRNs has been previously identified, [23]. In consequence, we have used Pearson correlation between time series data of gene pairs, to enhance solution space exploration. Based on absolute values of the correlation to gene g, each gene i is assigned segments on the WOF:

$$CORR_{gi} = \begin{cases} 0 & \text{if } |r_{gi}| < \text{1st decile} \\ 1 & \text{if 1st decile} < |r_{gi}| < \text{3rd decile} \\ 4 & \text{if 3rd decile} < |r_{gi}| < \text{7th decile} \\ 6 & \text{otherwise} \end{cases} \tag{2}$$

where r_{gi} is the Pearson coefficient between genes i and g. The deciles are based on all correlation values obtained. In this way, genes that show high correlation with the current gene will be more likely to be selected as possible regulators.

Knockout(KO) Experiments. Gene expression data from KO experiments can also be used to enhance the search for network models. Absolute values of log-ratios between wild-type and knockout samples can be used to allocate segments on the WOF to those genes that display a large effect on other genes. The number of segments (KO_{gi}) allocated for each gene i on the WOF of gene g depends on the magnitude of the log-ratio:

$$KO_{gi} = \begin{cases} 0 & \text{if } |\text{log-ratio}_{gi}| < 0.2 \\ 1 & \text{if } 0.2 < |\text{log-ratio}_{gi}| < 0.5 \\ 4 & \text{if } 0.5 < |\text{log-ratio}_{gi}| < 0.8 \\ 6 & \text{if } 0.8 < |\text{log-ratio}_{gi}| < 1.1 \\ 8 & \text{otherwise} \end{cases} \tag{3}$$

Gene Ontology (GO) Annotations. The GO database contains annotations of which gene products have been observed to have a specific function, and annotations of transcriptional regulator activity can be included in the EGIA framework. These genes will be allocated additional segments (4 in our experiments) on all the wheels of fortune of the genes in the network. In this way, known transcription factors become more likely to be selected as regulators:

$$ANNOT_{gi} = \begin{cases} 0 & \text{if gene } i \text{ is not annotated as TF} \\ 4 & \text{otherwise} \end{cases} \tag{4}$$

Binding Site Affinities. Binding site (BS) affinities can be integrated in a similar manner. To compute the affinity between a regulator and a gene, the position specofic weight matrix (PSWM) associated with the regulator is required, as well as promoter sequences for the gene. Using these two pieces of information, BS affinity values for each TF i and target gene g are retrieved. For each regulator i, the average ($\overline{A}$) and maximum affinity (A_{max}), over all target genes g, is computed, and segments on the WOF are allocated as follows:

$$BS_{gi} = \begin{cases} 0 & \text{if } A_{gi} < \overline{A} \\ 6 & \text{if } \overline{A} < A_{gi} < \overline{A} + \frac{\overline{A} - A_{max}}{2} \\ 8 & \text{otherwise} \end{cases} \tag{5}$$

where A_{gi} represents the affinity of gene i for binding to a promoter of gene g.

Once all the segments, corresponding to the different type of data, are allocated for all possible regulators, these are summed (Equation 6) and the segment distribution is normalised to represent a probability mass function (Equation 7).

$$WOF_{gi} = CORR_{gi} + KO_{gi} + BS_{gi} + ANNOT_{gi} \tag{6}$$

$$f_g(i) = \frac{WOF_{gi}}{\sum_i WOF_{gi}} \tag{7}$$

This probability mass function defines the probability that a gene i will be selected as regulator for gene g during mutation and initialisation. Each target gene g is associated with such a probability mass function. All data types mentioned can be integrated or omitted, depending on availability. When no additional data are available, the WOF mutation and initialisation are equivalent to the random assignment from the basic algorithm.

2.2.4 Extending Evaluation

The original algorithm uses a fitness function based on the RSS between data and simulation. This has been extended to include also the correlation between simulation and gene patterns [19]. However, this only considers time-series data for evaluation. Using additional data during model evaluation, which might provide information on possible structure, is one way of addressing the noise and under-determination problem, inherent in time-series data. This changes the fitness landscape, so that models which have a plausible topology as well as ability to simulate the time-series data, correspond to better fitness.

The WOF mechanism presented in Section 2.2.3 can be thus also used for model evaluation, by computing an average of all probabilities assigned to the model interactions by the WOF. This, used in combination with the previous fitness function discussed [19], enables construction of a fitness landscape that helps the optimisation algorithm find more plausible structures, as well as models that can simulate continuous behaviour. The final fitness function to be minimised is:

$$F = \frac{1}{2}\sum_i (o_i - t_i)^2 - cP - w\frac{1}{n}\sum_{(i,j)\in INT} f_j(i) \tag{8}$$

where the first term on the right hand side represents the squared error typical for ANN backpropagation (o_i is the expression level simulated by the model, while t_i is that observed in the data), the second the correlation term from [19], while the last term is an average, over all pair-wise interactions present in the model, of the probabilities obtained by the WOF mechanism. INT is the set of interactions predicted by the model ((i,j) is an inferred regulatory effect of i on j), while $f_j(i)$ represents the fraction of the WOF allocated to that interaction (Equation 7). This term is weighted by w, a parameter which needs to be provided by the user. This evaluation criterion is used both at the single-gene and complete-model optimisation stage.

3 Results

The customised evolutionary operators have been implemented using all data types available and models obtained compared to the original algorithm. In order to identify which type of data is more useful, different variants of WOF and evaluation have also been employed, by eliminating one data type at a time.

3.1 Performance on Synthetic Networks

For the synthetic datasets (DREAM networks), only correlation patterns and log-ratios for knockout experiments are available, so three versions of the algorithm

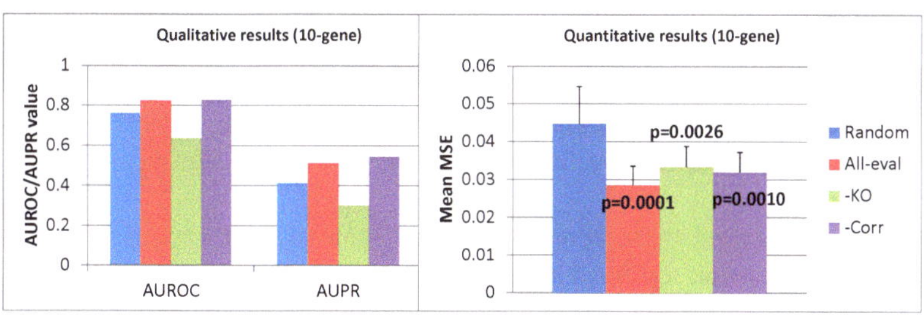

Fig. 1 Performance of WOF and extended evaluation for the 10-gene synthetic dataset

were compared to the basic one (Random). These three variants are denoted by *All-eval* (including all data available in WOF mutation, initialisation and evaluation), *-KO* (all data excluding knockout experiments) and *-Corr* (all data excluding correlation patterns).

Figure 1 displays AUROC and AUPR values obtained after 10 runs of each algorithm on the 10-gene synthetic network. It also includes average MSE over 10 runs for dual knockout simulations, and corresponding p-values of differences observed (compared to the basic algorithm - *Random*). As the figures show, extending the evaluation criterion appears to produce both qualitative and quantitative improvement when compared to the basic algorithm. The set of predicted interactions is slightly improved when knockout experiments only are used (*-Corr*), but quantitative behaviour is best (lowest MSE values) when both data types are integrated. However, when knockout experiments are excluded, AUROC/AUPR values decrease significantly. This suggests that knockout data are very important for extracting direct interactions.

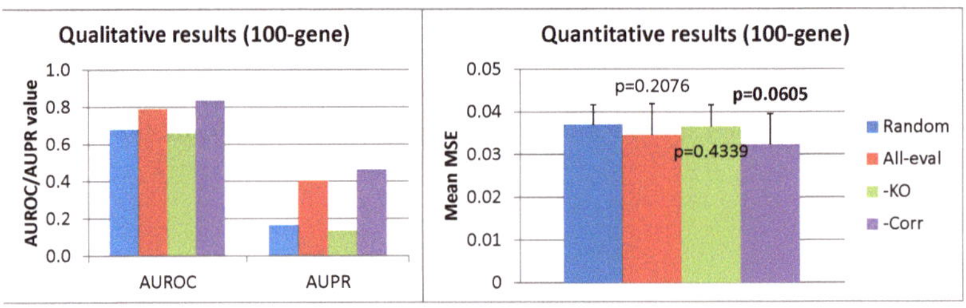

Fig. 2 WOF and extended evaluation for the 100-gene synthetic dataset

Similarly, for the 100-gene network, qualitative and quantitative results are displayed in Figure 2. Introducing the enhanced evaluation criterion markedly increases the number of correct interactions discovered, as shown by the AUROC and AUPR values. The best results are obtained after excluding correlation patterns from the data types used, indicating again that these are not particularly useful in this context, (as found also for the 10-gene network). On the other hand, if knockout experiments are excluded, AUROC/AUPR values decrease significantly, showing that these data are very important in predicting a good set of interactions. From the quantitative point of view, the novel evaluation criterion yields models with low MSE in dual knockout simulations, (minimum values under 0.025), with best results obtained for exclusion of correlation patterns. However, although minimum and average MSE are lower compared to the basic algorithm, the overall quantitative results from multiple experiments are only statistically significant at the 10% level (*-Corr*).

We have compared these results to those obtained by the participants in the DREAM4 competition, on the same networks used in this analysis. The top three teams, which submitted *quantitative* and *qualitative* results for *both network sizes*, have been selected for comparison. For these, AUROC/AUPR and MSE values are given in Table 1, with best performances outlined in bold font. EGIA has obtained the *best predicted interactions for the large scale network*, while for the small scale it scored 3rd. This indicates that our method is more scalable compared to the others. From the quantitative simulation point of view, EGIA has obtained models with lower MSE than the other methods on dual knockouts for both network sizes although, on average, behaviour is comparable to other methods. Nevertheless, given the good qualitative results, we conclude that this framework has something to contribute for extracting models with correct interactions, while it can also simulate unseen behaviour.

Table 1 Comparison of EGIA with DREAM4 results. For the dual knockout MSE values of EGIA, both the minimum and the average values obtained in repeated runs are provided.

	10-gene $\sqrt{AUROC} * \sqrt{AUPR}$	10-gene dual-KO MSE	100-gene $\sqrt{AUROC} * \sqrt{AUPR}$	100-gene dual-KO MSE
EGIA	0.6735	**0.019**/0.028	**0.624**	**0.0229**/0.0324
Team 548	0.654	0.038	0.544	0.0349
Team 532	**0.733**	0.020	0.505	0.0303
Team 498	0.702	0.029	0.28	0.0327

3.2 *Performance on the Drosophila Network*

For the real dataset, five variants of the algorithm have been analysed: *All-eval* (evaluation and WOF operators using all data available), *-Corr* (all data excluding correlation patterns), *-KO* (excluding knockout experiments), *-BS* (excluding binding site affinities), *-Annot* (excluding GO annotations), enabling assessment of the error structure in these data and how this influences the models obtained. Figure 3 displays AUROC and AUPR values for the five algorithm variants. These indicate that integrating all types of data yields the best prediction for interactions. The largest effect is from the binding site affinity data. However, all data types seem to contribute, unlike the synthetic data where correlation patterns disimproved performance compared to the basic algorithm.

Quantitative evaluation was performed again by computing the RMSE with the test dataset (DC), and Figure 3 also shows average results obtained by each of the algorithm variants in 10 runs, with p-values of observed differences from the basic

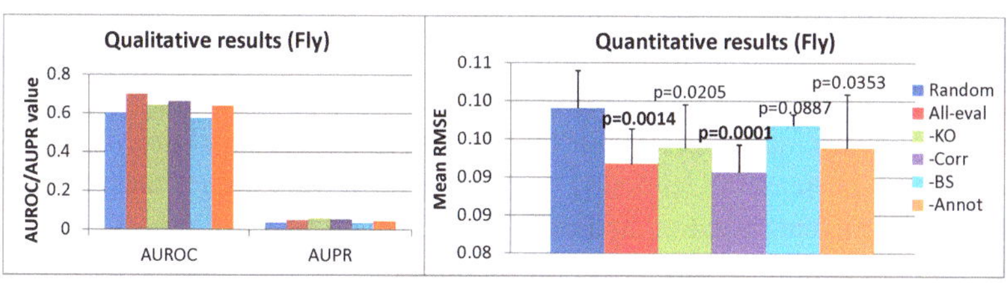

Fig. 3 Performance of WOF and extended evaluation for the 27-gene real dataset

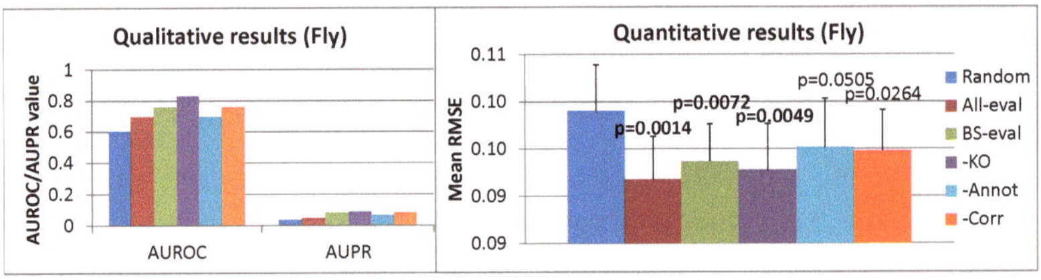

Fig. 4 WOF and binding site extended evaluation for the 27-gene real dataset

algorithm. Our algorithm improves quantitative behaviour, with RMSE values significantly lower than the basic algorithm (at the 1% level for *All-eval* and *-Corr*, and the 5% level for *-KO* and *-Annot*). This improvement means that models not only contain more valid interactions, but also simulate test data better, i.e. improvement in both qualitative and quantitative performance. The error structure analysis also indicates that correlation patterns are once again not particularly useful for improving quantitative performance, while binding site affinities seem to be crucial.

While WOF is a *weak* integration method, as it drives the algorithm only towards promising areas of the search space, without forcing it to choose one model or another, extended evaluation is a *strong* integration criterion, having the final say in which model is better. So, while the WOF operators can be resilient to some level of noise in the data, the evaluation criterion must include more specific data types. Given the results from the error structure analysis for the real dataset, correlation patterns, knockout experiments and GO annotation are more suitable for WOF alone, as they provide *guideline* information only on potential interactions. Binding site affinities are, however, suitable for formal model evaluation, as they have proved to be crucial for obtaining good quantitative performance (Figure 3). For the rest of this section, therefore, we present a similar analysis for different algorithm variants employing only binding site affinities in evaluation, but using various forms of WOF operators: *BS-eval* (using all data types for WOF), *-Corr* (excluding correlation patterns from WOF), *-KO* (excluding knockout experiments), *-Annot* (excluding GO annotations).

Figure 4 displays the performance for all four algorithm variants above, compared to *All-eval* (evaluation and WOF using all data types) and *Random*, the basic algorithm (no meta-data used). *BS-eval* produces models with better connections compared to *All-eval*, while RMSE on test data is maintained at a low level (*BS-eval* and *-KO* significantly different from *Random* at the 1% level).

On extending evaluation, RMSE values for training data display a slight increase, both for synthetic and real data. One explanation for this is that the *generalisation* ability of models is increased (RMSE on test data decreases), and the *over-fitting* of training data is decreased. Generally, machine learning techniques need to obtain a balance between generalisation and over-fitting, which was made possible here by the inclusion of additional data types for training.

4 Conclusion

This paper presented an analysis of data integration for GRN modelling. Two integration mechanisms have been analysed, namely *customised mutation and initialisation* (WOF) and *extended evaluation*. The *error structure* of available data has been studied, to identify which data type has larger effect on the networks analysed. WOF and extended evaluation led to both quantitative and qualitative improvement. This supports the hypothesis that optimisation with time-series data alone is not powerful enough, and that additional information from other data types is needed to aid further selection of GRN models.

The error structure analysis suggested that not all data types are useful for inference, however, and that great caution needs to be taken when integrating these. For synthetic data, knockout experiments proved to be highly important to improve predictions of regulatory interactions. For real data, binding site affinities seemed to have the largest impact. Correlation patterns, on the other hand, were of some help when integrated in WOF mutation with other data types, but had less individual importance. This might be due to the fact that correlation does not indicate only direct interaction, but also indirect effects, which can be captured by the models.

WOF proved to be a flexible integration tool, while evaluation provided additional rigour. For best results, only very reliable data should be used for the latter, while noisy data can be integrated into the former, following an error structure analysis. In our experiments, best performance on real data was found by using only binding affinities for evaluation, and all data types for WOF. This suggests that other data types can provide only general guidelines for possible structures. For instance, log ratios in knockout experiments, or correlations between gene expression patterns can sometimes be misleading, due to the existence of feedback loops, related to alternative regulatory paths or indirect interactions in the real network. The results presented here apply for the *Drosophila melanogaster* embryo development network and associated datasets available for this system. In analysing other systems, e.g. different processes or organisms, data types and quality available will vary, so performing an initial error analysis is crucial to determining the best integration strategy.

Acknowledgements. We would like to thank the Irish Centre for High-End Computing for the provision of computational resources for the experiments. This work was supported by the Irish Research Council for Science, Engineering and Technology, under the EMBARK scheme and partly by the EveryAware project funded by the EC under the EU RD contract IST-265432.

References

1. Baeck, T., Fogel, D.B., Michalewicz, Z.: Evolutionary Computation 1: Basic Algorithms and Operators. Institute of Physics Publishing, Bristol and Philadelphia (2000)

2. Bergman, C.M., Carlson, J.W., Celniker, S.E.: Drosophila DNase I footprint database: a systematic genome annotation of transcription factor binding sites in the fruitfly, Drosophila melanogaster. Bioinformatics 21(8), 1747–1749 (2005)
3. modENCODE Consortium, T.: Identification of functional elements and regulatory circuits by drosophila modencode. Science (2010)
4. Elgar, S.J., Han, J., Taylor, M.V.: Mef2 activity levels differentially affect gene expression during drosophila muscle development. Proceedings of the National Academy of Sciences of the United States of America 105(3), 918–923 (2008)
5. Estrada, B., Choe, S.E., Gisselbrecht, S.S., Michaud, S., Raj, L., Busser, B.W., Halfon, M.S., Church, G.M., Michelson, A.M.: An integrated strategy for analyzing the unique developmental programs of different myoblast subtypes. PLoS Genetics 2(2), e16 (2006)
6. Fox, R.M., Hanlon, C.D., Andrew, D.J.: The CrebA/Creb3-like transcription factors are major and direct regulators of secretory capacity. The Journal of Cell Biology 191(3), 479–492 (2010)
7. Heath, A., Kavraki, L.: Computational challenges in systems biology. Computer Science Review 3(1), 1–17 (2009)
8. Huttenhower, C., Mutungu, K.T., Indik, N., Yang, W., Schroeder, M., Forman, J.J., Troyanskaya, O.G., Coller, H.A.: Detailing regulatory networks through large scale data integration. Bioinformatics 25(24), 3267–3274 (2009)
9. Keedwell, E., Narayanan, A.: Discovering gene networks with a neural-genetic hybrid. IEEE/ACM Transactions on Computational Biology and Bioinformatics 2(3), 231–242 (2005)
10. Lee, W.P., Tzou, W.S.: Computational methods for discovering gene networks from expression data. Briefings in Bioinformatics 10(4), 408–423 (2009)
11. Liu, J., Ghanim, M., Xue, L., Brown, C.D., Iossifov, I., Angeletti, C., Hua, S., Negre, N., Ludwig, M., Stricker, T., Al-Ahmadie, H.A., Tretiakova, M., Camp, R.L., Perera-Alberto, M., Rimm, D.L., Xu, T., Rzhetsky, A., White, K.P.: Analysis of *Drosophila* Segmentation Network Identifies a JNK Pathway Factor Overexpressed in Kidney Cancer. Science 323(5918), 1218–1222 (2009)
12. Marbach, D., Prill, R.J., Schaffter, T., Mattiussi, C., Floreano, D., Stolovitzky, G.: Revealing strengths and weaknesses of methods for gene network inference. Proceedings of the National Academy of Sciences of the United States of America 107(14), 6286–6291 (2010)
13. Murali, T., Pacifico, S., Yu, J., Guest, S., Roberts, G.G., Finley, R.L.: DroID 2011: a comprehensive, integrated resource for protein, transcription factor, RNA and gene interactions for Drosophila. Nucleic Acids Research 39(suppl. 1), D736–D743 (2011)
14. Ontology, G.: `http://www.geneontology.org/` (accessed December 11, 2013)
15. Pollard, D.: Drosophila sequence specific transcription factor binding site matrices (2011), `http://www.danielpollard.com/matrices.html` (accessed March 2011)
16. Przytycka, T.M., Singh, M., Slonim, D.K.: Toward the dynamic interactome: it's about time. Briefings in Bioinformatics 11(1), 15–29 (2010)
17. Shabalin, A.A., Tjelmeland, H., Fan, C., Perou, C.M., Nobel, A.B.: Merging two gene-expression studies via cross-platform normalization. Bioinformatics 24(9), 1154–1160 (2008)
18. Sîrbu, A., Ruskin, H.J., Crane, M.: Comparison of evolutionary algorithms in gene regulatory network model inference. BMC Bioinformatics 11(59) (2010)

19. Sîrbu, A., Ruskin, H.J., Crane, M.: Regulatory network modelling: Correlation for structure and parameter optimisation. In: Karim, M., Lee, K., Ling, H., Maroudas, D., Sobh, T. (eds.) Proceedings of The IASTED Technology Conferences (International Conference on Computational Bioscience), Cambridge, Massachusetts (2010)
20. Toledano-Katchalski, H., Nir, R., Volohonsky, G., Volk, T.: Post-transcriptional repression of the drosophila midkine and pleiotrophin homolog miple by how is essential for correct mesoderm spreading. Development 134(19), 3473–3481 (2007)
21. Tomancak, P., Beaton, A., Weiszmann, R., Kwan, E., Shu, S., Lewis, S., Richards, S., Ashburner, M., Hartenstein, V., Celniker, S., Rubin, G.: Systematic determination of patterns of gene expression during Drosophila embryogenesis. Genome Biology 3(12) (2002)
22. Wessels, L.F.A., Reinders, M.J.T., Backer, E.: Robust genetic network modeling by adding noisy data. In: IEEE - EURASIP Workshop on Nonlinear Signal and Image Processing (2001)
23. Xulvi-Brunet, R., Li, H.: Co-expression networks: graph properties and topological comparisons. Bioinformatics 26(2), 205–214 (2010)

Designing a Social Network Survey for Cancer Care Coordination

Andrew Vakarau Levula, Kon Shing Kenneth Chung, and Kate White

Abstract. In this paper, we propose the use of social network analytics for investigating the effect of aggregate complexity on health care coordination for people with cancer. Here, we highlight the social networks data collection procedures, its benefits and limitations, and measures of relational data specific to aggregate complexity. Firstly, we suggest that collection and analysis of relational and attribute data offer richer insights to the health care coordination experience of cancer patients. Secondly, drawing from theoretical and methodological strength of previous social network studies conducted in health care, we describe the phases of design undertaken to develop our data collection instrument as well as challenges and solutions associated with the design phases. Thirdly, we discuss the sampling aspect of the study in the context of cancer patients at the Sydney Cancer Centre, New South Wales (NSW), Australia along with results and implications from our pre-pilot study.

Keywords: Social Networks, Complexity, Aggregate Complexity, Care Coordination.

1 Introduction

Social network concepts and measures have been widely adopted across a range of discipline such as organizational studies, diffusion of innovation, information

Andrew Vakarau Levula · Kon Shing Kenneth Chung
Complex System Research Group, Project Management Program,
The University of Sydney, NSW 2006, Australia
e-mail: {Andrew.levula,Kenneth.chung}@Sydney.edu.au

Kate White
Cancer Nursing Research Unit, Sydney Nursing School,
The University of Sydney, NSW 2006, Australia
e-mail: {kate.white}@sydney.edu.au

P. Contucci et al. (eds.), *Complex Networks V*,
Studies in Computational Intelligence 549,
DOI: 10.1007/978-3-319-05401-8_22, © Springer International Publishing Switzerland 2014

management, research, education and healthcare [1-4]. Social network studies focuses on the structure of relationships linking individuals (or aggregate social units, such as groups, teams and organizations) and the interdependencies in behavior or attitudes related to their social context [3, 4]. For instance, Healthcare systems may be viewed as a socio-technical network that comprise of diverse individuals (e.g. patients, nurses, physicians and clinicians) or technological units (e.g. computer systems, medical software) that have strong interdependencies on one another [2, 3, 5]. Healthcare systems are therefore complex systems – a collection of individual agents with the freedom to act in ways that are not always predictable, and whose actions are interconnected so that one agent's action affects the context of other agents systems[6, 7]. Examining the structural patterns and the interconnections within these structures is therefore important to unpacking complexity from a system-wide approach. In other words, we view complexity from an aggregate complexity perspective, which refers to the study of multiple concurrent and dynamic interactions amongst components within a social system [8, 9]. According to Kannampallil, Schauer [10], aggregate complexity can be characterized in terms of the number of components and the degree of interrelatedness amongst the components. Here, we argue that these variables can be captured through social network measures such as degree centrality and density [5], as will be explained in Section 3. In the following sections, we provide an overview of social network analysis, focus on the data collection procedure and describe the challenges in designing a data collection instrument, particularly in the context of patients diagnosed with cancer. We will also describe the findings from our pre-pilot study, along with conclusion and implications of the study.

2 Social Network

Social network relationships indicate connections between one or more units (also known as "actors", "nodes" or "vertices"). These units or actors are usually individual persons e.g. patients or clinicians. They can also be other social unit such as hospitals or objects such as text. Pairs of actors who maintain a particular relationship are said to be linked by that relationship (e.g. two people who are friends are linked by their friendship relation). Relationships often represent influence, communication, trust or friendship in the form of ties and can also represent conflicts or disputes [4]. The strength of ties can range from weak to strong depending on the number and type of resources they exchange, the frequency of exchanges and the intimacy of the exchanges [3, 4]. Furthermore, social ties consist of multiple relations (as in the case of cancer patients and his/her General Practitioner (GP) where a tie could constitute a patient-doctor relationship as well as a friendship relationship) and therefore are called "*multiplex ties*".

Table 1 Summary of Social Network Components

Components	Description of social network components
Ego	The focal person whose social network is being analyzed.
Alter	People who are connected to the ego.
Relationships	The tie(s) that connect the ego with alters.

Social Network Analysis (SNA) is a method for capturing the complexity of social relationships [3, 11]. O'Malley and Marsden [3] predict that health related applications of social network analysis will grow rapidly during the coming decades, since interpersonal relationships and support networks are crucial to the well-being of most persons and because appropriate methods for addressing the difficult analytic problems posed by social network data are increasingly available. Chambers, Wilson [12] also highlight that little evidence exists for the potential of SNA to be realized in the healthcare settings. Future work needs to move beyond the descriptive approach towards SNA-based interventions [12]. In the following section we examine the methodology and measures proposed in this study.

3 Measures for Social Network Data Analysis

This study is currently in the pilot phase to a small random sample of patients. Data from the survey will be stored in a MySQL database to allow for maintenance and flexible retrieval of data. Factor analyses and hypothesis testing will be conducted using SPSS. Density and degree centrality measures can be computed using UCINet and Netdraw [13]. These social network measures will be used as measures for the Aggregate Complexity Framework (ACF) explained in Levula, Chung [14]. In this paper however, we will only focus on the designing of an ego-centric social network instrument required for conducting this study.

The simplest and most obvious notion of centrality is degree centrality. The degree of a point p_i, is simply the count of the number of other points, p_j ($i \neq j$), that are adjacent to it and with which it is, in direct contact. With respect to communication an actor with relatively high degree is somehow in the "thick of things". To measure degree centrality we use the following equation.

$$C_D(p_k) = \sum_{i=1}^{n} a(p_i, p_k)$$

where $a(p_i, p_k) = 1$ if and only if p_i and p_k are connected by a line, 0 otherwise.

The density of a network is a commonly analyzed network property within social network analysis. Density is defined as *"the ratio of existing ties within the network as a proportion of all possible number of ties within the network"* [4]. Density in a directed network is defined using the following equation.

$$\frac{l}{n(n-1)/2}$$

where l is the number of lines present and n is the number of nodes in the graph[4]. The density value for any social network ranges from 0 to 1.The higher the density within a network the more connections there would be between the actors in that network [4].

4 Limitations of Social Network Analysis

A significant challenge in social network analysis (SNA) is associated with recall and recognition bias when respondents identify components of their social network in response to the name generator item. Researchers have argued that name generators elicit only a fraction of those persons having a criterion relationship to a respondent [15]. However, studies have shown that people are able to remember long term events and patterns of interaction fairly well. Data gathered through self-reports could explore the influences of other personal variables that would not have been possible through observational data [3, 16].

In addition, SNA only captures the social network at a given timestamp. Social network analysis according to O'Malley and Marsden [3] is not dynamic, however social systems consists of human actors who are always moving and frequently changing (dynamic systems) their behaviors due to interactions with their environment. Such social dynamic modeling are being explored by scientists such as [17] and [18] using computational and mathematical models to predict the interaction patterns and behavior of individuals and groups. Finally, further applied work needs to be undertaken in the domain of healthcare especially from a patient's perspective [3, 12]. The current social network instruments need to be modified and adjusted to meet people with complex cases in healthcare organizations such as cancer [3].

5 Context of the Study

In this study we examine the aggregate complexity involved in cancer care coordination from a patient's perspective using aggregate social network measures [14]. We will be collecting primary data from the Royal Prince Alfred Hospital (RPAH), Sydney Cancer Centre, in New South Wales (NSW), Australia. Cancer care is a high priority area in Australia and managing cancer is complex [19-21]. Cancer care often requires multiple interventions provided by a variety of health professionals over prolonged periods [9, 22]. Problems such as poor care coordination resulting from socio-demographic factors, physical demographic factors, lack of association with peers, size of the social health professional networks, treatment types and infrastructure (e.g. transportation, technology and hospital services) provide the motivation for an understanding of the interplay between aggregate complexity using social network concepts and measures and care coordination for patients diagnosed with cancer care.

6 Network Data Collection

There are two main social network approaches for data collection. These are egocentric and sociocentric network approaches. The egocentric network approach views the network from the perspective of an actor in the network and the sociocentric network approach views the network as a whole [4, 16]. In this paper we will only focus on an egocentric network approach because it is not practical to apply a sociocentric network approach given the context of this study. In terms of sampling, we utilize the purposive sampling technique. Purposive sampling is one of the most commonly used sampling strategies that involve participants who have been pre-selected according to criteria relevant to a particular research question [20]. The pre-selection criteria for this study is to include patients that have been diagnosed with head and neck, breast, gynecological (surgical and medical) and lung and brain and have had at least 3 – 6 months of treatment. The recruitment for this study will take place at the oncology and outpatient clinics of the Sydney Cancer Centre. Patients will be informed about this research by the medical staff and via posters and flyers that will be placed in selected locations e.g. lifts and waiting areas. These will point the patients to the research investigators that will be collecting the data onsite.

6.1 Design Phase 1: Survey

In phase 1 of this study we developed a social network survey instrument to be completed by cancer patients. It is an egocentric social network instrument in which the cancer patients are the ego and those within their social network are alters. In this section we refer to the social network for patients as their "social health professional network". By social health professional network we mean "social (e.g. family member, friend) and professional (e.g. Doctors, care nurses etc) people who the patient would communicate, interact or consult for matters related to cancer care and the provision of care". An egocentric network instrument contains two main questions - 1) a name generator to identify the respondent's alters and 2) a name interpreter aimed towards obtaining the associations between alters [16, 23]. Name generators are free to recall questions that delineate network boundaries. However, name interpreters are used to elicit data about both the ego-alter and alter-alter relationships.

The name generator question that we used to elicit cancer patients social health professional network is: *"Looking back over the last six months, please identify people (up to 15 maximum) who are or were important in providing you with information or advice related to cancer care. Please also identify their occupation or role and their proximity in their involvement with you"*. The name interpreter question that we used is *"In this section we would like to determine how the members of your professional network relate to each other. This is most essential for conducting an analysis of your network"*. The answer for this question was to be completed using an adjacency matrix as shown in figure 1. The patients are required to select an answer from one of six possible answers - unsure (missing data), 1 – do not know each other, 2 – distant, 3 – less than close, 4 – close and 5 – especially close.

	Person 1	Person 2	Person 3	Person 4	Person 5	Person 6	Person 7	Person 8	Person 9	Person 10	Person 11	Person 12	Person 13	Person 14	Person 15
Person 1	X														
Person 2	X	X													
Person 3	X	X	X												
Person 4	X	X	X	X											
Person 5	X	X	X	X	X										
Person 6	X	X	X	X	X	X									
Person 7	X	X	X	X	X	X	X								
Person 8	X	X	X	X	X	X	X	X							
Person 9	X	X	X	X	X	X	X	X	X						
Person 10	X	X	X	X	X	X	X	X	X	X					
Person 11	X	X	X	X	X	X	X	X	X	X	X				
Person 12	X	X	X	X	X	X	X	X	X	X	X	X			
Person 13	X	X	X	X	X	X	X	X	X	X	X	X	X		
Person 14	X	X	X	X	X	X	X	X	X	X	X	X	X	X	
Person 15	X	X	X	X	X	X	X	X	X	X	X	X	X	X	X

Fig. 1 Response matrix of formal data elicited through the name interpreter item

This social network survey instrument was tested and reviewed by several individuals (including two subject matter experts in cancer research, two clinicians and five PhD research students) before it was circulated to the healthcare professionals (e.g. medical oncologists, hematologists and nurse practitioners) in the Sydney Cancer Centre to evaluate and provide feedback. The feedback from the healthcare professionals was anonymous. They felt that the instrument was too complicated and difficult for the patients to complete especially the name generator item – to elicit up to fifteen people. They thus proposed that we undertake a pre-pilot study so we could better understand the patient's situation.

Given these feedbacks we conducted a pre-pilot study at the Sydney Cancer Centre. The pre-pilot study was conducted by a healthcare expert in the area of cancer research. Patients were firstly asked to identify up to 15 people who they interacted with, communicated and/or consulted for information relating to their cancer care treatment. The responses from the patients were unclear and it was evident that most of the patients were unable to understand the question. The interviewer then rephrased the questions so that the patients could understand and answer them. It was evident that the patients were mostly already stressed and some were suffering from physical impairment such as hearing disability, fatigue and diminished physical activities. Another issue that was identified from the pre-pilot was due to the sequential structure of the instrument. It created an obstacle for the patients such that the patients were unable to tell their cancer care journey story naturally.

However, an interesting finding from this pre-pilot study was that patients were able to express their views of the clinician's role and what they would consult them for. Given the feedback from the pre-pilot study and that from the healthcare professionals we redesigned our instrument by developing a more semi-structured and qualitative interview instrument. This would enable patients to "tell their story" without being interrupted in a natural way. These findings were incorporated in the second design phase which is discussed in section 6.2 below.

6.2 Design Phase 2: Interview

The feedback and results from the pre-pilot study led to the development of a semi-structured qualitative interview instrument. This instrument would enable patients to tell their cancer care journey story naturally and the questions were simplified so that cancer patients could easily understand them. With the new instrument the interviewer would simply prompt and insert probes to help guide the discussion. Furthermore, the name generator item was slashed from eliciting up to 15 individuals to up to 5 individuals as alluded to by the healthcare professionals [24].

Another critical change that was identified by the healthcare professionals was the six month duration in the name generator question. This was reduced to three months and the name generator was rephrased to "*Looking back over the last three months, who has been* <u>*important*</u> *in providing you with information, advice or support related to cancer care?*" In order to elicit up to five alters, the interviewer would simply probe the respondent through the follow-up question "Anyone else?". This would enable the investigator to collect relational data pertaining to the ego-alter network. The key with this approach was that it would enable patients to answer questions naturally in a conversational manner. Another reason for the change in duration was that it would enable patients to clearly recall their experiences and their relational attributes as they journeyed through their cancer treatment. We also changed the name interpreter question as such "*Prompt: XXX (insert name or initials) was the 1st person, you listed. Does XXX know any of the other members (prompt: unsure, don't know, distant, less than close, close and especially close).*" This would be repeated for all five or so alters depending on the number of alters identified by the patients. Furthermore, the patients will not have to fill out anything as the interviewer will be simultaneously asking the questions and filling in the form on their behalf.

Given the feedback from the healthcare professional (i.e. medical oncologists, hematologists, administrators, nurses etc.) and from the pre-pilot study it became clear that the quantitative survey instrument that had been developed initially was too complex for the patients to complete on their own. This led to changes to the design of the instrument so that it would be more qualitative. This would allow the patients to share their experience in a natural and conversational manner.

7 Summary

This paper provides an overview of the concepts of social network in terms of understanding the aggregate complexity associated with the coordination of care for cancer patients in the Sydney Cancer Centre, Royal Prince Alfred Hospital (RPAH), in New South Wales (NSW), Australia. In this study, we focused specifically on egocentric network approach to elicit the ego-alter and alter-alter network for cancer diagnosed patients. We also discuss two design phases for the development of our survey instrument. In design phase 1, patients were required to complete a quantitative survey. The instrument was pre-piloted to a sample of patients at the Sydney Cancer Centre by a senior cancer researcher. In design phase 2, we developed a semi-structured qualitative interview instrument based on the response from the healthcare specialists and the issues identified in the pre-pilot study. The instrument would enable patients to tell their story in a more natural and conversational manner.

The contributions that this study makes to the field are novel insights into the design of social network instrument for complex diseases such as cancer. This study also contributes to the theoretical aspect of aggregate complexity and care coordination. These insights can be leveraged by practitioners and researchers working on similar projects or on complex diseases such as cancer care. The next step for this study is to start data collection which would provide more insights into how the social networks for each patient contributes toward their treatment journey in the healthcare system.

References

1. Barabasi, A.-L.: Network Science, 1–81 (November 2012)
2. Valente, T.: Social Networks and Health: Models, Methods, and Applications 2010. Oxford University Press, USA (2010)
3. O'Malley, A.J., Marsden, P.V.: The Analysis of Social Networks. Health Services and Outcomes Research Methodology 8(4), 222–269 (2008)
4. Scott, J.: Social Network Analysis: A Handbook/John Scott, x, 210 p. : ill. ; 22 cm. Sage, London (1991)
5. Chung, K.S.K., Young, J., White, K.: Towards a Network-enabled Complexity Profile for Examining Responsibility for Decision-making by Healthcare Professionals. In: International Symposium on Network Enabled Health Informatics, Biomedicine and Bioinformatics, Niagara Falls, Canada (2013)
6. Zimmerman, B., Lindberg, C., Plsek, P.: Edgeware: Lessons From Complexity Science for Health Care Leaders 1998. VHA Inc., Dallas (1998)
7. Plsek, P.E., Greenhalgh, T.: Complexity Science: The Challenge of Complexity in Health Care. British Medical Journal 323(7313), 625–628 (2001)
8. Manson, S.M.: Simplifying Complexity: A Review of Complexity Theory. Geoforum 32(3), 405–414 (2001)
9. Litaker, D., et al.: Using Complexity Theory to Build Interventions that improve Health Care Delivery in Primary Care. J. Gen. Intern. Med. 21(2), S30–S34 (2006)
10. Kannampallil, T.G., et al.: Considering Complexity in Healthcare Systems. Journal of Biomedical Informatics 44(6), 943–947 (2011)
11. Hawe, P., Ghali, L.: Use of Social Network Analysis to Map the Social Relationships of Staff and Teachers at School. Health Educ. Res. 23(1), 62–69 (2008)

12. Chambers, D., et al.: Social Network Analysis in Healthcare Settings: A Systematic Scoping Review. PloS One 7(8), 3 (2012)
13. Borgatti, S.P., Everett, M.G., Freeman, L.C.: UCINET IV: Network Analysis Software; User's Guide 1996: Analytic Technology (1996)
14. Levula, A.L., et al.: Envisioning Complexity in Healthcare Systems through Social Networks. In: International Symposium on Network Enabled Health Informatics, Biomedicine and Bioinformatics 2013, Niagara Falls, Canada (2013)
15. Mick, S., Wyttenback, M.: Advances in Health Care Organization Theory 2003. Jossey-Bass, San Francisco (2003)
16. Carrington, P.J., Scott, J., Wasserman, S.: Models and Methods in Social Network Analysis. Cambridge University Press, Cambridge (2005)
17. Carley, K.: Computational Modeling for Reasoning about the Social Behavior of Humans. Computational and Mathematical Organization Theory 15(1), 47–59 (2009)
18. Newman, M.E.J.: Communities, Modules and Large-Scale Structure in Networks. Nature Physics 8(1), 25–31 (2012)
19. Young, J., et al.: Measuring Cancer Care Coordination: Development and Validation of a Questionnaire for Patients. BMC Cancer 11(1), 298 (2011)
20. Walsh, J., et al.: What is Important in Cancer Care Coordination? A Qualitative Investigation. European Journal of Cancer Care 20(2), 220–227 (2011)
21. Bickell, N.A., Young, G.J.: Coordination of Care for Early-Stage Breast Cancer Patients. J. Gen. Intern. Med. 16(11), 737–742 (2001)
22. Walsh, J., et al.: What are the Current Barriers to Effective Cancer Care Coordination? A Qualitative Study. BMC Health Services Research 10(1), 132 (2010)
23. Burt, R.S.: Network Items and the General Social Survey. Social Networks 6(4), 293–339 (1984)
24. Marsden, P.V.: The Reliability of Network Density and Composition Measures. Social Networks 15(4), 399–421 (1993)

Dynamic Contact Network Analysis in Hospital Wards

Lucie Martinet, Christophe Crespelle, and Eric Fleury

Abstract. We analyse a huge and very precise trace of contact data collected during 6 months on the entire population of a rehabilitation hospital. We investigate the graph structure of the average daily contact network. Our main results are to unveil striking properties of this structure in the considered hospital, and to present a methodology that can be used for analysing any dynamic complex network where nodes are classified into groups.

The MOSAR project aims at examining the factors determining the dynamics of AMRB (AntiMicrobial Resistant Bacteria) spread within healthcare facilities. To further reduce transmission, in addition to classical prevention measures (such as admission controls, isolation of carriers and hand hygiene), changing contacts within the hospital is considered as the next step [1]. Indeed, contacts strongly influence how transmission occurs [2]. Yet, contacts are difficult to measure efficiently in practice, and they may even be harder to change. Recently, however, advances in communication technologies have made it possible to record person-to-person interactions with unprecedented detail, allowing an in depth view of the structure of contacts in real-life settings [3]. If such contacts actually support transmission, it may open the way to further improvement in hospital hygiene.

Lucie Martinet · Eric Fleury
ENS de Lyon, DANTE/INRIA, LIP UMR CNRS 5668,
Université de Lyon

Christophe Crespelle
Université Claude Bernard Lyon 1,
DANTE/INRIA, LIP UMR CNRS 5668,
ENS de Lyon, Université de Lyon
e-mail: `christophe.crespelle@inria.fr`

P. Contucci et al. (eds.), *Complex Networks V*,
Studies in Computational Intelligence 549,
DOI: 10.1007/978-3-319-05401-8_23,

In this article, we analyse the contact trace recorded on the entire population of a rehabilitation hospital during 6 months between June and November 2009, within the MOSAR project. We focus on a period of 8 weeks of the measurement, from July 6th to September 2nd involving 492 individuals, 253 patients and 239 staffs. We describe the methodology we used to uncover the key characteristics of this dynamic contact network and the main results we obtained: we point out big differences in the contact profiles of services (Sec. 1), as well as in contact patterns of patients and staffs (Sec. 2), and we reveal the structure of interconnections between the mainly introverted services of the hospital (Sec. 3).

Related Works. There have been some works using sensor devices in order to unfold contact patterns among individuals in environments involving patients or children, which present critical risks for spreading of diseases. The measurement analysed in [4] was made on an entire primary school during 3 days. Two similar experiments, described in [5, 6], were both conducted during one week in some paediatric ward. Compared to those works, our analyses present two important advantages. Firstly, the measurement we use was made on a much longer period of time (6 months), which allows to assess the generality of the conclusions we can derive on shorter period of times (like one day or one week). Secondly, our measurement is not limited to a specific part of the hospital, it involves all patients and all staffs of all services of the hospital, which is a key point to have an accurate view of the actual possibility of spreading into a given service. Indeed, these possibilities highly depend on the contacts occurring outside the service under study.

Preliminaries. The contact data was recorded using sensor devices carried by the participants and that send signals every 30s. Those signals include the ID of their source device which is recorded together with a time stamp by devices that are close enough from this source (typically 1 to 2 meters). The sending time of the different sensors are not synchronised but their internal clocks are. Afterwards, time is sliced in slots of 30s and we keep, for each slot, the list of pairs A, B of sensors such that at least one (possibly both) recorded the signal of the other. Each of these pairs is unordered (we do not keep track of which node receives the signal and which one sends it) and appears at most once in a given time slot. Finally, in all this article, we manipulate intervals of contacts instead of punctual contacts, i.e. a *contact* is a quadruplet (A, B, t_s, t_e) where A and B are two nodes of the network and t_s and t_e are respectively the time slots where starts and ends the interval of contact between A and B, the *length* of the contact being $t_e - t_s$.

Throughout the article, we analyse sets of contacts over a specified time period (typically one day) using three parameters: number of contacts, cumulated length of contacts and number of adjacency pairs. (A, B) is an adjacency pair on a given time period iff there is at least one contact between A and B during this period. A contact (A, B, t_s, t_e) between A and B gives rise to two *semi-contacts*: one attached to A, denoted $(A, t_s, t_e)^B$, and one attached to B, denoted $(B, t_s, t_e)^B$. And similarly, every adjacency pair gives rise to two *adjacency semi-pairs*. In the rest of the article, for sake of simplicity of vocabulary, we use the term contact (resp. pair) instead of semi-contact (resp. semi-pair), but all statistics are actually made using semi-contacts (resp. semi-pairs). The reason is that it gives a straightforward meaning to mean statistics per individual.

In the rest of the article, for sake of comparison, we make extensive use of a uniformised version of the network of the hospital, which we call the *full-uniform network* and which is defined as follows. The full-uniform network is a complete weighted graph where each pair of nodes receives 1) a weight equal to the density of the real network (i.e.the mean number of adjacency pairs per pair of nodes in the real network), 2) a number of contacts equal to the mean number of contacts per pair of nodes, and 3) a cumulated length of contacts equal to the mean cumulated length per pair of nodes.

General Organisation of the Hospital. Over the period of study, the mean number of people present in the hospital in one day is about 103 patients and 64 staffs. The patients and staffs are divided into 9 services (see repartition on Fig. 1), only the first five of which (S1 to S5) contain both patients and staffs, the other four (S6 to S9) containing only staffs. Each of the services S1 to S5, containing both patients and staffs, occupies one floor in one of the two wings of the building: S1, S2 and S3 occupy respectively the 1st, 2nd and 3rd floor of the 1st wing, while services S4 and S5 occupy the 2nd and 3rd floor of the 2nd wing. Services S7 to S9 contain rehabilitation staffs and S6 is the night service, regrouping people replacing staffs from services S1 to S5 during nights. S7 and S8 are located in two distinct places between the two wings of the buildings, but S6 and S9 do not have a unique location in the hospital. It must be clear that the division of the hospital into services is not meaningful only from an administrative point of view but has also a strong impact on the structure of the network: in average in one

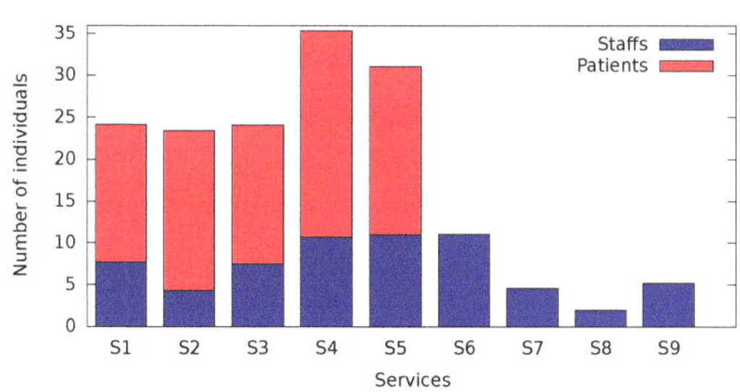

Fig. 1 Number of individuals per day for each service, distinguishing between patients (light red) and staffs (deep blue)

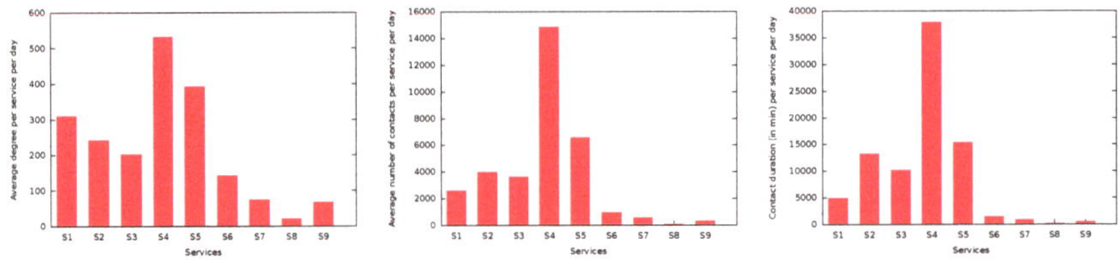

Fig. 2 Mean activity per day in each service. Left: number of adjacency pairs. Centre: number of contacts. Right: cumulated length of contacts.

day, 66% of the adjacency pairs of the hospital occur inside services, and 92% of the cumulated length of contacts, while these values are only 25% in the full-uniform network.

1 Different Levels of Activity of Services

Figure 2 shows the repartition of contacts among the 9 services of the hospital, in terms of total number of adjacency pairs (left), number of contacts (centre) and cumulated length of contacts (right). It reveals some big differences between services. The 5 services including patients seem to be more active than the 4 others, for each of the three criteria. But there are also clear differences between these 5 services as well. As one may guess, one reason for this is that services have different sizes (see Fig. 1). For adjacency pairs, this is confirmed by the fact that the number of mean adjacency pairs per individual per day varies only little between two different services (Fig. 3 left). On the other hand, the number of contacts and the cumulated length of contacts per day remain very different from one service to another even when computed in average for one individual (Fig. 3 centre and right). This indicates that for these two criteria, the sizes of services cannot be hold for entirely responsible of the disparities between global activity of services appearing on Fig. 2.

Services S6 to S9, which do not include any patients, have a mean number of contacts and cumulated length of contacts per individual which is far less than those of services S1 to S5, which do include patients (Fig. 3 centre and right). Moreover, among these latter services, it appears that services S4, S5 and S2 present a higher mean individual activity, for these two parameters, than services S1 and S3; and it turns out that S4, S5 and S2 are the 3 services that contain the greater number of patients (see Fig. 1). These observations suggest that the individual activity of patients wrt. number of contacts and cumulated length of contacts may be much higher than the one of staffs.

Another interesting fact revealed by Fig. 2 and 3 is that the number of contacts and the cumulated length of contacts per service behave very similarly. We conducted more analyses (not presented here) which showed that

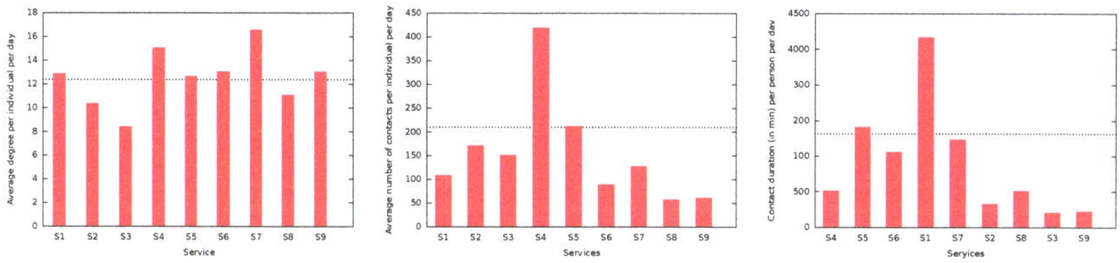

Fig. 3 Mean individual activity per day in each service. Left: number of adjacency pairs. Centre: number of contacts. Right: cumulated length of contacts. The doted lines depicts the mean values per individual in the hospital.

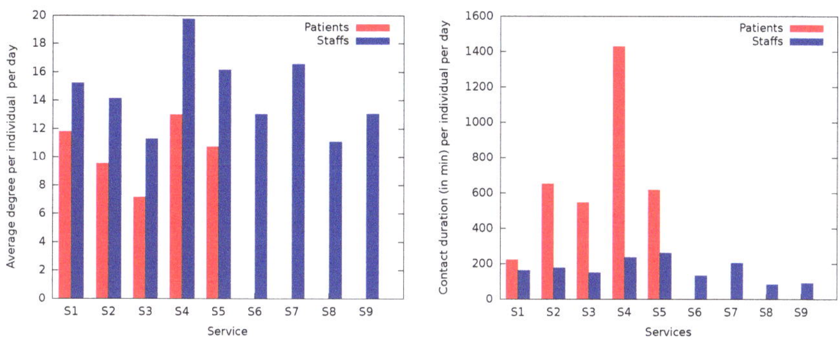

Fig. 4 Mean individual activity per day in each service, distinguishing between patients (light red) and staffs (deep blue). Left: number of adjacency pairs. Right: cumulated length of contacts.

this is a more general fact, not only visible for services: for one node over the whole period of study, these two parameters appear to be strongly correlated. Therefore, as they give very similar results in all the experiments we conducted, we chose to keep only one of them in the rest of the paper, namely the cumulated length of contacts.

2 Different Behaviours of Patients and Staffs

As pointed out above, patients and staffs seem to have a very different activity. We then refine our analysis of the mean activity per individual and per day by separating patients from staffs in the 5 concerned services (Fig. 4). It turns out that patients are a bit less active than staffs (about 20% to 30% less) in terms of adjacency pairs, but are much more active in terms of cumulated length of contact (between 2 and 6 times more, except for service S1 where cumulated length of contacts of patients and staffs are comparable). This explains why the differences between services that appeared on Fig. 2 left for the whole service disappear when considering the adjacency pairs per person (Fig. 3 left), while this difference does not disappear for cumulated length (see Fig. 2 right and Fig. 3 right).

This rises an even more accurate question: what is the role of patients and staffs in the global contact pattern of the hospital? Where is located the majority of contacts? between patients, between staffs or between patients and staffs? Table 1 shows that a vast majority of the cumulated length of contacts in the hospital (80%) occurs between two patients, while only 12% of this length involve one patient and one staff, and 8% involve two staffs. Nevertheless, the picture for adjacency pairs is quite different: those between patients represent only 24% of all pairs, which is about 35% less than in the full-uniform network. The majority of adjacency pairs (56%) involves one patient and one staff, and 20% of them involve two staffs. Both of these values are about 20% higher than in the full-uniform network, suggesting that the contacts of staffs and in particular the contacts between staffs and patients are very important for the structure of the network, and may then play a key role regarding the possibility of spreading in the hospital.

Table 1 Repartition of contacts between patients and staffs in the hospital

	PA-PA	PA-ST	ST-ST
Pairs	0.24	0.56	0.20
Length	0.80	0.12	0.08

(a) Global repartition

PA vs	PA	ST
Pairs	0.46	0.54
Length	0.93	0.07

(b) Patients centred

ST vs	PA	ST
Pairs	0.60	0.40
Length	0.42	0.58

(c) Staffs centred

Table 1 centre and right give the repartition of contacts respectively for an average patient and an average staff. They show that the majority of the adjacency pairs of a patient (54%) occurs with a staff, and that the majority of the adjacency pairs of a staff (60%) occurs with a patient. Note that, opposite to the the case of patients whose cumulated length of contacts is strongly unbalanced in favour of contacts with patients (93%), staffs share much more equitably their length of contacts between patients (42%) and staffs (58%). This confirms that staffs present a more open pattern of contacts than the one of patients, which may result for them in particular spreading abilities.

3 Introversion and Interconnection of Services

In the introduction, we mentioned that most of the activity of the network takes place inside services. Here we investigate further this question by examining the *deviation* of introversion of each service with regard to adjacency pairs, number of contacts and cumulated length of contacts. The introversion of a service S with regard to one of these 3 parameters, denoted α, is defined as $\alpha_{int}(S)/\alpha_{ext}(S)$, where $\alpha_{int}(S)$ is the value of parameter α (e.g. number of adjacency pairs) inside S and $\alpha_{ext}(S)$ is the value of parameter α between S and the rest of the hospital. In all the rest of the article, we qualify contacts and adjacency pairs as *internal* or *external* depending whether they take place inside a service or between two distinct services. We define the *factor of deviation* of introversion of service S as the ratio between the introversion of S in the real network and the introversion of S in some specifically defined uniform network. For adjacency pairs, we use for comparison the full-uniform network defined in the preliminaries. For number of contacts, we use the *contact-uniform network*, which has exactly the same adjacency pairs as the real network, each of which receives a number of contacts equal to the mean number of contacts per adjacency pair in the real network. And finally, for cumulated length of contacts we use the *length-uniform network*, which has the same adjacency pairs as the real network, each of which has the same number of contacts as in the real network, but each of this contacts receives a length equal to the mean cumulated length per contact in the real network. The rational behind these definitions is that for the number of contacts, we compute its deviation knowing the adjacency pairs of the real network, and for the cumulated length of contacts, we compute its deviation knowing both the adjacency pairs and the number of contacts of the real network.

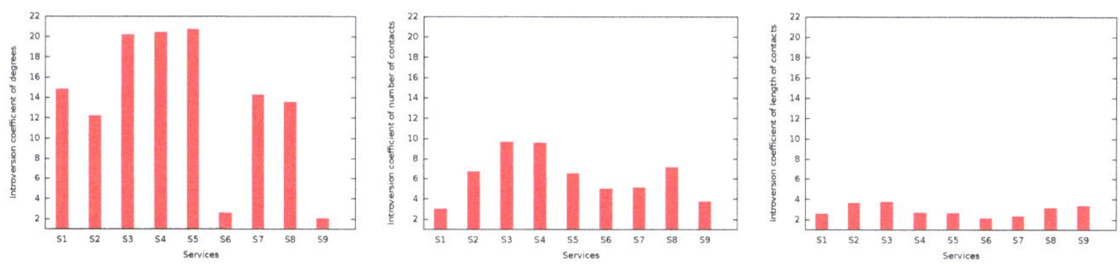

Fig. 5 Factor of deviation of the introversion per day for each service. Left: deviation of the introversion wrt. number of adjacency pairs. Centre: deviation of the introversion wrt. number of contacts, knowing adjacency pairs. Right: deviation of the introversion wrt. cumulated length of contacts, knowing adjacency pairs and number of contacts.

The results are depicted on Fig. 5. They show that services are strongly introverted in terms of adjacency pairs: most of them have a factor of deviation of introversion between 9 and 18, except two services S6 and S9 having factors 2 and 3. Note that these two services are those that do not have a single location in the building of the hospital. Going further, even knowing this structure of the adjacency pairs, services are still clearly introverted in terms of number of contacts (factors between 3 and 10). This means that services do not have only a strong preference for making adjacency pairs inside rather than outside, but they are also much more likely to repeat contacts for their internal adjacency pairs. For cumulated length, the factor of deviation of introversion is between 2 and 4 for all services. The fact that these values

Table 2 Repartition of adjacency pairs between patients and staffs, distinguishing between internal and external pairs

	PA-PA	PA-ST	ST-ST	Total
External	0.05	0.23	0.06	0.34
Internal	0.19	0.33	0.14	0.66
Total	0.24	0.56	0.20	1.00

(a) Global repartition.

PA vs	PA	ST	Total
Ext.	0.10	0.22	0.32
Int.	0.36	0.32	0.68
Total	0.46	0.54	1

(b) Patients centred.

ST vs	PA	ST	Total
Ext.	0.24	0.12	0.36
Int.	0.34	0.30	0.64
Total	0.58	0.42	1

(c) Staffs centred.

are lower than the previous ones is a consequence of the correlation between cumulated length of contacts and number of contacts (see Section 1). But still, they indicate that services not only favour internal adjacency pairs and internal repetition of contacts, but also prefer longer contacts between their members rather than outside.

Table 2 gives some global statistics distinguishing both between internal and external contacts and between patients and staffs. It reveals a strong bipartite-like structure of the network between the staffs divided into services on one side (9 classes), and the patients divided into services on the other side (5 classes). Indeed, more than 83% of the links between these 14 classes occur between one patient and one staff. In addition, links between patients and staffs represent more than 67% of the external links between services of the hospital (18% of these links occur between staffs and 15% between patients). This shows that the contacts between patients and staffs play a prevalent role in connecting the introverted services of the hospital. These observations are confirmed from an individual centred point of view (see Tab. 2 centre and right): an individual (either patient or staff) has only few external adjacency pairs with his own side of the bipartition, while the repartition between its external and internal pairs with the other side are more balanced than internal/external pairs in the whole network.

Perspectives

The main perspective of our work is to determine the impact of the specific structure of contacts we highlighted on spreading processes. In this context, it is still to establish whether there is a correlation between the contaminations contained in the MOSAR dataset (which also includes biological data) and the pattern of contacts in the dynamic network. The second perspective is to take into account the variation of the contact pattern of the hospital along time and determine its impact on the possibility of spreading in the network.

References

1. Wernitz, M., Swidsinski, S., Weist, K., Sohr, D., Witte, W., Roloff, K.F.D., Ruden, H., Veit, S.: Effectiveness of a hospital-wide selective screening programme for methicillin-resistant staphylococcus aureus (MRSA) carriers at hospital admission to prevent hospital-acquired MRSA infections. Clinical Microbiology and Infection 11(6), 457–465 (2005)
2. Stehle, J., Voirin, N., Barrat, A., Cattuto, C., Colizza, V., Isella, L., Regis, C., Pinton, J.F., Khanafer, N., Van den Broeck, W., Vanhems, P.: Simulation of an SEIR infectious disease model on the dynamic contact network of conference attendees. BMC Medicine 9(1), 87 (2011)
3. Lucet, J.C., Laouenan, C., Chelius, G., Veziris, N., Lepelletier, D., Friggeri, A., Abiteboul, D., Bouvet, E., Mentré, F., Fleury, E.: Electronic sensors for assessing interactions between healthcare workers and patients under airborne precautions. PLoS ONE 7(5), e37893 (2012)
4. Stehle, J., Voirin, N., Barrat, A., Cattuto, C., Isella, L., Pinton, J.F., Quaggiotto, M., den Broeck, W.V., Régis, C., Lina, B., Vanhems, P.: High-resolution measurements of face-to-face contact patterns in a primary school. PLoS ONE 6(8), e23176 (2011)
5. Isella, L., Romano, M., Barrat, A., Cattuto, C., Colizza, V., den Broeck, W.V., Gesualdo, F., Pandolfi, E., Ravà, L., Rizzo, C., Tozzi, A.E.: Close encounters in a pediatric ward: Measuring face-to-face proximity and mixing patterns with wearable sensors. PLoS ONE 6(2), e17144 (2011)
6. Hornbeck, T., Naylor, D., Segre, A., Thomas, G., Herman, T., Polgreen, P.: Using sensor networks to study the effect of peripatetic healthcare workers on the spread of hospital-associated infections. The Journal of Infectious Diseases 206, 1549–1557 (2012)

Social Network Analysis Metrics and Their Application in Microbiological Network Studies

Juliana Saragiotto Silva, Nancy de Castro Stoppe, Tatiana Teixeira Torres, Laura Maria Mariscal Ottoboni, and Antonio Mauro Saraiva

Abstract. In the last decade, several researchers have been using interaction networks resources to investigate of the role of biologic interactions in biodiversity maintenance. The conceptual foundations are the same as in Social Networks (such as Facebook), that have brought a set of metrics to study the network structure and the function of each node in the network. Thus, the aim of this work was to assess the application of Social Network Analysis (SNA) concepts and metrics in microbiological interaction networks, to identify patterns of cohesive subgroups, besides discovering new knowledge regarding the underlying structure of

Juliana Saragiotto Silva · Nancy de Castro Stoppe ·
Tatiana Teixeira Torres · Antonio Mauro Saraiva
Research Center on Biodiversity and Computing (BioComp-USP),
Universidade de São Paulo (USP), São Paulo, SP, Brazil

Juliana Saragiotto Silva
Instituto Federal de Educação, Ciência e Tecnologia de Mato Grosso (IFMT),
Cuiabá, MT, Brazil
email: juliana.silva@cba.ifmt.edu.br

Nancy de Castro Stoppe · Laura Maria Mariscal Ottoboni
Centro de Biologia Molecular e Engenharia Genética (CBMEG),
Universidade Estadual de Campinas (UNICAMP), Campinas, SP, Brazil
email: ncstoppe@gmail.com, ottoboni@unicamp.br

Tatiana Teixeira Torres
Departamento de Genética e Biologia Evolutiva, Instituto de Biociências,
Universidade de São Paulo (USP), São Paulo, SP, Brazil
email: tttorres@ib.usp.br

Antonio Mauro Saraiva
Departamento de Engenharia de Computação e Sistemas Digitais, Escola Politécnica,
Universidade de São Paulo (USP), São Paulo, SP, Brazil
email: saraiva@usp.br

P. Contucci et al. (eds.), *Complex Networks V*,
Studies in Computational Intelligence 549,
DOI: 10.1007/978-3-319-05401-8_24,

subgroups. We built a bipartite microbiological interaction database containing frequency of phylogenetic subgroups in water bodies and applied the following SNA metrics: dependence distribution, strength, betweenness centrality and clique. The *sna* package for the *R* program, *Pajek*, *Dieta* and *Ucinet* programs were the tools used. Among the results, we found that SNA concepts and metrics are extremely useful in microbiological studies to understand the correlation between each node in the network (the generalist and the predominant nodes), as well as to analyze the co-occurrence pattern of microorganisms in the network (cohesive subgroups).

1 Introduction

The concepts, metrics and tools commonly used in interaction network studies have their foundations in Social Network Analysis (SNA), which uses the graph theory concepts, computing techniques and resources, to analyze the network structure and its interactions [26]. The SNA is considered one of the broadest applications in the network science field to research human relationships and connections [16]. Among the challenges that posed in this area are: (i) how networks improve, weaken or transform; and (ii) how to tackle the dynamic process of changes in the network structure. These same challenges arise in several biological contexts, such as in pollination studies [3, 29], that use interaction network analysis as a resource for studying the factors that have contributed and influenced biodiversity maintenance.

In the ecological field, interaction network is defined as a set of species (pollinators and plants) which are connected by means of links, which represent interactions [3]. Thus, researchers apply the algorithms, metrics and software commonly available in the SNA area to identify the role of each species in the network, as well as the properties of the network structure (e.g. connectance, number of species at each trophic level, among others). Furthermore, in the microbiological context, network analysis has become increasingly common in the last decade, and has been used in protein interactions with other macromolecules, such as carbohydrates, nucleic acids, lipids and other chemical molecules (metabolites and drugs) [31]. Besides, network analysis can also be used with microorganisms in more complex data, as phylogenetic relationships [22], microbial source tracking [21], soil bacterial diversity [25], health and disease variants [10, 19], and microbiota diversity [2, 22] .

Concerning microbial indicators, they have been used to survey water quality as surrogates to detect the presence of pathogens of fecal origin. However, these indicators do not provide any information about the origin of fecal pollution, i.e., whether the host source is human, cattle or birds, or even a combination of these [14]. Phylogenetic groups were proposed as a screening tool in source tracking due to rapidity and simplicity [9]. Nevertheless, new approaches should be used to evaluate pollution sources relationships. Considering this limitation, we deem that network analysis is an important tool for addressing this issue, particularly as it has not been used yet in interactions between phylogenetic subgroups and environmental sites. This proof-of-principle study will also provide a framework to evaluate the potential use of network metrics in more complex biological problems.

Therefore, our aim is to assess the application of SNA concepts and metrics in microbiological interaction networks, to identify patterns of cohesive subgroups, besides discovering new knowledge regarding the underlying structure of subgroups.

2 Material and Methods

2.1 Sample Description

Phylogenetic subgroups (A_0, A_1, B1, $B2_2$, $B2_3$, D_1 and D_2) described by [11] and [12] were identified from *Escherichia coli* strains isolated from twelve water bodies with different pollution levels in the State of São Paulo as previously described by [27]. The phylogenetic subgroups were used as biomarkers in the rivers and reservoirs studied, which are Tietê River (TIET2050 and TIET3120), Paratei River (PTEI2900), Ipiranga River (IPIR0018), Pau de Bala Stream (PBAL0014), Aguapei River (AGUA2800), Jaguari Mirim River (JAMI2100), Tanque Grande Reservoir (TGDE0900), Billings Reservoir (BILL2801 and BILL2251) and Guarapiranga Reservoir (GUAR0502 and GUAR0601).

These data were organized in a bipartite microbiological interaction database (water bodies x phylogenetic subgroups) – available as part of a project of the Research Center on Biodiversity and Computing (BioComp-USP), under development by the authors. This dataset is composed of a weighted matrix (phylogenetic group abundance), instead of a binary matrix, in which rows were represented by water bodies, and columns by phylogenetic subgroups; therefore, each cell contains a positive integer representing the frequency of occurrence of a phylogenetic subgroup in the corresponding water body.

2.2 SNA Metrics

The SNA metrics was the method used to analyze the role of each node on the microbial interaction network (species level); it was chosen because it allows identifying the following characteristics of the network nodes: *dependence*, *strength*, *betweenness centrality* and *w-clique* (cohesive subgroups) – which are described below.

In a weighted network, the *dependence* of i node on j node (d_{ij}) is considered the proportion of all interactions of the i node with j node [17]. Therefore, the formula used is $d_{ij} = N_{ij}/N_i$, where N_{ij} represents the number of interactions observed between node i and j, and Ni the total number of interactions identified to node i [18].

The interaction *strength* of the *j* node comprises the sum of all dependencies of that *j* node with each *i* node ($strength_{(j)} = \Sigma d_{ij}$) and represents the effect of one *j* node with the total population of the network [30, 28]. For instance, in plant-pollinator mutualism networks, it is usually measured as the contribution of a pollinator to the maintenance of the plant species [28]. In our network, is the effect of one *j* phylogenetic subgroup in the network structure.

The *betweenness centrality* metric allows analyzing how vital a node is in the network; values near 1 indicate the node that has the bridge function in the network (if it is removed, any node would be disconnected from the other nodes in the network), and 0 indicate nodes with low betweenness centrality (if the node is removed, the network structure would not be altered) [16]. The calculation of this metric is based on the geodesic distances (the shortest paths between a given pair of nodes in the network) and it is described in [13].

The last metric applied was the *w-clique* that allows identifying the cohesive subgroups (clusters) in the network structure; it is composed of a set of vertices (nodes) that are connected to each other by strong interactions, i.e., the weights of which are higher than the average network weight [1]. This metric is an alternative approach to the *clique*[1] metric, because it takes into account the interaction abundance (weighted matrices) to identify the network clusters.

2.3 Tools Applied to Calculate the Metrics

To calculate the SNA metrics in the network, we used the following tools: *sna* package for *R* program (The R Foundation for Statistical Computing), *Pajek*, *Dieta* and *Ucinet* programs.

The *sna* package [8] has a set of tools for Social Network Analysis that allows calculating the metrics related to the nodes level in the network, such as *dependence* and *strength*; from this metrics we can analyze the role of each node in the interaction networks.

The *Pajek*[2] program [4] – Slovene word for spider – is a computational tool used for performing large network analysis and drawing the network structure in graph form [24]. In this context, it was used to calculate the *betweenness centrality* metric and its representation in graph form.

The *Dieta*[3] program [1] allows analyzing individual specialization and identifying cohesive subgroups (*w-cliques*) in weighted networks; it is based on the complex network theory. It is used in combination with *Ucinet*[4] [7] – a software that

---

[1] A clique is composed by a set of three or more vertices (nodes) totally connected to each other on the network [24].

[2] Available online for free download at <`http://pajek.imfm.si/doku.php`>

[3] Available online for free download at <`http://esapubs.org/archive/ecol/E089/115/Dieta1.exe`>

[4] Available online for free download in <`http://www.analytictech.com/downloaduc6.htm`>. A Ucinet tutorial is available at [15].

has incorporated a range of SNA metrics, such as cohesive subgroups and measure of centrality [6]. We firstly used the *Dieta* program to verify the nodes that have been connected to each other by strong interactions and, in sequence, we applied the resulting matrix in the *Ucinet* program to identify the *w-cliques*.

2.4 Data Analysis

From the bipartite microbiological database (phylogenetic subgroups frequency in water bodies), available in an excel spreadsheet, we organized the data, sorted the rows and columns in descending order of degree (number of connections of each node), as recommended by [23], in order to draw the standard graphs. Following these spreadsheets, we prepared two matrices formats and saved in text files: *(i)* the first, in which labels of columns and rows are represented, besides the values of frequencies – to be used in the *R* program (*matrix_r.txt*); and *(ii)* the second, in which only the row labels and values of frequencies remain to be used in the *Dieta* program (*matrix_dieta.txt*).

Firstly, we use the matrix (prepared to the R program) and applied the *sna* package to calculate the node level metrics, specially the *dependence* distribution and the *strength* of each node in the network. Furthermore, this matrix was also used in the *Pajek* program to draw the network graphs, and to calculate and plot the betweenness centrality metric values in graph form.

Another application performed in these networks was the use of the *Dieta* program, in order to identify niche overlap (cohesive subgroups). More details about the parameters used in the *Dieta* program can be seen in its manual, available in the supplementary section of [1]. The output file generated by *Dieta* corresponds to a binary matrix representing the interactions with strong connections, i.e., interactions whose values were greater than or equal to the network average (*w-clique*). This matrix was opened with the *Pajek* program to transform it into an undirected network, because it was a niche overlap network. Finally, we used the *Ucinet* program for the identification of cliques and for drawing the clusters by means of a dendrogram.

3 Results and Discussion

Among the results, by analyzing *dependence* distribution and *strength* metrics of the network nodes (Table 1), it was possible to identify the generalist nodes (e.g., phylogenetic subgroups occurring in most of the water body nodes) and the predominant nodes (e.g., which phylogenetic subgroups were important for certain water body groups). In this case, Table 1 represents the *dependence* distribution of each water body by phylogenetic subgroups, and the phylogenetic subgroups strength metric (sum of the dependencies of all *i* water bodies by each *j* phylogenetic subgroup). Thus, it can be observed that although the BILL2801 reservoir has a positive value of dependence for all phylogenetic subgroups, it has the highest *dependence* of A_1 (0.4314), because this phylogenetic subgroup is the most frequent in this reservoir.

Table 1 Dependence distribution of water bodies by phylogenetic subgroups, and phylogenetic subgroups strength metric

Dependence of water bodies by phylogenetic subgroups[(a)]	A_0	B1	$B2_3$	D_2	A_1	D_1	$B2_2$
BILL2801	0.2549	0.0980	0.1373	0.0392	0.4314	0.0196	0.0196
GUAR0601	0.2245	0.1020	0.1837	0.0408	0.3469	0.0612	0.0408
TIET2050	0.3913	0.0870	0.2391	0.0217	0.1739	0.0652	0.0217
TIET3120	0.3269	0.1154	0.1731	0.0577	0.2115	0.0769	0.0385
BILL2251	0.2453	0.0943	0.3396	0.0189	0.2642	0.0000	0.0377
GUAR0502	0.2692	0.0769	0.2885	0.0385	0.2692	0.0577	0.0000
TGDE0900	0.3725	0.0392	0.0784	0.0000	0.3922	0.0588	0.0588
PTEI2900	0.6136	0.0682	0.0455	0.0227	0.2273	0.0000	0.0227
AGUA2800	0.4630	0.2222	0.0741	0.1296	0.0000	0.1111	0.0000
IPIR0018	0.6667	0.0513	0.0513	0.1538	0.0000	0.0769	0.0000
PBAL0014	0.3913	0.0000	0.0435	0.1739	0.0000	0.3913	0.0000
JAMI2100	0.7727	0.0909	0.0000	0.0000	0.1364	0.0000	0.0000
Phylogenetic subgroup strength[(b)]	4.9920	1.0455	1.6539	0.6969	2.4529	0.9188	0.2399

Notes:

[(a)] The sum total of the dependence values of each water body should be equal to 1. High value of water body dependence (closer to 1) comprises higher association with the phylogenetic subgroup; whilst low value of water body dependence (closer to 0) consists of lower association with the phylogenetic subgroup.

[(b)] The strength metric represents the abundance of phylogenetic subgroups (based on their frequency values) in each water bodies; thus, high strength values can demonstrate the nodes that are more generalists or more predominant in the microbiological network. Strength (j) = $\sum$ d(ij).

In our network, A_0 was the phylogenetic subgroup with the highest *strength* (4.9920). This subgroup has the highest abundance in the water bodies sampled and showed the highest interaction frequency values. However, not necessarily, the most frequent and ubiquitous node displays the highest strength. For instance, the phylogenetic subgroup A_1 which is not present in some reservoirs (AGUA2800, IPIR0018 and PBAL0014) has the second highest *strength* in the network (2.4529). This higher *strength* resulted from the highest *dependence* values in those water bodies in which it is present (as in BILL2801, TGDE0900 and GUAR0601). From this analysis, we can conclude that the *strength* metric is directly related to the presence of association (interaction) and with the *dependence* value. Moreover, by means of the *dependence* and *strength* metrics, it is possible to note the role of each network node and then analyze the network tolerance regarding to the extinction/inclusion of a node – similarly to analyzing the potential of an actor in Social Networks studies, as performed by [20].

Another result found in this study refers to the *betweenness centrality* metric. As it can be seen in Figure 1, this metric allows verifying that the phylogenetic subgroups have a central role in the network – specially A_0, B1, $B2_3$ and D_2 –, the most centralized in the graph and with the largest circles). An analogy to Social

Networks would be to know the most popular people in the network (those that have many friends). In our context, by means of this metric we can verify that the water body interactions have been established through the phylogenetic subgroups, as expected. An additional result to support this kind of analysis is the graphical representation of the metric values, as can be done with the *Pajek* program.

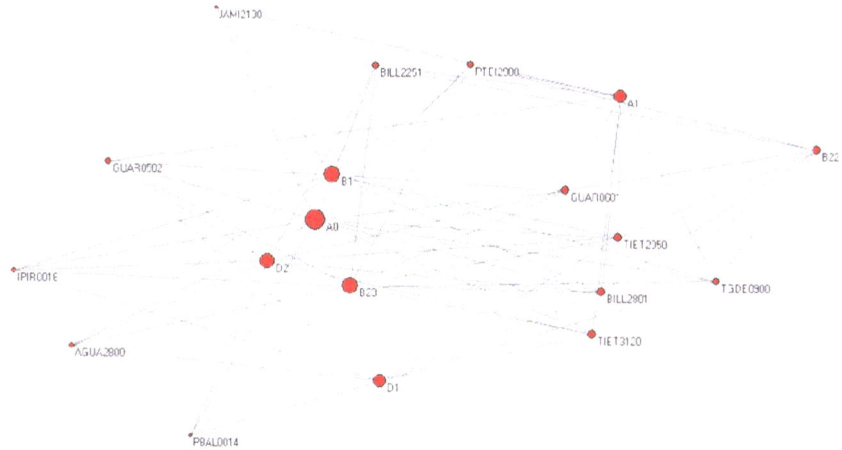

Fig. 1 Betweenness *centrality* graph – phylogenetic subgroups (A_0, A_1, B1, $B2_2$, $B2_3$, D_1, D_2) and water bodies (AGUA2800, BILL2251, BILL2801, GUAR0502, GUAR0601, IPIR0018, JAMI2100, PBAL0014, PTEI2900, TGDE0900, TIET2050, TIET3120)

Once the relationship between water bodies and phylogenetic subgroups was confirmed (by means of *dependence*, *strength* and *betweenness centrality* metric), we found a metric that can support the identification of phylogenetic subgroup co-occurrence patterns in water bodies, from a clustering method. The first alternative was the use of statistical methods – commonly applied for cluster identification – such as Correspondence Analysis (CA) and Unweight Par Group Method with Arithmetic mean (UPGMA). As described by [27], the CA did not recover any consistent grouping. The UPGMA clustering of the dissimilarities split the nodes into two groups, but these groups did not reflect features that are of interest for water studies, such as geographic location, source of pollution or abiotic factors.

Finally, by means of *cliques* identification, it was possible to discover new knowledge regarding a simple interactions database, such as grouping patterns of water bodies (in protected environments and contaminated environments), based on the abundance of phylogenetic subgroups. From the identification of network *cliques* (nodes connected by stronger interactions – greater than or equal to the network average, as showed in Table 2), we could find the cohesive subgroups in the network and it was an innovative contribution for this research area.

Table 2 The output file 'B[*file*_name].mat' generated by Dieta program that corresponds to a binary matrix representing the presence (1) or absence (0) of the interactions with strong connections

*Vertices	12	*Matrix	1	2	3	4	5	6	7	8	9	10	11	12
1	BILL2801	1	0	1	1	1	1	1	1	1	0	0	0	0
2	GUAR0601	2	1	0	1	1	1	1	1	0	0	0	0	0
3	TIET2050	3	1	1	0	1	1	1	1	1	1	0	0	0
4	TIET3120	4	1	1	1	0	1	1	1	1	1	0	0	0
5	BILL2251	5	1	1	1	1	0	1	1	1	0	0	0	0
6	GUAR0502	6	1	1	1	1	1	0	1	1	0	0	0	0
7	TGDE0900	7	1	1	1	1	1	1	0	1	0	0	0	0
8	PTEI2900	8	1	0	1	1	1	1	1	0	0	1	0	1
9	AGUA2800	9	0	0	1	1	0	0	0	0	0	1	1	0
10	IPIR0018	10	0	0	0	0	0	0	0	1	1	0	1	1
11	PBAL0014	11	0	0	0	0	0	0	0	0	1	1	0	0
12	JAMI2100	12	0	0	0	0	0	0	0	1	0	1	0	0

(a) (b)

Note: The left side of this table (a) shows the vertices (nodes) of the network and the right side of this table (b) demonstrate the binary matrix. This matrix gets the result of processing of Dieta program from the bipartite microbiological interaction database (containing the distribution of phylogenetic subgroups in these water bodies). In this case, the nodes between 1 and 7 represent the "contaminated environments" and, the others, the "protected environments".

It is noteworthy that there is a set of metrics (measures) for network analysis. However, the choice should take into account the specific problem and the dataset. Besides, it is possible to analyze at the network level (network structure) or at the node level (the role of one node in the network) – this also depends on the context analysis. More details about a set of metrics can be seen in [5, 29].

Another finding in this research concerning computational tools is that there is a trend towards having graphical and user-friendly interfaces, based on open source codes and having extensive documentation, supporting data import/export to other software, metrics calculation and interactions representation in graph form. As an example, we can mention the *Pajek* program that allows a set of tools to be accessed only through menu options (such as *betweenness centrality*) and also provides a set of graphs for visualizing the results (networks).

4 Conclusion

From this research, we found that SNA concepts and metrics can be used as a tool in microbiological studies, allowing the visualization of phylogenetic subgroups and water bodies associations. For instance, they can help to identify pollution sources (*w-clique*), higher/lower association between water bodies and phylogenetic subgroups (*dependence*), abundance of phylogenetic subgroups in each water bodies (*strength*) and, moreover, the node that has the bridge function in the network. However, is important to emphasize the support of experts in Biodiversity Informatics and microbiological fields, to identify which kind of new knowledge would be generated with biological significance.

In conclusion, a considerable part of the microbiological Interaction Networks studies requirements, especially microbial genetics, concerns the need to identify subgroups (clusters) and co-occurring microorganisms – that have already been exploited in the Social Networks and Ecological Networks field. Thus, these findings will guide further developments of the application of SNA tools and metrics to microbiological studies.

Acknowledgments. This research was developed in the Research Center on Biodiversity and Computing, of the University of São Paulo (BioComp-USP). JSS is a PhD student supported by CAPES, FAPEMAT and IFMT. NCS is a fellowship FAPESP, TTT and LMMO are researcher fellowship CNPq. The authors thank Dr. Marco A. R. Mello (UFMG) and his Guide for Ecological Networks Analysis (http://marcomello.casadosmorcegos.org/Redes.html) for helping with the data analysis.

References

1. Araújo, M.S., Guimarães, P.R., Svanbäck, R., Pinheiro, A., Guimarães, P., Reis, S.F., Bol-nick, D.I.: Network analysis reveals contrasting effects of intraspecific competition on indi-vidual vs. population diets. Ecology 89, 1981–1993 (2008)
2. Bapteste, E., Bicep, C., Lopez, P.: Evolution of genetic diversity using networks: the human gut microbiome as a case study. Clin. Microbiol. Infect. 18, 40–43 (2012)
3. Bascompte, J., Jordano, P.: Plant-Animal Mutualistic Networks: The Architecture of Biodi-versity. Annu. Rev. Ecol. Evol. Syst. 38, 567–593 (2007)
4. Batagelj, V., Mrvar, A.: Pajek – program for large network analysis. Connections 21, 47–57 (1998)
5. Blüthgen, N., Fründ, J., Vázquez, D.P., Menzel, F.: What do interaction network metrics tell us about specialization and biological traits. Ecology 89, 3387–3399 (2008), http://dx.doi.org/10.1890/07-2121.1
6. Borgatti, S.P., Everett, M.G., Freeman, L.C.: UCINET 5 for Windows: Software for Social Network Analysis (USER'S GUIDE) (1999), http://www.analytictech.com/ucinet6/Ucinet_Guide.doc (accessed in November 15, 2012)
7. Borgatti, S.P., Everett, M.G., Freeman, L.C.: Ucinet for Windows: Software for Social Network Analysis. Analytic Technologies, Harvard (2002)
8. Butts, C.T.: Social Network Analysis with sna. Journal of Statistical Software 24, 1–51 (2008)
9. Carlos, C., Pires, M.M., Stoppe, N.C., Hachich, E.M., Sato, M.I.Z., Gomes, T.A.T., Amaral, L.A., Ottoboni, L.M.M.: Escherichia coli phylogenetic group determination and its application in the identification of the major animal source of fecal contamination. BMC Microbiol. 10, 161 (2010)
10. Chen, M., Cho, J., Zhao, H.: Incorporating biological pathways via a Markov random field model in genome-wide association studies. PLoS Genet. 7, e1001353 (2011)
11. Clermont, O., Bonacorsi, S., Bingen, E.: Rapid and simple determination of the Escherichia coli phylogenetic group. Appl. Environ. Microbiol. 66, 4555–4558 (2000)

12. Escobar-Páramo, P., Grenet, K., LeMenac'h, A., Rode, L., Salgado, E., Amorin, C., Gouriou, S., Picard, B., Rahimy, C., Andremont, A., Denamur, E., Ruimy, R.: Large-scale population structure of human commensal Escherichia coli isolates. Appl. Environ. Micro-Biol. 70, 5698–5700 (2004)
13. Freeman, L.C.: Centrality in social networks: conceptual clarification. Social Networks 1(3), 215–239 (1979)
14. Hagedorn, C., Blanch, A.R., Harwood, V.J.: Microbial source tracking: methods, application, and case studies, 642 p. Springer, New York (2011)
15. Hanneman, R.A., Riddle, M.: Introduction to social network methods. University of California, Riverside, Riverside, CA (2005), published in digital form at `http://faculty.ucr.edu/~hanneman/`
16. Hansen, D.L., Shneiderman, B., Smith, M.A.: Analysing social media networks with NodeXL: insights from a connected world, 284 p. Morgan Kaufmann, Amsterdan (2011)
17. Jordano, P.: Patterns of mutualistic interactions in pollination and seed dispersal: connec-tance, dependence asymmetries, and coevolution. American Naturalist 129, 657–677 (1987)
18. Jordano, P., Vázquez, D., Bascompte, J.: Redes complejas de interacciones mutualistas planta-animal. In: Mendel, R., Aizen, M.A., Zamora, R. (eds.) Ecología y Evolución de Interacciones Planta-Animal, 1[a] ed., pp. 17–41. Universitaria, Santiago de Chile (2009)
19. Jostins, L., Ripke, S., Weersma, R.K., Duerr, R.H., McGovern, D.P., Hui, K.Y., Lee, J.C., Schumm, L.P., Sharma, Y., et al.: Host-microbe interactions have shaped the genetic architecture of inflammatory bowel disease. Nature 491, 119–124 (2012)
20. Kumar, K., Novak, J., Tomkins, A.: Structure and evolution of online social networks. In: Proc. of ACM SIGKDD Intl. Conf. of Knowledge Discovery and Data Mining, New York, pp. 611–617 (2006)
21. Lee, J.E., Lee, S., Sung, J., Ko, G.P.: Analysis of human and animal fecal microbiota for microbial source tracking. The ISME J. 5, 362–365 (2011)
22. Ley, R.E., Hamady, M., Lozupone, C., Turnbaough, P.J., Ramey, R.R., Bircher, S., Schlegel, M., Tucker, T.A., Schrenzel, M.D., Knight, R., Gordon, J.I.: Evolution of mammals and their gut microbes. Science 320, 1647–1651 (2008)
23. Mello, M.A.R.: Guia para Análise de Redes Ecológicas. Version March 3 (2013), `http://marcomello.casadosmorcegos.org/Redes.html` (accessed in April 1, 2013)
24. Nooy, W., Mrvar, A., Batagelj, V.: Exploratory Network Analysis with Pajek, 334 p. Cambridge University Press (2005)
25. Roesch, L.F.W., Fulthorpe, R.R., Pereira, A.B., Pereira, C.K., Lemos, J.N., Barbosa, A.D., Suleiman, A.K.A., Gerber, A.L., Pereira, M.G., Loss, A., Costa, E.M.: Soil bacteria community abundance and diversity in ice-free areas of Keller Peninsula, Antarctica. Appl. Soil Ecol. 61, 7–15 (2012)
26. Scott, J.: Social Network Analysis: A handbook, 2nd edn. Sage, London (2000)
27. Stoppe, N.C., Silva, J.S., Torres, T.T., Carlos, C., Hachich, E.M., Sato, M.I.Z., Saraiva, A.M., Ottoboni, L.M.M.: Clustering of water bodies in unpolluted and polluted environments based on Escherichia coli phylogroup abundance using a simple interaction database. BMC Microbiology (unpublished)
28. Vázquez, D.P., Morris, W.F., Jordano, P.: Interaction frequency as a surrogate for the total effect of animal mutualists on plants. Ecology Letters 8, 1088–1094 (2005)
29. Vázquez, D.P., Blüthgen, N., Cagnolo, L., Chacoff, N.P.: Uniting pattern and process in plant–animal mutualistic networks: a review. Annals of Botany 103, 1445–1457 (2009)
30. Wootton, J.T., Emmerson, M.: Measurement of interaction strength in nature. Annual Review of Ecology, Evolution and Systematics 36, 419–444 (2005)
31. Yamada, T., Bork, P.: Evolution of biomolecular networks – lessons from metabolic and protein interactions. Nature Rev. Mol. Cell Biol. 10, 791–803 (2009)

Inducing Language Networks from Continuous Space Word Representations

Bryan Perozzi, Rami Al-Rfou', Vivek Kulkarni, and Steven Skiena

Abstract. Recent advancements in unsupervised feature learning have developed powerful latent representations of words. However, it is still not clear what makes one representation better than another and how we can learn the ideal representation. Understanding the structure of latent spaces attained is key to any future advancement in unsupervised learning. In this work, we introduce a new view of continuous space word representations as language networks. We explore two techniques to create language networks from learned features by inducing them for two popular word representation methods and examining the properties of their resulting networks. We find that the induced networks differ from other methods of creating language networks, and that they contain meaningful community structure.

Keywords: Language Networks, Word Embeddings, Natural Language Processing, Unsupervised Learning, Distributed Representations.

1 Introduction

Unsupervised feature learning (*deep learning*) utilizes huge amounts of raw data to learn representations that model knowledge structure and disentangle the explanatory factors behind observed events. Under this framework, symbolic sparse data is represented by lower-dimensional continuous spaces. Integrating knowledge in this format is the secret behind many recent breakthroughs in machine learning based applications such as speech recognition, computer vision, and natural language processing (NLP) [3].

We focus here on word representations (*word embeddings*) where each word representation consists of a dense, real-valued vector. During the pre-training stage, the representations acquire the desirable property that similar words have lower

Bryan Perozzi · Rami Al-Rfou' · Vivek Kulkarni · Steven Skiena
Department of Computer Science, Stony Brook University, Stony Brook, NY
e-mail: {bperozzi,ralrfou,vvkulkarni,skiena}@cs.stonybrook.edu

P. Contucci et al. (eds.), *Complex Networks V*,
Studies in Computational Intelligence 549,
DOI: 10.1007/978-3-319-05401-8_25,

distance to each other than to unrelated words [13]. These representations have been used successfully in supervised learning applications such as part-of-speech tagging, named entity recognition, language modeling, and sentiment analysis [10, 11, 15].

Several methods and algorithms have been proposed to learn word representations using different benchmarks for evaluation [9]. However, these evaluations are hard to comprehend as they squash the analysis of the representation's quality into abstract numbers. To enable better understanding of the actual structure of word relationships which have been captured, we have to address the problems that come with analyzing high-dimensional spaces (typically between 50-1000 dimensions). We believe that network induction and graph analysis are appropriate tools to give us new insights.

In this work, we seek to induce meaningful graphs from these continuous space language models. Specifically, our contributions include:

- **Analysis of Language Network Induction** - We propose two criteria to induce networks out of word embeddings. For both methods, we study and analyze the characteristics of the induced networks. Moreover, the networks generated lead to easy to understand visualizations.
- **Comparison Between Word Representation Methods** - We evaluate the quality of two well known words embeddings. We contrast between their characteristics using the analysis developed earlier.

The remainder of this paper is set up as follows. First, in Section 2, we describe continuous space language models that we consider. In Section 3, we discuss the choices involved with inducing a network from these embeddings and examine the resulting networks. Finally, we finish with a discussion of future work and our conclusions.

2 Continuous Space Language Models

The goal of a language model is to estimate the likelihood of observing any given sequence of words. The training objective usually maximizes the joint probability of the training corpus. A continuous space probabilistic language model aims to estimate such probability distribution by, first, learning continuous representations for the words and phrases observed in the language. Such mapping is useful to cope with the curse of dimensionality in cases where data distribution is sparse as in natural language.

More precisely, given a sequence of words $S = [w_1 \ldots w_k]$, we want to maximize $P(w_1, \ldots, w_k)$ and learn representations for words. During the training process the continuous space language model learns a mapping of words to points in $\mathbb{R}^d$, where d usually ranges between $20 - 200$. Prior to training we build a vocabulary V that consists of the most frequent $|V|$ words, and we map each word to a unique identifier that indexes an embeddings matrix C that has a size of $|V| \times d$. The sequence S is now represented by a matrix $\left[C[w_1]^T \; \ldots \; C[w_k]^T\right]^T$, enabling us to compose a new representation of the sequence using one of several compositional functions. The simplest is to concatenate all the rows in a bigger vector with size kd. Another option is to sum the matrix row-wise to produce a smaller representation of size d. While the first respects the order of the words, it is more expensive to compute.

We will focus our investigations, here, on two embeddings which are trained with different tasks and compositional functions; the Polyglot and SkipGram embeddings.

2.1 Polyglot

The Polyglot project offers word representations for each language in Wikipedia [18]. For large enough Wikipedias, the vocabulary consists of the most frequent 100,000 words. The representations are learned through a procedure similar to the one proposed by Collobert et al. [10]. For a given sequence of words $S_t = [w_{t-k} \dots w_t \dots w_{t+k}]$ observed in the corpus T, a corrupted sequence S'_t will be constructed by replacing the word in the middle w_t with a word w_j chosen randomly from the vocabulary V. Once the vectors are retrieved, we compose the sequence representation by concatenating the vectors into one vector called the projection layer S_t. The model is penalized through the hinge loss function,

$$\frac{1}{T}\sum_{t=1}^{t=T} |1 - score(S'_t) + score(S_t)|_+$$

where *score* is calculated through a hidden layer neural network

$$score(S_t) = W_2(tanh(W_1 S_t + b_1)) + b_2.$$

For this work, we use the Polyglot English embeddings[1] which consist of the 100,000 most frequent words in the English Wikipedia, each represented by a vector in $\mathbb{R}^{64}$.

2.2 SkipGram

While the Polyglot embeddings consider the order of words to build the representation of any sequence of words, the SkipGram model proposed by Mikolov et al. [14] maximizes the average log probability of the context words independent of their order

$$\frac{1}{T}\sum_{t=1}^{T}\Big[\sum_{j=-k}^{k} \log p(w_{t+j}|w_t)\Big]$$

where k is the size of the training window. This allows the model to scale to larger context windows. In our case, we train a SkipGram model[2] on the English Wikipedia

[1] Polyglot embeddings and corpus available at `http://bit.ly/embeddings`

[2] SkipGram training tool available at `https://code.google.com/p/word2vec/`

corpus offered by the Polyglot project for the most frequent 350,000 words with context size k set to 5 and the embeddings vector size set to 64.

2.3 Random

In order to have a baseline, we also generate random embeddings for the most frequent 100,000 words. The initial position of words in the Polyglot embeddings were sampled from a uniform distribution, therefore, we generate the random embedding vectors by sampling from $\mathscr{U}(\bar{m}-\sigma, \bar{m}+\sigma)$, where $\bar{m}$ and σ are the mean and standard deviation of the trained Polyglot embeddings' values respectively. This baseline allows us to see how the language networks we construct differ from networks induced from randomly initialized points.

3 Word Embedding Networks

We now consider the problem of constructing a meaningful network given a continuous space language model. As there are a variety of ways in which such a network could be induced, we start by developing a list of desirable properties for a language network. Specifically, we are seeking to build a network which:

1. **Is Connected** - In a connected graph, all the words can be related to each other. This allows for a consistent approach when trying to use the network to solve real-world problems.
2. **Has Low Noise** - Minimizing the spurious correlations captured by our discrete representation will make it more useful for application tasks.
3. **Has Understandable Clusters** - We desire that the community structure in the network reflects the syntactic and semantic information encoded in the word embeddings.

We also require a method to compute the distance in the embedding space. While there are a variety of metrics that could be used, we found that Euclidean distance worked well. So we use:

$$dist(x,y) = ||x-y||_2^2 = (\sum_{i=1}^{m}(x_i - y_i)^2)^{(1/2)} \tag{1}$$

where x and y are words in an d-dimensional embedding space ($x, y \in \mathbb{R}^d$). With these criteria and a distance function in hand, we are ready to proceed. We examine two approaches for constructing graphs from word embeddings, both of which seek to link words together which are close in the embedding space. For each method, we induce networks for the 20,000 most frequent words for each embedding type, and compare their properties.

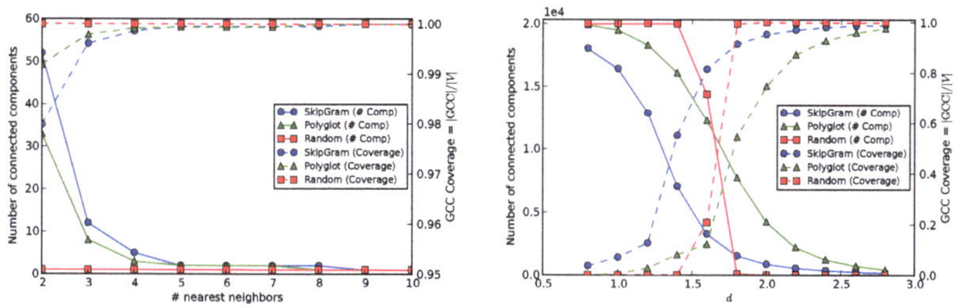

(a) k-NN Coverage **(b)** d-Threshold Coverage

Fig. 1 Graph Coverage. The connected components and relative size of the Giant Connected Component (GCC) in graphs created by both methods. We see that very low values of k quickly connect the entire network (1a), while relatively large values of d are required before a a GCC emerges (1b).

3.1 k-Nearest Neighbors

The first approach we will consider is to link each word to the k closest points in the embedding space. More formally, we induce a set of directed edges through this method:

$$E_{knn} = \{(u,v) : \min_x dist(u,v)\} \quad \forall u,v \in V, x \leq k \tag{2}$$

where $\min_x$ denotes the rank of the x-th number in ascending sorted order (e.g. $\min_0$ is the minimum element, $\min_1$ the next smallest number). After obtaining a directed graph in this fashion, we convert it to an undirected one.

The resulting undirected graph does not have a constant degree distribution. This is due to the fact that the nearest-neighbor relation may not be symmetric. Although all vertices in the original directed graph have an out-degree of k, their orientation in the embedding space means that some vertices will have higher in-degrees than others.

Results from our investigation of basic network properties of the k-NN embedding graphs are shown in Figures 1 and 2. In Figure 1a we find that the embedding graphs have few disconnected components, even for small values of k. In addition, there is an obvious GCC which quickly emerges. In this way, the embeddings are similar to the network induced on random points (which is fully connected at $k = 2$). We performed an investigation of the smaller connected components when k was low, and found them to contain dense groupings of words with very similar usage characteristics (including ordinal values, such as Roman numerals (`II, III, IV`)).

In Figure 2a we see that the clustering coefficient initially grows quickly as we add edges to our network ($k \leq 6$), but has leveled off by ($k = 20$). This tendency to bridge new clusters together, rather than just expand existing ones, may be related to the *instability* of the nearest neighbor [5] in high dimensional spaces.

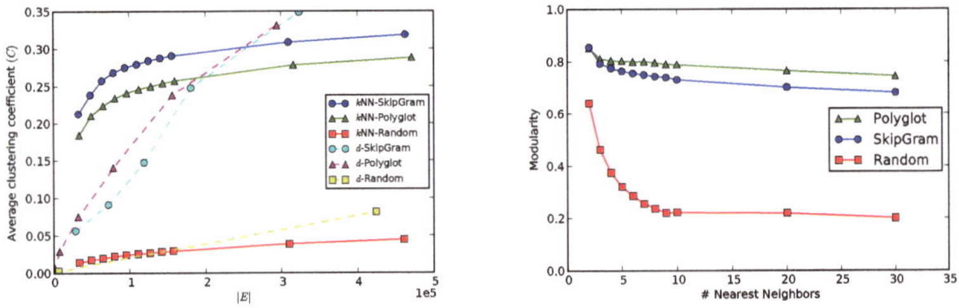

(**a**) Clustering Coeff., C (**b**) k-NN Modularity, Q_{knn}

Fig. 2 Community Metrics. In (2a), C shown for $k = [2,30]$ and $d = [0.8,1.6]$ against number of edges in the induced graph. When the total number of edges is low ($|E| < 150,000$), networks induced through the k-NN method have more closed triangles than those created through d-Proximity. In (2b), Q_{knn} starts high, but slowly drops as larger values of k include more spurious edges.

In Figure 2b, we see that the networks induced by the k-NN are not only connected, but have a highly modular community structure.

3.2 *d*-Proximity

The second approach we will consider is to link each word to all those within a fixed distance d of it:

$$E_{proximity} = \{(u,v) : dist(u,v) < d\} \quad \forall u,v \in V \tag{3}$$

We perform a similar investigation of the network properties of embedding graphs constructed with the d-Proximity method. The results are shown in Figures 1 and 2. We find that networks induced through this method quickly connect words that are near each other in the embedding space, but do not bridge distant groups together. They have a large number of connected components, and connecting 90% of the vertices requires using a relatively large value of d (Fig. 1b).

The number of connected components is closely related to the average distance between points in the embedding space (around d =(3.25, 3.80, 2.28) for (SkipGram, Polyglot, Random)). As the value of d grows closer to this average distance, the graph quickly approaches the complete graph. Figure 2a shows that as we add more edges to the network, we add triangles at a faster rate than using the k-NN method.

3.3 Discussion

Here we discuss the differences exposed between the methods for inducing word embeddings, and the differences between the embeddings themselves.

Comparison of Network Induction Methods. Which method then, provides the better networks from word embeddings? To answer this question, we will use the properties raised at the beginning of this section:

1. **Connectedness** - Networks induced through the k-NN method connect much faster (as a function of edges) than those induced through d-Proximity (Fig. 1). Specifically, the network induced for $k = 6$ has nearly full coverage (Fig. 1a) with only 100K edges (Fig. 2a). This inability to create connected graphs is a serious limitation of using the d-Proximity approach.
2. **Spurious Edges** - We desire that our resulting networks should be modular. As such we would prefer to add edges between members of a community, instead of bridging communities together. For low values of $|E|$, the k-NN approach creates networks which have more closed triangles than d-Proximity (Fig. 2a). However this does not hold in networks with more edges.
3. **Understandable Clusters** - In order to qualitatively examine the quality of such a language network, we induced a subgraph with the k-NN of the most frequent 5,000 words in the Polyglot embeddings for English (Fig. 3). We find the identified clusters to be highly meaningful. The lack of a connected graph precludes us from reasoning about how well d-Proximity preserves global relationships.

According to our three criteria, k-NN seems better than d-Proximity. In addition to the reasons we already listed, we note that k-NN has the additional advantage of requiring less parameterization (d-Proximity has a different optimal d for each embedding type).

Comparison of Polyglot and SkipGram. Having chosen to use k-NN as our preferred method for inducing language networks, we now examine the difference between the Polyglot and SkipGram networks.

Clustering Coefficient. We note that in Figure 2a, the SkipGram model has a consistently higher clustering coefficient than Polyglot in k-NN networks. A larger clustering coefficient denotes more triangles, and this may indicate that points in the SkipGram space form more cohesive local clusters than those in Polyglot. Tighter local clustering may explain some of the interesting regularities observed in the SkipGram embeddings [16].

Modularity. In Figure 2b, we see that Polyglot modularity is consistently above the SkipGram modularity. SkipGram's embeddings capture more semantic information about the relations between words, and it may be that causes a less optimal community structure than Polygot whose embeddings are syntactically clustered.

Clustering Visualizations. In order to understand the differences between the language networks better, we conducted an examination of the clusters found using the Louvain method [7] for modularity maximization. Figure 4 examines communities from both Polyglot and SkipGram in detail.

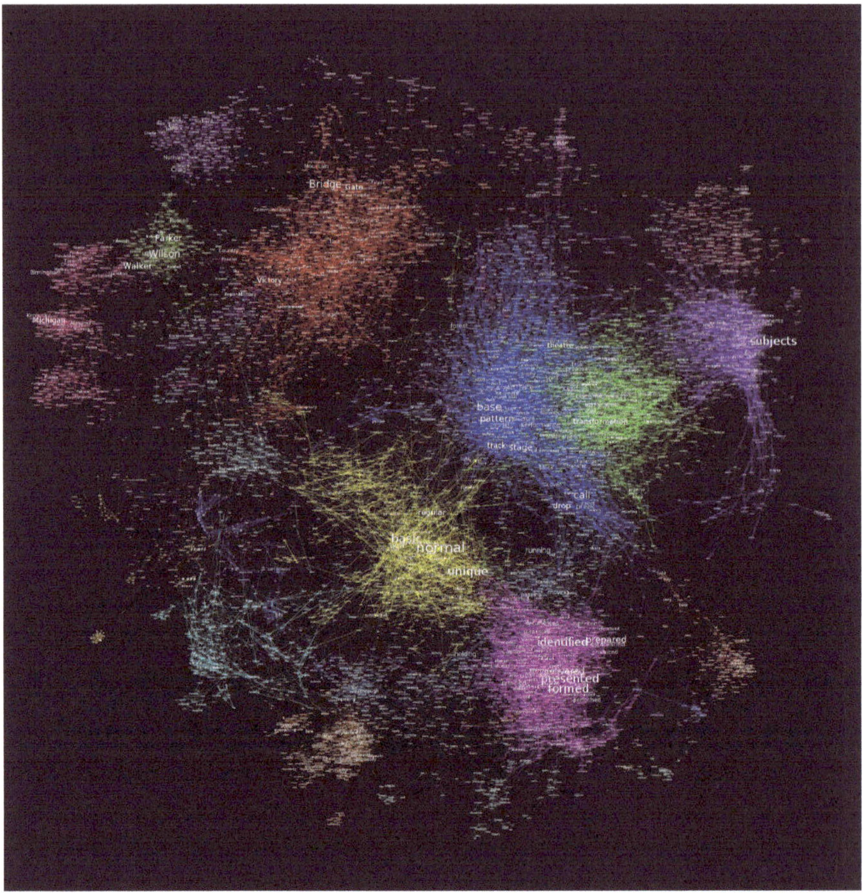

Fig. 3 Polyglot Nearest Neighbor Graph. Here we connect the nearest neighbors ($k = 6$) of the top 5,000 most frequent words from the Polyglot English embeddings. Shown is the giant connected component of the resulting graph ($|V| = 11,239$; $|E| = 26,166$). Colors represent clusters found through the Louvain method (modularity $Q = 0.849$). Vertex label size is determined by its PageRank. Best viewed in color.

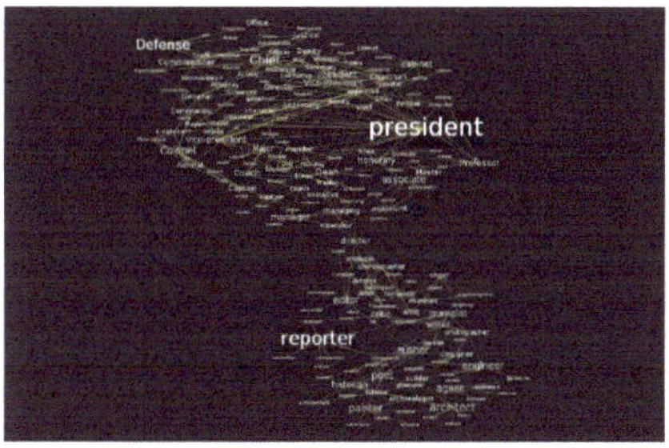

(a) Professions (SkipGram)

(b) Professions (Polyglot)

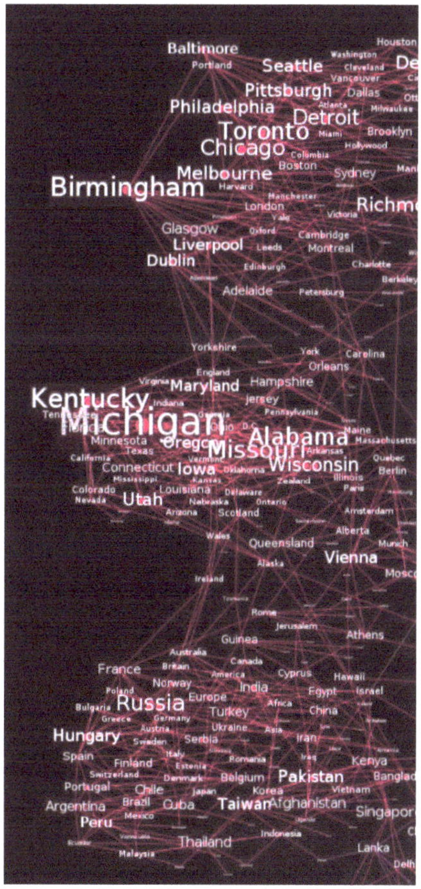

(c) Locations (SkipGram)

(d) Locations (Polyglot)

Fig. 4 Comparison of clusters found in Polyglot and SkipGram language networks. Polyglot clusters viewed in context of the surrounding graph, SkipGram clusters have been isolated to aide in visualization. SkipGram's bag-of-words approach favors a more semantic meaning between words, which can make its clusters less understandable (Note how in Figure 4c `Petersburg` is included in a cluster of religious words, because of `Saint`.) Images created with Gephi [2].

4 Related Work

Here we discuss the relevant work in language networks, and word embeddings.

Language Networks. One branch of the study of language as networks seeks to build networks directly from a corpus of raw text. Cancho and Solé [8] examine word co-occurrence graphs as a method to analyze language. In their graph, edges connect words which appear below a fixed threshold ($d \leq 2$) from each other in sentences. They find that networks constructed in this manner show both small world structure, and a power law degree distribution. Language networks based on word co-occurrence have been used in a variety of natural language processing tasks, including motif analysis of semantics [6], text summarization [1] and resolving disambiguation of word usages [20].

Another approach to studying language networks relies on studying the relationships between words exposed by a written language reference. Motter et al. [17] use a thesaurus to construct a network of synonyms, which they find to find to exhibit small world structure. In [19], Sigman and Cecchi investigate the graph structure of the Wordnet lexicon. They find that the semantic edges in Wordnet follow scale invariant behavior and that the inclusion of polysemous edges drastically raises the clustering coefficient, creating a small world effect in the network.

Much of the previous work in language networks build networks that are prone to noise from spurious correlations in word co-occurrence or infrequent word senses [8, 19]. Dimensionality reduction techniques have been successful in mitigating the effects of noise in a variety of domains. The word embedding methods we examine are a form of dimensionality reduction that has improved performance on several NLP tasks and benchmarks.

The networks produced in our work are considerably different from language networks created by previous work that we are aware of. We find that our degree distribution does appear to follow a power-law (like [8, 17, 19]) and we have some small world properties like those present in those works (such as $C \gg C_{random}$). However, the average path length in our graphs is considerably larger than the average path length in random graphs with the same node and edge cardinalities. Table 1 shows a comparison of metrics from different approaches to creating language networks.[3]

Word Embeddings. Distributed representations were first proposed by Hinton [12], to learn a mapping of symbolic data to continuous space. These representations are able to capture fine grain structures and regularities in the data [16]. With the recent advancement in hardware performance, Bengio et al. [4] used the distributed representations to produce a state-of-the-art probabilistic language model. More applications followed, Collobert et al. [10] developed SENNA, a system that offers part of speech tagger, chunker, named entity recognizer, semantic role labeler and discriminative syntactic parser using the distributed word representations. Al-Rfou' et al. [18] trained word embeddings for more than a hundred languages and

[3] Our induced networks available at `http://bit.ly/inducing_language_networks`

Table 1 A comparison of properties of language networks from the literature against those induced on the 20,000 most frequent words in the Polyglot and SkipGram Embeddings. (C clustering coefficient, pl average path length, γ exponent of power law fits to the degree distribution) '*' denotes values which have been estimated on a random subset of the vertices.

	$\|V\|$	$\|E\|$	C	C_{random}	pl	pl_{random}	γ
Cancho and Solé [8](UWN)	478,773	1.77×10^7	0.687	1.55×10^{-4}	2.63*	3.03	-1.50,-2.70
Cancho and Solé [8](RWN)	460,902	1.61×10^7	0.437	1.55×10^{-4}	2.67*	3.06	-1.50,-2.70
Motter et al. [17]	30,244	–	0.53	0.002	3.16	–	–
Polyglot, 6-NN	20,000	96,592	0.241	0.0004	6.78*	4.62*	-1.31
SkipGram, 6-NN	20,000	94,172	0.275	0.0004	6.57*	4.62*	-1.32

showed that the representations help building multilingual applications with minimal human effort. Recently, SkipGram and Continuous bag of words models were proposed by Mikolov et al. [14] as simpler and faster alternatives to neural network based models.

5 Conclusions

We have investigated the properties of recently proposed distributed word representations, which have shown results in several machine learning applications. Despite their usefulness, understanding the mechanisms which afford them their characteristics is still a hard problem.

In this work, we presented an approach for viewing word embeddings as a language network. We examined the characteristics of the induced networks, and their community structure. Using this analysis, we were able to develop a procedure which develops a connected graph with meaningful clusters. We believe that this work will set the stage for advances in both NLP techniques which utilize distributed word representations, and in understanding the properties of the machine learning processes which generate them.

Much remains to be done. In the future we would like to focus on comparing word embeddings to other well known distributional representation techniques (e.g. LDA/LSA), examining the effects of different vocabulary types (e.g. topic words, entities) on the induced graphs, and the stability of the graph properties as a function of network size.

Acknowledgments. This research was partially supported by NSF Grants DBI-1060572 and IIS-1017181, with additional support from TASC Inc, and a Google Faculty Research Award.

References

[1] Antiqueira, L., Oliveira Jr., O.N., da Fontoura Costa, L., das Graças Volpe Nunes, M.: A complex network approach to text summarization. Information Sciences 179(5), 584–599 (2009), http://dx.doi.org/10.1016/j.ins.2008.10.032, http://www.sciencedirect.com/science/article/pii/S0020025508004520 ISSN 0020-0255

[2] Bastian, M., Heymann, S., Jacomy, M.: Gephi: An open source software for exploring and manipulating networks (2009), http://www.aaai.org/ocs/index.php/ICWSM/09/paper/view/154

[3] Bengio, Y., Courville, A., Vincent, P.: Representation learning: A review and new perspectives (2013)

[4] Bengio, Y., Schwenk, H., Senécal, J.-S., Morin, F., Gauvain, J.-L.: Neural probabilistic language models. In: Holmes, D.E., Jain, L.C. (eds.) Innovations in Machine Learning. STUDFUZZ, vol. 194, pp. 137–186. Springer, Heidelberg (2006)

[5] Beyer, K., Goldstein, J., Ramakrishnan, R., Shaft, U.: When is "nearest neighbor" meaningful? In: Beeri, C., Bruneman, P. (eds.) ICDT 1999. LNCS, vol. 1540, pp. 217–235. Springer, Heidelberg (1998)

[6] Biemann, C., Roos, S., Weihe, K.: Quantifying semantics using complex network analysis. In: Proceedings of COLING 2012, Mumbai, India, pp. 263–278. The COLING 2012 Organizing Committee (December 2012), http://www.aclweb.org/anthology/C12-1017

[7] Blondel, V.D., Guillaume, J.-L., Lambiotte, R., Lefebvre, E.: Fast unfolding of communities in large networks. Journal of Statistical Mechanics: Theory and Experiment 2008(10), P10008 (2008)

[8] Cancho, R.F.I., Solé, R.V.: The small world of human language. Proceedings of the Royal Society of London. Series B: Biological Sciences 268(1482), 2261–2265 (2001)

[9] Chen, Y., Perozzi, B., Al-Rfou', R., Skiena, S.: The expressive power of word embeddings. In: ICML 2013 Workshop on Deep Learning for Audio, Speech, and Language Processing, Atlanta, USA, vol. abs/1301.3226 (2013)

[10] Collobert, R., Weston, J., Bottou, L., Karlen, M., Kavukcuoglu, K., Kuksa, P.: Natural language processing (almost) from scratch. The Journal of Machine Learning Research 12, 2493–2537 (2011)

[11] Glorot, X., Bordes, A., Bengio, Y.: Domain adaptation for large-scale sentiment classification: A deep learning approach, vol. 27, pp. 97–110 (June 2011)

[12] Hinton, G.E.: Learning distributed representations of concepts. In: Proceedings of the Eighth Annual Conference of the Cognitive Science Society, Amherst, MA, pp. 1–12 (1986)

[13] Hinton, G.E., Salakhutdinov, R.R.: Reducing the dimensionality of data with neural networks. Science 313(5786), 504–507 (2006)

[14] Mikolov, T., Chen, K., Corrado, G., Dean, J.: Efficient estimation of word representations in vector space. arXiv preprint arXiv:1301.3781 (2013)

[15] Mikolov, T., Kombrink, S., Burget, L., Cernocky, J.H., Khudanpur, S.: Extensions of recurrent neural network language model. In: 2011 IEEE International Conference on Acoustics, Speech and Signal Processing (ICASSP), pp. 5528–5531. IEEE (2011)

[16] Mikolov, T., Yih, W.-T., Zweig, G.: Linguistic regularities in continuous space word representations. In: Proceedings of NAACL-HLT, pp. 746–751 (2013)

[17] Motter, A.E., de Moura, A.P.S., Lai, Y.-C., Dasgupta, P.: Topology of the conceptual network of language. Phys. Rev. E 65, 065102 (2002), http://link.aps.org/doi/10.1103/PhysRevE.65.065102, doi:10.1103/PhysRevE.65.065102

[18] Al-Rfou', R., Perozzi, B., Skiena, S.: Polyglot: Distributed word representations for multilingual nlp. In: Proceedings of the Seventeenth Conference on Computational Natural Language Learning, Sofia, Bulgaria, pp. 183–192. Association for Computational Linguistics (August 2013), http://www.aclweb.org/anthology/W13-3520

[19] Sigman, M., Cecchi, G.A.: Global organization of the wordnet lexicon. Proceedings of the National Academy of Sciences 99(3), 1742–1747 (2002)

[20] Véronis, J.: HyperLex: lexical cartography for information retrieval. Computer Speech & Language 18(3), 223–252 (2004) ISSN 0885-2308, http://dx.doi.org/10.1016/j.csl.2004.05.002, http://www.sciencedirect.com/science/article/pii/S0885230804000142

Network Differences between Normal and Shuffled Texts: Case of Croatian

Domagoj Margan, Sanda Martinčić-Ipšić, and Ana Meštrović

Abstract. This paper is an initial attempt to study the properties of the Croatian word order via complex networks. We present network properties of normal and shuffled Croatian texts for different co-occurrence window sizes and different linkage boundaries. The results of network analysis show that the text shuffling causes the decrease of the network diameter, due to the establishment of previously non-existing links. This indicates that the syntax does play a significant role in the Croatian language, although it is a mostly free word-order language.

Keywords: complex networks, linguistic co-occurrence networks, Croatian corpus, shuffled text, randomized text.

1 Introduction

The complex networks sub-discipline tasked with the analysis of language has been recently associated with the term of linguistic's network analysis. The linguistic network can be based on various language constraints: structure, semantics, syntax dependencies, etc. It has been shown that language networks share various non-trivial topological properties and may be characterized as small-world networks and scale-free networks which are well-known and studied classes of complex networks. Small-world networks [14] have a small average shortest path length and a large clustering coefficient; scale-free networks [4] have power law degree distribution.

Domagoj Margan · Sanda Martinčić-Ipšić · Ana Meštrović
Department of Informatics,
University of Rijeka,
Radmile Matejčić 2, 51000 Rijeka, Croatia
e-mail: {dmargan,smarti,amestrovic}@uniri.hr

P. Contucci et al. (eds.), *Complex Networks V*,
Studies in Computational Intelligence 549,
DOI: 10.1007/978-3-319-05401-8_26,

In the linguistic co-occurrence complex networks properties are derived from the word order in texts. The open question is how the word order itself is reflected in topological properties of the linguistic network. One approach to address this question is to compare networks constructed from normal texts with the networks from randomized or shuffled texts. Since the majority of linguistic network studies have been performed for English, it is important to check whether the same properties hold for Croatian language as well. In this context the study of the Croatian language is notably behind other European languages [1]. So far, there have been only sporadic efforts to model the phenomena of the Croatian language through complex networks. Croatian is a highly flective Slavic language and words can have 7 different cases for singular and 7 for plural, genders and numbers. The Croatian word order is mostly free, especially in non-formal writing. These features are positioning Croatian among morphologicaly rich and free word-order languages.

In this paper we address the problem of Croatian text complexity by constructing the linguistic co-occurrence networks form two types of corpora: a) Croatian original texts, b) Croatian word-level shuffled texts. For the construction of the networks we varied two different criteria: a) the co-occurrence window size, b) the delimiters for limiting the linkage of the words only to the borders of a sentence.

Section 2 presents an overview of related work on complex network analysis of randomized texts. In Section 3 we define measures for the network structure analysis. In Section 4 we present the construction of eight different co-occurrence networks. The network measurements are in Section 5. In the final Section, we elaborate the obtained results and make conclusions regarding future work.

2 Related Work

Some of the early work related to the analysis of random texts dates to 1992, when Li [8] showed that the distribution of word frequencies for randomly generated texts is very similar to Zipf's law observed in natural languages such as in English. Thus, the feature of being a scale-free network does not depend on the syntactic structure of the language. Watts and Strogatz [14] showed that the network formed by the same amount of nodes and links but only establishing links by choosing pairs of nodes at random has a similar small network distance measures as in the original one. Caldeira *et al.* [5] analyzed the role played by the word frequency and sentence length distributions to the undirected co-occurrence network structure based on shuffling. Shuffling procedures were conducted either on the texts or on the links. Liu and Hu [9] discussed whether syntax plays a role in the complexity measures of a linguistic network. They built up two random linguistic networks based on syntax dependencies and compared the complexity of non-syntactic and syntactic language networks. Masucci and Rodgers showed [11, 12] that the power law distribution holds when they randomized the words in the text. Thus, they showed that degree distribution is not the best measure of the self-organizing nature of weighted linguistic networks. Due to the equivalence between frequency and strength of a node, shuffled texts obtain the same degree distribution, but lose all the syntactic structure. They have analyzed the differences between the statistical properties of a real and a shuffled weighted network and showed that the scale-free degree distribution and the scale-free weight distribution are induced by the scale-free strength distribution. They defined a measure, the node selectivity, that can distinguish a real network from a shuffled network. Krishna *et al.* [7] studied the effect of linguistic constraints on the large scale organization of language. They described the properties of linguistic co-occurrence networks with the randomized words. These properties were compared to those obtained for a network built over the original text. It is observed that the networks from randomized texts also exhibit small-world and scale-free characteristics.

Preliminary results on Croatian co-occurrence networks presented in [10] point out that the increase of the co-occurrence window size is followed by a decrease in diameter, average path shortening and, expectedly, the condensing of the average clustering coefficient. The stopwords removal causes the same effect. When comparing Croatian literature networks to networks from other languages such as English and Italian [3] some expected universalities such as small-world properties are shown, but there are still some differences. The Croatian language exhibits a higher path length than English and Italian language which can be caused by the mostly free word order nature of Croatian.

3 The Network Structure Analysis

In the network, N is the number of nodes and K is the number of links. In weighted networks every link connecting two nodes has an associated weight $w \in R_0^+$. The co-occurrence window m_n of size n is defined as n subsequent words from a text. The number of network components is denoted by ω.

For every two connected nodes i and j the number of links lying on the shortest path between them is denoted as d_{ij}, therefore the average distance of a node i from all other nodes is:

$$d_i = \frac{\sum_j d_{ij}}{N}. \tag{1}$$

And the average path length between every two nodes i, j is:

$$L = \sum_{i,j} \frac{d_{ij}}{N(N-1)}. \tag{2}$$

The maximum distance results in the network diameter:

$$D = max_i d_i. \tag{3}$$

For weighted networks the clustering coefficient of a node i is defined as the geometric average of the subgraph link weights:

$$c_i = \frac{1}{k_i(k_i-1)} \sum_{ij} (\hat{w}_{ij}\hat{w}_{ik}\hat{w}_{jk})^{1/3}, \tag{4}$$

where k_i is the degree of the node i, and the link weights $\hat{w}_{ij}$ are normalized by the maximum weight in the network $\hat{w}_{ij} = w_{ij}/\max(w)$. The value of c_i is assigned to 0 if $k_i < 2$.

The average clustering of a network is defined as the average value of the clustering coefficients of all nodes in a network:

$$C = \frac{1}{N} \sum_i c_i. \tag{5}$$

If $\omega > 1$, C is computed for the largest network component.

An important property of complex networks is degree distribution. For many real networks this distribution follows power law [13], which is defined as:

$$P(k) \sim k^{-\alpha}, \tag{6}$$

where the distribution parameter α is typically in range between 2 and 3.

4 Methodology

4.1 Data

For the construction and analysis of co-occurrence networks, we used two corpora. First is the original text of literature (C1), and second is the shuffled version of the same text (C2). In C2, the content of the original corpus is randomized by shuffling the words and punctuation marks, so C2 has the same quantity and frequency of words as the original corpus, but the text itself is meaningless.

Corpus C1 contains 10 books written in or translated into the Croatian language: I. Andrić "The Bridge on the Drina", M. Krleža "On the Edge of Reason" and "The Return of Philip Latinowicz", A. Šenoa "Branka", M. Jergović "Mama Leone", C. Collodi "Pinocchio", U. Eco "The Name of the Rose", E. Hemingway "The Old Man and the Sea", S. King "The Mist", and H. Lee "To Kill a Mockingbird".

The C1 has 895547 words, of which 91714 are unique, in 59128 sentences. The shuffling algorithm randomized words and punctuation marks which raised the new structure of sentences in the C2. The C2 has the same number of words in 58896 sentences.

4.2 The Construction of Co-occurrence Networks

Text can be represented as a complex network of linked words: each individual word is a node and interactions amongst words are links. From C1 and C2 we constructed eight different co-occurrence networks, all weighted and directed. Words are nodes linked within the co-occurrence window and according to the usage of the delimiters (punctuation marks).

The co-occurrence window m_n of size n is defined as a set of n subsequent words from a text. Within a window the links are established between the first word and $n-1$ subsequent words. In the networks where the linkage is limited to the sentence borders during the construction, we consider the sentence boundary as the window boundary too. Three steps in the network construction for a sentence of 6 words, with usage of the delimiters, for the co-occurrence window sizes $n=2$ and $n=6$ are shown in Fig. 1.

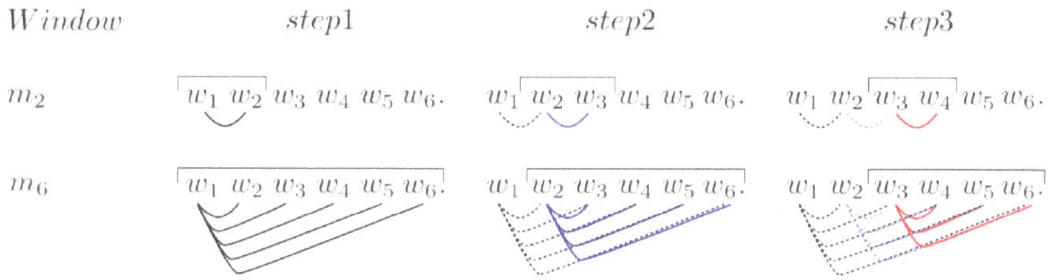

Fig. 1 An illustration of 3 steps in a network construction with a co-occurrence window m_n of sizes $n=2$, and $n=6$. $w_1 \ldots w_6$ are words within a sentence

The weight of the link is proportional to the overall co-occurrence frequencies of the corresponding words within a co-occurrence window. Network construction and analysis was implemented with the Python programming language using the NetworkX software package developed for the creation, manipulation, and study of the structure, dynamics, and functions of complex networks [6]. Numerical analysis of power law distributions was made with the 'powerlaw' software package [2] for the Python programming language.

5 Results

The comparison of the properties for networks differing in the co-occurrence window sizes (m_2, m_6) and the usage of delimiters are shown in Tables 1 and 2. The results show that the networks constructed with larger co-occurrence window emphasize small-world properties in both networks: from original

Table 1 Networks constructed with delimiters: the *rand* subscript is for the networks from C2

	m_2	m_6
N	91647	91647
N_{rand}	91526	91535
K	464029	2009187
K_{rand}	598519	2233643
L	3.10	2.38
L_{rand}	2.998	2.40
D	23	7
D_{rand}	9	5
C	0.32	0.71
C_{rand}	0.35	0.73
ω	22	22
ω_{rand}	15	8

Table 2 Networks constructed without delimiters: the *rand* subscript is for the networks from C2

	m_2	m_6
N	91714	91714
N_{rand}	91714	91714
K	513297	2459706
K_{rand}	636748	2666998
L	3.05	2.30
L_{rand}	2.95	2.29
D	17	6
D_{rand}	7	4
C	0.34	0.68
C_{rand}	0.38	0.70
ω	1	1
ω_{rand}	1	1

and shuffled texts. More precisely, in networks built with m_6, values of the average path length L and network diameter D are smaller, and the average clustering coefficient C is larger in comparison to the same measures from networks built with m_2.

Furthermore, in Tables 1 and 2 we compare the characteristics of networks constructed for co-occurrence window limited within or across the sentence boundaries. In the networks without delimiters, words are linked within a given co-occurrence window regardless of being in different sentences.

All of the networks constructed without the usage of delimiters show smaller network distance measures. Also, the clustering coefficient becomes larger only in the case of m_2, while the larger co-occurrence window m_6 decreases its value.

The number of nodes N ($N_{rand} < N$) is different from the number of words in C1, due to the used co-occurrence criteria. Our approach (Table 1) limits the co-occurrence window size within the sentence delimiters. This causes sentences with exactly one word to be isolated from the network, which reduces the number of nodes N. This is the reason why we considered the co-occurrence window across sentence boundaries (Table 2). $\omega_{rand} < \omega$ indicates that the number of connected components is smaller in the shuffled text C2. Therefore, when co-occurrence window disregarded the sentence boundaries networks have only 1 connected component (Table 2).

Fig. 2. shows the comparison of the plots of the clustering coefficient against the node degree for four different networks. Each plot shows clustering coefficient values spread on a log-log scale. The difference between plots constructed for networks based on original (a, b) and shuffled text (c, d) is that the clustering coefficients are more dispersed for the C1 than for the C2. It is especially emphasized in the case of small window size (m_2). The

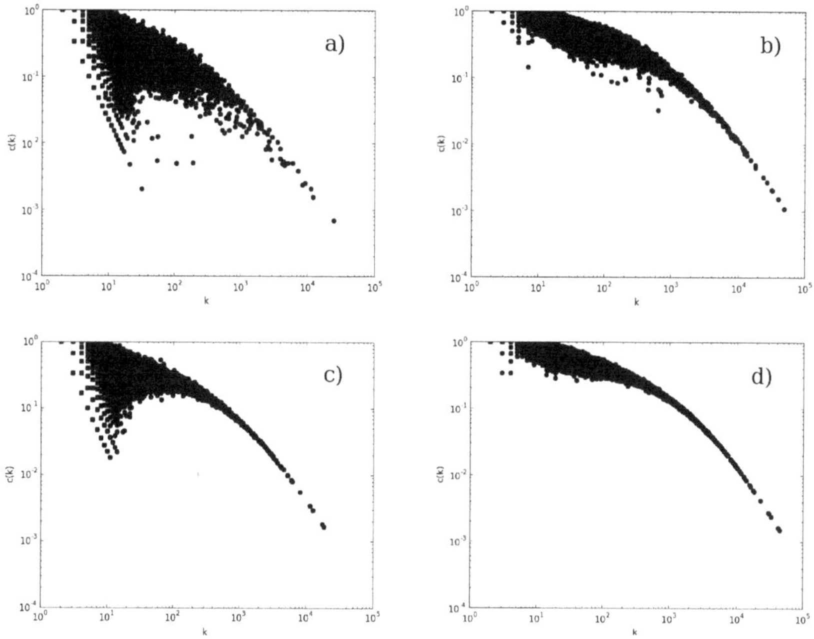

Fig. 2 Plots of the clustering coefficient against the degree of the vertices for four networks: (a) network based on the original text with m_2, (b) network based on the original text with m_6, (c) network based on the shuffled text with m_2, (d) network based on the shuffled text with m_6; always with delimiters used

dispersion of the clustering coefficient values associated with the properties of the word neighborhood reflects the complex organization of words [11]. Therefore, the more dispersed plots for the networks from the original texts, may indicate the more complex structure of original texts in comparison to the shuffled texts.

The clustering coefficient, as a local measure, is calculated considering the links' weights (Eq. 4). The results shown in Fig. 2 indicate that clustering coefficient of the weighted networks should be considered in the further study of the syntax structure.

Numerical results of power law distribution analysis indicate the presence of the power law distribution. The numeric values of α for the power law distributions are: 2.167 for m_2, C1; 2.090 for m_2, C2; 2.158 for m_6, C1; 2.137 for m_6, C2.

The global network measures: average shortest path length, diameter, clustering coefficient and degree distribution may not be well-suited properties for fine-grained network analysis. This may be explained by the fact that the syntax is a local language property. Therefore, it is necessary to include local network measures such as clustering coefficient of a node.

6 Conclusion

We studied the topologies of the linguistic networks constructed from normal and shuffled Croatian texts. As expected, the text shuffling causes the decrease of the network diameter, due to the establishment of previously non-existing links. This indicates that syntax does play a significant role in the Croatian language, although it is a free word-order language.

We have shown that the Croatian language networks have similar properties as language networks from English and other languages. Firstly, all Croatian language co-occurrence networks, based on normal and shuffled texts, have a power law degree distribution. That means that text shuffling has no influence on the degree distribution, which has already been shown for English [11, 12], English and Portuguese [5] and English, French, Spanish and Chinese [7]. Furthermore, all eight networks constructed for the Croatian language have small-world properties. There is a slight difference in the average clustering coefficient which is higher for the networks based on shuffled text. Distance measures (average shortest path length and diameter) show that each of the four networks based on normal texts have a greater L and D value than the corresponding network based on shuffled text. The same relations for average clustering coefficient, average shortest path length and diameter are shown in [7] for all studied languages (English, French, Spanish and Chinese). Similar results are shown for English and Portuguese in [5], although the authors used different shuffling procedures.

Our results imply that the syntax structure of the Croatian language has impact on the network properties, which needs further detailed analysis in order to find which network measures perform a fine-grained differentiation between an original and shuffled text. This should be thoroughly examined in the future work, which will cover: a) the comparison of the topological properties of the networks constructed from shuffled texts with preserved sentence length frequencies, b) shuffling of each book separately, c) using the node selectivity measure, and d) the analysis of the syntax dependencies in the Croatian linguistic networks.

References

1. Meta-net white paper series: Key results and cross-language comparison (2012), `http://www.meta-net.eu/whitepapers/key-results-and-cross-language-comparison`
2. Alstott, J., Bullmore, E., Plenz, D.: Powerlaw: a python package for analysis of heavy-tailed distributions. arXiv preprint arXiv:1305.0215 (2013)
3. Ban, K., Martinčić-Ipšić, S., Meštrović, A.: Initial comparison of linguistic networks measures for parallel texts. In: 5th International Conference on Information Technologies and Information Society (ITIS), pp. 97–104 (2013)
4. Barabási, A., Albert, R.: Emergence of scaling in random networks. Science 286(5439), 509–512 (1999)

5. Caldeira, S., Lobao, P., Andrade, R., Neme, A., Miranda, V.: The network of concepts in written texts. The European Physical Journal B-Condensed Matter and Complex Systems 49(4), 523–529 (2006)
6. Hagberg, A., Swart, P., Chult, D.: Exploring network structure, dynamics, and function using networkx. Technical report, Los Alamos National Laboratory (LANL) (2008)
7. Krishna, M., Hassan, A., Liu, Y., Radev, D.: The effect of linguistic constraints on the large scale organization of language. arXiv preprint arXiv:1102.2831 (2011)
8. Li, W.: Random texts exhibit zipf's-law-like word frequency distribution. IEEE Transactions on Information Theory 38(6), 1842–1845 (1992)
9. Liu, H., Hu, F.: What role does syntax play in a language network? EPL (Europhysics Letters) 83(1), 18002 (2008)
10. Margan, D., Martinčić-Ipšić, S., Meštrović, A.: Preliminary report on the structure of Croatian linguistic co-occurrence networks. In: 5th International Conference on Information Technologies and Information Society (ITIS), Slovenia, pp. 89–96 (2013)
11. Masucci, A., Rodgers, G.: Network properties of written human language. Physical Review E 74(2), 026102 (2006)
12. Masucci, A., Rodgers, G.: Differences between normal and shuffled texts: structural properties of weighted networks. Advances in Complex Systems 12(1), 113–129 (2009)
13. Newman, M.: Power laws, pareto distributions and zipf's law. Contemporary Physics 46(5), 323–351 (2005)
14. Watts, D., Strogatz, S.: Collective dynamics of small-world networks. Nature 393(6684), 440–442 (1998)

Application of Text Mining to Analysis of Social Groups in Blogosphere

Bogdan Gliwa, Anna Zygmunt, Jarosław Koźlak, and Krzysztof Cetnarowicz

Abstract. The paper concerns analysis of social groups in blogosphere using text mining methods to discover additional knowledge about groups and users. Two methods to distinguish messages (the first one - between messages from main and secondary thread, the second one - between facts and opinions) in blogosphere were proposed and their quality was assessed on manually annotated dataset. Both tasks are very important and proposed methods deal with them in a fully automatic way. The results were obtained on real-world data from Polish blogosphere.

Keywords: social network analysis, group topics, subjectivity detection, blogosphere, text mining.

1 Introduction

Nowadays, more and more elements of our everyday life are transferred to the virtual reality, especially communication with other people: we participate in discussions on forums, comment on blogs, chat and express our opinions using social media. Many companies are interested in automatic way of extraction information from users messages left in forums, blogs etc.

For analysis of user activity in social media, the application of methods of social network analysis is very popular. Discussions between people in blogs or forums can be modeled as social network and in such a network there are formed some groups of users that are more strongly connected between themselves than with the rest of network. This approach lets us analyse groups of people at different angles. Analysing groups in the context of written messages is the main goal of the paper.

Bogdan Gliwa · Anna Zygmunt · Jarosław Koźlak · Krzysztof Cetnarowicz
AGH University of Science and Technology
Al. Mickiewicza 30, 30-059 Kraków, Poland
e-mail: {bgliwa,azygmunt,kozlak,cetnar}@agh.edu.pl

P. Contucci et al. (eds.), *Complex Networks V*,
Studies in Computational Intelligence 549,
DOI: 10.1007/978-3-319-05401-8_27,

2 Related Work

Many methods for groups detection were proposed [3], [8]. They can find overlapping or non-overlapping groups, changing in time or not, etc. One of the most popular representative of algorithms finding overlapping groups is CPM method (Clique Percolation Method) [10].

In different methods regarding dynamics of groups, many events in groups lifecycle (also called groups evolution) were proposed. Palla et al. [11] described some events that can be identified during groups evolution: growth, contraction, merging, splitting, birth and death.

Topic Modeling [9] is a statistical technique that detects abstract "topics" existing in a collection of documents. "Topic" can be defined as a set of words that tend to co-occur in multiple documents, and, therefore, they are expected to have similar semantics. One of the biggest advantage of this method is that similar texts can be discovered even if they use different vocabulary, which is hard to achieve using other methods. Latent Dirichlet Allocation (LDA) [1] is one of the most popular methods in topic modeling and aims to reduce dimensionality by grouping words with similar semantics together.

In literature most applications of Text Mining in the field of Social Network Analysis regard some specific cases [2]. In [7] the authors showed usefulness of topic modeling to analysis of groups dynamics in social networks in blogosphere.

3 Analysis of Text Messages and Groups in Blogs

This section provides the concept of methods used to further analysis. In 3.1 and 3.2 we describe methods used to find out whether a message is a fact or an opinion and whether given message relates to the main topic of discussion thread (called *in the main thread*) or not (called *in the secondary thread*). Next, we depict methods used to analyse groups in dynamic social network.

3.1 Finding Messages in the Main and the Secondary Thread for Comments

Distinction between messages in the main and the secondary thread is based on topics uncovered by LDA method (and manually labelled) for given comment and post in analysed conversation thread. Additionally, one from LDA topics was labelled as *various* (it was hard to annotate as one particular topic), so in this method during comparison of topics in the case when post has topic *various* and comments has topic *various*, we assumed that they are different ones.

Let us define c as analysed comment, $post_c$ - post in thread where comment c was written and $topics(m)$ as topics for message m. Method is quite simple and can be described in the following way (*MS* is a function assigning label for a comment whether such comment is in the main thread or in the secondary one):

$$MS(c, post_c) = \begin{cases} \text{main}, & \text{if } topics(c) = \emptyset \vee |topics(c) \cap topics(post_c)| > 0, \\ \text{secondary}, & \text{otherwise.} \end{cases} \tag{1}$$

3.2 Finding Facts and Opinions in Comments

To distinguish messages containing only facts from messages containing opinions we also employed detection of topics (by means of LDA method) for a comment and a post in the same discussion thread.

Method consists of 2 steps:

Step 1. Analysing content of message to find out some symptoms of opinions, which we defined as occurence one of opinion words (manually defined, about 20 words such as (translated to English) *think*, *convince*, *respect*...), containing exclamation sign or have one topic from annotated as *opinions* and *critics* (LDA method uncovered such clusters). If any of above mentioned conditions are fulfilled, then message is annotated as *opinion* and second step is omitted.

Step 2. Analysing similarity of topics for given comment and post in thread. If they are similar i.e. $|topics(c) \cap topics(post_c)| > 0$, then we assumed that message c is a fact. When there are no topics for given post and comment then such comment is treated as *opinion*, but when there are no topics for post and comment has some topics – the comment is labelled as *fact*. Otherwise, we marked comment as *opinion*.

Above conditions can be expressed also as an assumption that when topic of comment and post matches then people discuss facts (except the case when we found some symptoms of opinions) and when they introduce new topic, then they express their opinions (or attacks personally between themselves).

3.3 Groups in Dynamic Social Network

Data from whole time range is divided into series of time slots and each time slot contains static snapshot of network from defined period of time.

In each slot we used the comments model for building graph, introduced by us in [4] - the users are nodes and relations between them are built in the following way: from user who wrote the comment to the user who was commented on (if the user whose comment was commented on is not explicitly referenced by using @ and name of author of comment in title of comment by commenting author, the target of the relation is the author of post).

In every static snapshot of social network groups were detected. Groups from adjacent time slots can be matched to find continuation of groups from different periods of time. For this goal, the SGCI (Stable Group Changes Identification) [5] method was applied. SGCI algorithm consists of the following steps: identification of short-lived groups in each time slot; identification of groups continuation, separation of the stable groups (lasting for a certain time interval) and the identification of types of group changes (transitions between stable groups).

Identification of group continuation is conducted using modified Jaccard measure with minimal threshold equals 0.5 (A and B are examined groups from the consecutive time slots):

$$MJ(A,B) = \begin{cases} 0, & \text{if } A = \emptyset \vee B = \emptyset, \\ max(\frac{|A \cap B|}{|A|}, \frac{|A \cap B|}{|B|}), & \text{otherwise.} \end{cases} \quad (2)$$

and ratio of group size with maximal threshold equals 50:

$$ds(A,B) = max(\frac{|A|}{|B|}, \frac{|B|}{|A|}). \quad (3)$$

4 Results

4.1 Description of Experiments

The experiments were conducted on data set containing data from the portal *salon24*[1]. The data set consists of 31 750 users (12 750 of them have their own blog), 380 700 posts and 5 703 140 comments within the period 1.01.2008 - 1.07.2013. The analysed period was divided into time slots, each lasting 7 days and neighboring slots overlap each other by 4 days. In the examined period there are 503 time slots.

For group detection we used CPM [10] method (directed version of CPM from CFinder[2]).

Topic for messages were extracted using LDA algorithm from mallet tool[3]. The method discovered 350 clusters of topics, which were manually annotated and some of them were manually joined. After this operation the number of clusters shortened to 67.

[1] Mainly focused towards politics, `www.salon24.pl`
[2] `http://www.cfinder.org/`
[3] `http://mallet.cs.umass.edu/`

4.2 Testing Quality of the Methods for Detection of Opinions, Facts, Messages in the Main and in the Second Thread

To assess quality of proposed methods we prepared set of discussion threads chosen in a random way from threads having at least 10 comments inside. Test dataset consists of 30 threads and 833 comments. Each comment was manually annotated whether contains only facts or contains opinions (possibly mixed with facts), and whether is related to the main topic in discussion thread or maybe is related to other one (e.g. personal messages between bloggers are annotated as messages not related to the main topic). The shortest thread has 11 comments and the longest one – 69 comments.

We evaluated F-measure (the harmonic mean of precision and recall) for each thread expressing quality of both methods. The results are presented in table 1. One can observe that results are quite good for both tasks. The lowest values (below or equal 0.5) in task determining whether message is fact or opinion are for 3 cases and in task assessing the fact that message belongs to the main topic of discussion thread or not are for 2 cases.

Table 1 Number of cases with given F-measure for methods detecting facts/opinions and main/secondary thread on manually annotated set of threads

range	main/secondary	opinion/fact
0-0.5	2	3
0.51-0.6	0	2
0.61-0.7	5	7
0.71-0.8	11	10
0.81-0.9	10	6
0.91-1	2	2

4.3 Discussion Threads with Messages Related to the Main and the Secondary Topic

We analysed the impact of the discussion thread topic on tendency to moving discussion to other topics. In figure 1a we can see topics of discussion threads, in which users most frequently discussed also other topics. Figure shows for topics the number of messages in the secondary thread divided by the number of messages in the main thread. We can observe that people often change topic of discussion in e.g. discussion threads with controversial content (like *abortion*) or in *philosophical* threads. Opposite situation is described in figure 1b – there are topics, in which users rarely change the subject of discussion. Among them we can find such topics as *sport* and *music*.

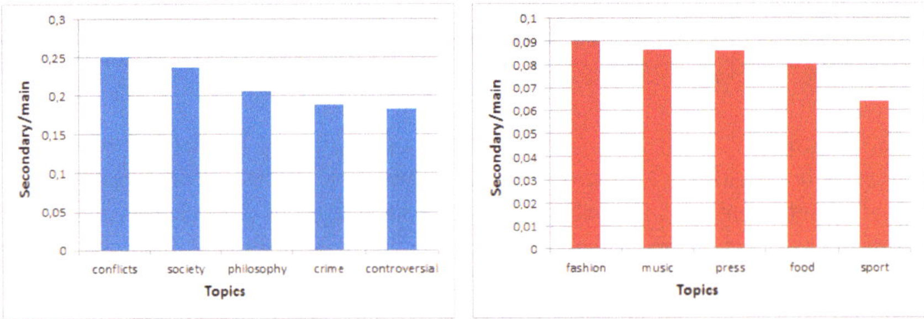

(a) Max secondary part. (b) Min secondary part.

Fig. 1 Top 5 topics of discussion threads with max and min secondary part

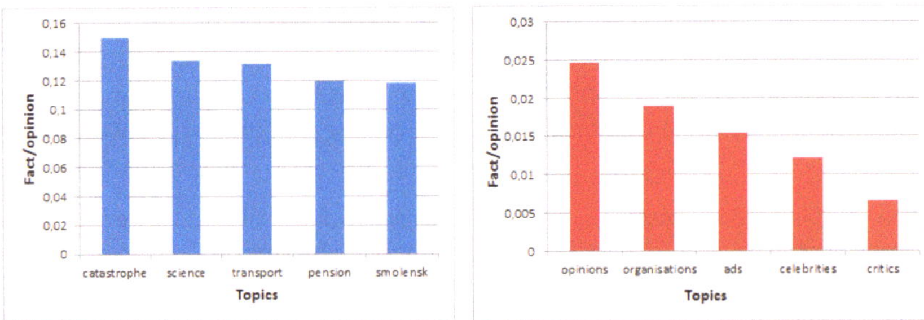

(a) Max fact part. (b) Min fact part.

Fig. 2 Top 5 topics of thread with max and min fact part

4.4 Discussion Threads with Facts and Opinions

We conducted similar analysis – we tried to find out the topics where users mostly express their opinions and where they discuss also about facts. Fig. 2a presents topics with the highest number of facts in comments in discussion threads. It is not surprise that we can find there *science* topic. On the other hand, fig. 2b shows topics with the lowest number of facts in comments in discussion threads. Among such topics, one can notice topics related with *critics*, *celebrities* and *opinions*.

4.5 Topics in Groups

In fig. 3 the most popular topics in groups with different size are shown. We can notice that the most popular topics in groups are *various* and *politics*. *Science* topic is dominated by groups of medium size (11-50 members). Another interesting observation is that the topic of *religion* mostly occurs in small and

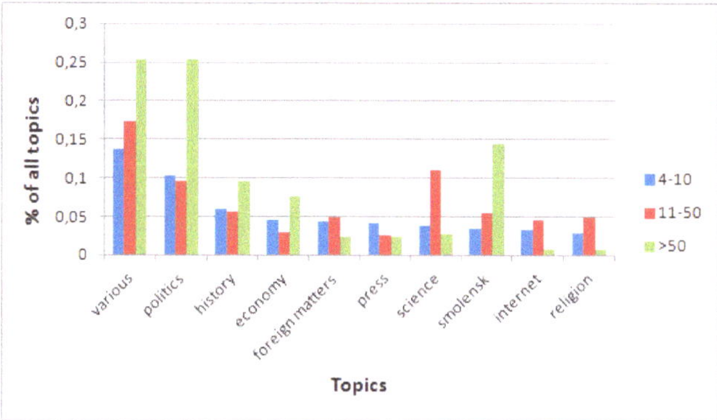

Fig. 3 Most popular topics in groups with different size

medium size groups. One can see also *smolensk* topic which is very popular. This topic concerns event of Polish President airplane crash in Smolensk and some other events related with investigation of this catastrophe.

4.6 Groups Formed Around Messages Deviating from the Main Topic

In fig. 4 we can observe what part of a group consitutes the part related with the main thread of discussion or, in other words, in how many groups the people during their discussions are stuck to the main topic. One can notice that for most groups the fraction of the main threads in discussions established inside them is very high, which means that people form groups to discuss particular topics. The highest variety can be noted for small groups (with 4-10 members) and it decreases when size of groups increases.

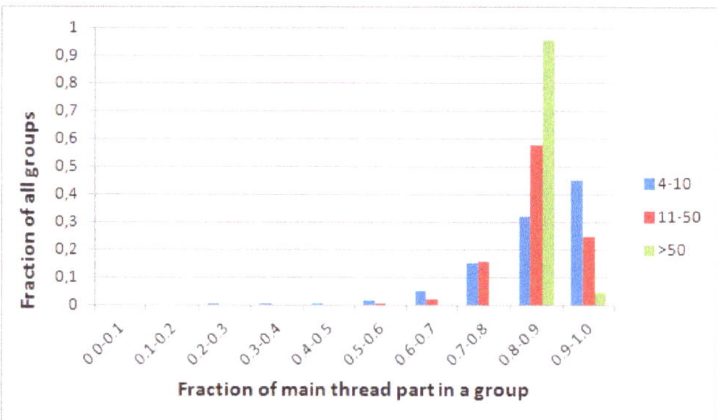

Fig. 4 Fraction of the main thread part in groups with different size

4.7 Groups Formed around Facts and Opinions

Fig. 5 presents how many groups with different size talk mostly about opinions. As we could anticipate, in blogs people in groups mostly share their opinions with others. However, we can notice that there are some small groups that talk almost completely about facts without expressing their own opinions. Moreover, in large groups (with more than 50 members) for most of them the part related with facts is quite high (about 20%) which is different from small and medium size groups.

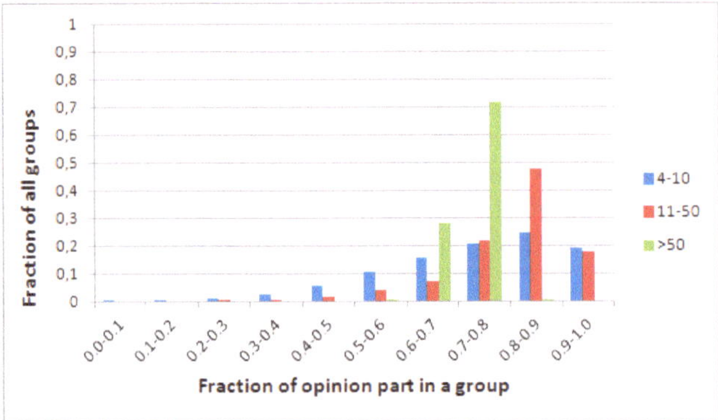

Fig. 5 Fraction of opinion part in groups with different size

5 Conclusion

In this paper we proposed 2 methods – the first for the distinction of messages in the main and in the secondary thread and the second one – detection opinions and facts, both in blogosphere. We assessed quality of these methods on manually annotated subset of whole analysed data and achieved results seem promising. Moreover, we analysed groups in social network under those angles. The obtained results allow us to better understand structure of groups.

Future work may follow in several directions. Firstly, there is a place to improve these methods (e.g. maybe some assumptions are not well suited for all types of topics). The second is to analyse roles of users in groups who e.g. change the main discussed topic or express mostly facts. For this purpose we want to employ our method of detecting roles of users [6]. Another interesting direction is to conduct experiments on other datasets including datasets in English language.

Acknowledgements. The research leading to these results has received funding from the dean grant.

References

1. Blei, D., Ng, A., Jordan, M.: Latent dirichlet allocation. Journal of Machine Learning Research 3, 993–1022 (2003)
2. Aggarwal, C., Wang, H.: Text mining in social networks. In: Aggarwal, C. (ed.) Social Network Data Analytics, pp. 353–378. Springer (2011)
3. Fortunato, S.: Community detection in graphs. In: Phys. Rep., ch. 486 (2010)
4. Gliwa, B., Koźlak, J., Zygmunt, A., Cetnarowicz, K.: Models of social groups in blogosphere based on information about comment addressees and sentiments. In: Aberer, K., Flache, A., Jager, W., Liu, L., Tang, J., Guéret, C. (eds.) SocInfo 2012. LNCS, vol. 7710, pp. 475–488. Springer, Heidelberg (2012)
5. Gliwa, B., Saganowski, S., Zygmunt, A., Bródka, P., Kazienko, P., Kozlak, J.: Identification of group changes in blogosphere. In: IEEE/ACM International Conference on Advances in Social Networks Analysis and Mining, ASONAM 2012, Istanbul, Turkey, August 26-29. IEEE Computer Society (2012)
6. Gliwa, B., Zygmunt, A., Kozlak, J.: Analysis of roles and groups in blogosphere. In: Burduk, R., Jackowski, K., Kurzynski, M., Wozniak, M., Zolnierek, A. (eds.) CORES 2013. AISC, vol. 226, pp. 299–308. Springer, Heidelberg (2013)
7. Gliwa, B., Zygmunt, A., Podgórski, S.: Incorporating text analysis into evolution of social groups in blogosphere. In: Federated Conf. on Computer Science and Information Systems, FedCSIS 2013, Krakow, Poland, September 8-11 (2013)
8. Greene, D., Doyle, D., Cunningham, P.: Tracking the evolution of communities in dynamic social networks. In: Proc. International Conference on Advances in Social Networks Analysis and Mining, ASONAM 2010. IEEE (2010)
9. Huang, Y.: Support vector machines for text categorization based on latent semantic indexing. Tech. rep., Electrical and Computer Engineering Department, The Johns Hopkins University (2003)
10. Palla, G., Derenyi, I., Farkas, I., Vicsek, T.: Uncovering the overlapping community structure of complex networks in nature and society. Nature 435, 814–818 (2005)
11. Palla, G., Barabási, A.L., Vicsek, T., Hungary, B.: Quantifying social group evolution. Nature 446, 2007 (2007)

The Role of the Shannon Entropy in the Identification of Acronyms

Marco Alberto Javarone

Abstract. Acronyms are linguistic signs composed of the initial components of other signs, therefore a codification process is needed to find their meaning. Usually, people are able to evaluate whether a sign is an acronym by considering its grammatical form and its phonic aspect. For instance, signs as "WWW" (i.e., World Wide Web) or "btw" (i.e., by the way) are easily identified as acronyms, although sometimes their meaning can be unknown. On the other hand, acronyms as "radar" and "laser" can be exchanged for simple nouns both from a grammatical and a phonetic perspective. We hypothesize the existence of a relation between the identification (i.e., the correct classification) of an acronym and its Shannon entropy. In order to investigate this hypothesis, we define an agent-based model to study the spreading dynamics of acronyms. Numerical simulations of the proposed model seem to confirm that the Shannon entropy has a central role in these dynamics. In particular, we found that the number of time steps to identify the solution of an acronym increases with its Shannon entropy.

1 Introduction

Nowadays, the human language and its evolution are investigated from different perspectives, as philosophy, psychology, classical linguistics, statistics and also statistical physics. In particular, many scientists have shown that several linguistic phenomena can be represented as complex dynamical systems [1][2][3][4][5]. In this scenario, it is worthy to note that Wittgenstein was the first to introduce an

Marco Alberto Javarone
DUMAS - Dept. Humanities and Social Sciences,
University of Sassari, Sassari 07100, Italy
DIEE - Dept. of Electrical and Electronic Engineering,
University of Cagliari
Cagliari 09123 Italy
e-mail: marco.javarone@diee.unica.it

P. Contucci et al. (eds.), *Complex Networks V*,
Studies in Computational Intelligence 549,
DOI: 10.1007/978-3-319-05401-8_28,

interdisciplinary approach in the studying of the human language, as he observed that it can be represented as a game [6]. Later on, his insightful observation has been used as reference in many language models, as the famous Naming Game [7]. This latter has been widely investigated by several authors. Just to cite a few, Baronchelli et al. [8] introduced a microscopic model of communicating autonomous agents, Dall'Asta et al. [9] studied the dynamics of the Naming Game on complex networks and Liu et al. [10] analyzed its behavior on small-world networks. The term language dynamics [1][11] identifies this research line, that makes use of the statistical physics to study phenomena as the emergence of a common vocabulary in communities of agents. In this work, we analyze a linguistic phenomenon related to the spreading dynamics of acronyms (see also [12]). Usually, people are able to evaluate whether a linguistic sign [13] (sign hereinafter) is an acronym. In particular, people consider its grammatical form and its phonic aspect. Notwithstanding, there are acronyms that can be exchanged for simple nouns, e.g., radar and laser. We hypothesize that a relation holds between how an acronym is considered (i.e., as an acronym or as a noun) and its Shannon entropy. The Shannon entropy [14][15] has a fundamental role in several branches of knowledge, as communications [16][17], data compression [18], network theory [19], genomics [20][21] and linguistics [22][23]. In order to investigate our hypothesis, we define an agent-based model where agents, interacting in a network, have to compute the solution of an acronym (i.e., the set of signs to generate it). Agents that compute a solution consider the acronym as such; otherwise, they consider it as a simple noun. Finally, we perform numerical simulations to study the proposed model.

2 Shannon Entropy of Acronyms

The Shannon entropy of acronyms depends on the degrees of freedom of signs that constitute a vocabulary. In this context, the term degree of freedom means the number of combinations that a sign can generate with other signs of a vocabulary. In principle, all signs can be used together to set up a phrase with a logic meaning. For instance, signs as "car" and "flower" can be combined in phrases as "A car is not a flower". Notwithstanding, if we consider our experience in the use of a language, we can observe that not all signs are combined together. This concept can be put into practice by representing a vocabulary as a network of signs, where signs are connected in the event they can be used together. Therefore, the topology of the network of signs is fundamental to compute the Shannon entropy of acronyms [12]. In general, if a network of signs has a fully-connected structure, there is a large set of possible solutions for each acronym. The cardinality of the set of solutions can be computed as:

$$|\Omega_{fc}| = \prod_{i=1}^{z} \omega_i \tag{1}$$

where z is the acronym length, $|\Omega_{fc}|$ is the cardinality of the set of all possible solutions considering a fully-connected network of signs, and ω_i is the number of signs that begin with the ith character. On the other hand, if we consider a different topology of the network of signs, the number of possible solutions can be widely reduced. For instance, in the event a network of signs has a scale-free structure, we can write $|\Omega_{sf}| \leq |\Omega_{fc}|$ (with $|\Omega_{sf}|$ cardinality of the set of all possible solutions considering a network of signs with a scale-free structure). The reported inequality holds because not all signs, in the scale-free network, are reciprocally connected. Considering a network of signs fully-connected, the Shannon entropy [24] of acronyms can be computed as follows:

$$H_{fc} = \sum_{i=1}^{z} h_i \tag{2}$$

where h_i is the entropy of the ith character of the acronym, computed as:

$$h_i = -\sum_{j=1}^{\omega_i} s_j \cdot \log_2 s_j \tag{3}$$

with s_j probability of occurrence of the jth sign. Instead, by using a network of signs with a scale-free structure, the Shannon entropy H_{sf} of acronyms becomes:

$$H_{sf} = -\sum_{j_1}\sum_{j_2}\cdots\sum_{j_z} s_{j_1,j_2,\ldots,j_z} \cdot \log_2 s_{j_1,j_2,\ldots,j_z} \tag{4}$$

with z acronym length and $s_{j_1,j_2,\ldots,j_z}$ probability that all signs occur together.

3 Identification of Acronyms

Let us now introduce our model to study the spreading dynamics of acronyms and their identification. In the proposed model, we consider N interacting agents which communicate by a linguistic convention, i.e., a common vocabulary. Relations among signs are mapped to a network [12], named *NetSigns*. The edges of *NetSigns* are generated between signs that can be used together, since their combination has a logic meaning. All agents know the rule for generating/codifying an acronym, which consists in the utilization of the first character of each sign. In so doing, more than one solution (i.e, meaning) can be associated to an acronym. At the time $t = 0$, a randomly chosen agent invents a new acronym with the aim to spread it in the population. The inventor communicates the acronym and the first sign of the solution to its neighbors. For example, it wants to spread the acronym laser, hence it communicates to its neighbors both the acronym and the sign *light*. In this case, to compute the Shannon entropy of acronyms, we have to consider that the first sign is known. Therefore, the value of H_{sf} can be computed as:

$$H_{sf} = -\sum_{j_2}\cdots\sum_{j_z} s_{j_1,j_2,\ldots,j_z|j_1} \cdot \log_2 s_{j_1,j_2,\ldots,j_z|j_1} \tag{5}$$

The first sign j_1 constitutes a partial solution that agents use to codify the acronym. In the event they compute the solution (i.e., the meaning) defined by the inventor, they add the acronym in their vocabulary and consider it as an acronym. In the opposite case, they add the acronym in their vocabulary but consider it as a simple noun, hence the acronym has not been identified. At each time step, all agents that receive an unknown acronym (with its first sign), try to compute the solution. On the other hand, all agents that know the acronym, but not its solution, can try again to compute the solution if they have neighbors that know both the acronym and its solution. Agents have a limited number of attempts to codify an acronym. The number of attempts depends on the knowledge of their neighbors. In particular, in the event there are more neighbors that know the acronym and its solution, the number of attempts is 4; otherwise, the number of attempts is 2. To summarize, the proposed model is composed by the following steps:

(1) In a network with N agents, a randomly selected agent invents a new acronym.
(2) Each agent, who does not know the acronym and/or its solution, computes the number of its neighbors that know the acronym:

    (a) in the event there are more neighbors that know the acronym and its solution, it tries to codify the acronym by 4 attempts.
    (b) in the event there are more neighbors that know only the acronym but not its solution, it tries to codify the acronym by 2 attempts.

    if the agent does not compute the correct solution, it saves the acronym but considers it as a simple noun. Otherwise, it saves the acronym and its solution, therefore it considers this sign as an acronym.
(3) Repeat from (2) until all agents knows both the acronym and its solution.

In order to compute the solution of an acronym, agents use the following algorithm:

(1) Identify the "list 1", that contains all signs connected to the first (known) sign.
(2) Randomly select one sign from the "list 1":

    (a) if the selected sign is correct, define the "list 2" that contains all signs connected to the 2nd sign;
    (b) else increase the number of failures and, if there are still attempts, restart from step (2).

(3) Randomly select one sign from the "list n", that contains all signs connected to the nth sign:

    (a) if the selected sign is correct, define the "list $n + 1$", that contains all signs connected to the n+*1*th sign and repeat from step (3) using the new list;
    (b) else increase the number of failures and, if there are still attempts, restart from step (3).

(4) Repeat, starting from the partial correct solution, until the acronym has been codified or until the number of failures is equal to the number of attempts.

In so doing, the task of codifying acronyms is simplified. For example, let us consider the acronym laser. At the first attempt, an agent computes the following partial solution: (Light, Amplification, Stimulated, *Element*). Since the 4th sign is wrong, the number of failures of the agent increases. Hence, if the agent has other available attempts, it tries again to compute a solution starting from the partial solution

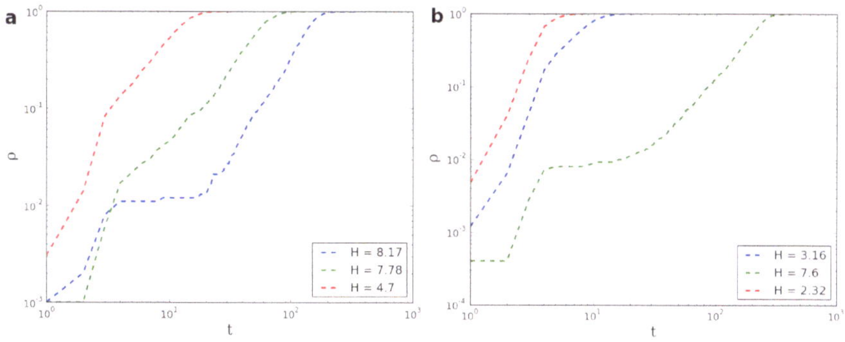

Fig. 1 Density of agents that add an acronym in their vocabulary over time, and know its solution. The legend indicates values of the Shannon entropy of each acronym. **a** Results achieved in a population with $N = 1000$. **b** Results achieved in a population with $N = 5000$. Values are averaged over 10 different realizations.

(Light, Amplification, Stimulated); otherwise it saves the acronym as a noun. It is worthy to recall that, agents that saved the acronym as a noun can change opinion by trying to compute again a solution (in the event they have at least one neighbor that considers the acronym as such).

4 Simulations

We performed numerical simulations of the proposed model with a community of language users composed by a number of agents in the range $N = [100, 10000]$. Agents have a common vocabulary, organized as a network (named *NetSigns*), with a number of signs in the range $S = [1000, 5000]$. The agent network and *NetSigns* have a scale-free structure achieved by the Barabasi-Albert model [25]. In so doing, both networks have a degree distribution $P(k) \sim k^{-\gamma}$, with a scaling parameter $\gamma \approx 3$. The number of attempts to codify an acronym has been set to 4 in the event an agent has a greater number of neighbors that know the acronym and its solutions than those of neighbors that know the acronym but not its solution; otherwise, in the opposite case, the number of attempts is 2. The first analysis is related to the density of agents (ρ) that add an acronym in their vocabulary over time, and know its solution –see Figure 1.

This analysis allows also to compare the length of acronyms with their Shannon entropy in terms of time steps to let all agents to identify the solution. Figure 2 illustrates this comparison, showing that the Shannon entropy affects this process more than the length of acronyms. These results highlight that the number of time steps, to let agents compute the correct solution of an acronym, increases as the Shannon entropy of the considered acronym increases. Another important measure is the scaling of the consensus time T_c [26], i.e., the time at which the system reaches the ordered phase, with the size of the population and with the Shannon entropy of acronyms. As shown in Figure 3 (panel **a**), the value of T_c increases as the size

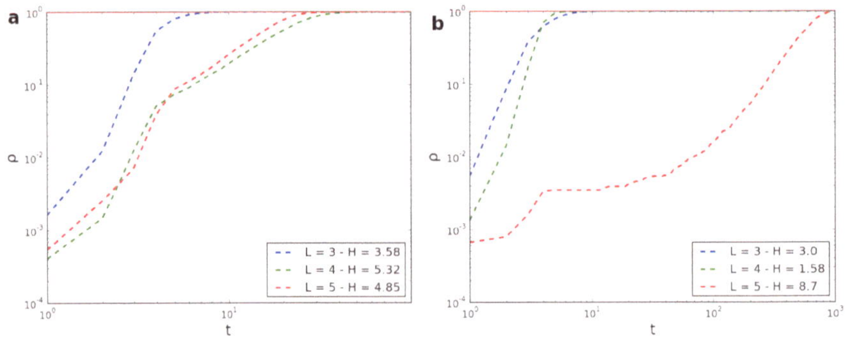

Fig. 2 Density of agents, in a population with $N = 2500$, that add an acronym in their vocabulary over time, and know its solution. The legend indicates the length of acronyms and their Shannon entropy. Values are averaged over 10 different realizations. **a** and **b** show results for two different sets of acronyms (see the Shannon entropy).

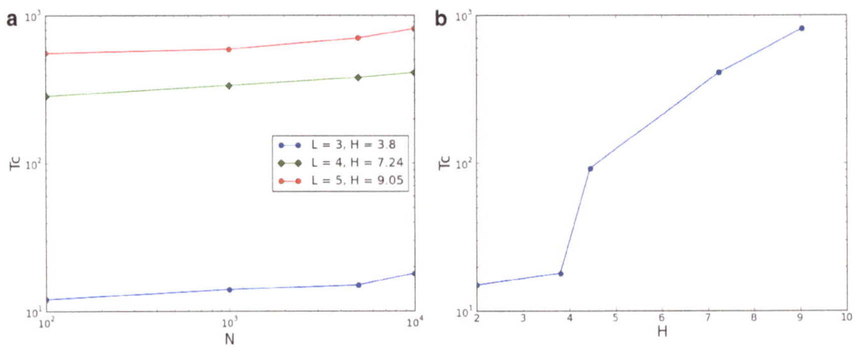

Fig. 3 **a** Scaling of the consensus time T_c with the size of the population. Each curves referes to a different acronym. **b** Scaling of the consensus time T_c with the Shannon entropy of acronyms, in a population of 10000 agents. Values are averaged over 10 different realizations.

of the population increases. Moreover, in the cases of 3 character acronyms and 4 character acronyms, it seems that the growth of T_c is almost linear. Results shown in Figure 3 (panel **b**) confirm that as the Shannon entropy of acronyms increases the T_c increases.

4.1 Discussion

In this work we investigate the relation between the Shannon entropy of acronyms and their identification in a population of language users. We propose a model where interacting agents have to compute the solution of an acronym, i.e., the set of signs to generate it. Agents, arranged in a network, communicate the acronym to their neighbors. In particular, they send both the acronym and the first sign of the solution. Agents that compute a solution add the new sign to their vocabulary and they consider it as an acronym. In the opposite case, agents save the acronym but they consider it as a simple noun. For each agent, the number of attempts to compute a solution depends on the amount of neighbors that consider the acronym as such or as a noun. The density of agents that compute the solution of an acronym allows to evaluate whether the Shannon entropy affects the identification process. We found that as the Shannon entropy increases the number of time steps to let all agents compute the solution increases –see Figure 1. Furthermore, we compared the Shannon entropy with the length of acronyms. Figure 2 illustrates that the entropy affects the identification process more than the acronyms length. Finally, we analyzed the scaling of the consensus time with the size of the population, and with the Shannon entropy of the acronyms. Results shown in panel **a** of Figure 3 highlights that the growth of T_c is almost linear while increasing the size of the population. Moreover, results shown in panel **b** of Figure 3 confirms that the number of time steps to reach the ordered phase depends on the entropy of acronyms. In particular, as the entropy increases the T_c increases. It is worthy to observe that all achieved results support our hypothesis about the existence of a relation between the probability of identifying acronyms and their Shannon entropy.

5 Conclusions

In this work, we propose a model for studying the identification of acronyms in a community of language users. This work originates by the observation that, from a linguistic perspective, there are acronyms that can be exchanged for simple nouns because of their grammatical and phonic aspect, e.g., radar, laser and sonar. We hypothesize that the Shannon entropy of acronyms affects their probability to be identified as such. We introduce an agent-based model to study the spreading dynamics of acronyms, with the aim to investigate the above hypothesis. In the proposed model, an acronym spreads in the agent network and each agent tries to compute a solution. In the event an agent finds a solution, it considers the acronym as such; otherwise it considers the acronym as a noun. Results of numerical simulations show that the Shannon entropy strongly affects the identification of acronyms. In particular, several time steps are needed for computing acronyms with a high Shannon entropy. This latter seems more important than the acronym length, since the codification of short acronyms with a high Shannon entropy requires more time steps than that of longer acronyms with a low entropy. Finally, we deem the proposed model useful also for more general problems related to the codification of sequences of symbols.

Acknowledgements. The author wishes to thank Fiorenzo Toso for his helpful comments. This work was supported by Fondazione Banco di Sardegna.

References

1. Loreto, V., Baronchelli, A., Mukherjee, A., Puglisi, A., Tria, F.: Statistical Physics of Language Dynamics. Journal of Statistical Mechanics: Theory and Experiment (2011)
2. Borge-Holthoefer, J., Arenas, A.: Semantic Networks: Structure and Dynamics. Entropy 12, 1264–1302 (2010)
3. Steyvers, M., Tenenbaum, J.B.: The Large-Scale Structure of Semantic Networks: Statistical Analyses and a Model of Semantic Growth. Cognitive Science 29, 41–78 (2005)
4. Hills, T.T., Maouene, M., Maouene, J., Sheya, A., Smith, L.: Categorical Structure among Shared Features in Networks of Early-learned Nouns. Cognition 112, 381–396 (2009)
5. Xavier Castello, X., Eguiluz, V.M., San Miguel, M.: Ordering dynamics with two non-excluding options: Bilingualism in language competition. New Journal of Physics 8, 308 (2006)
6. Wittgenstein, L.: Tractatus logico-philosophicus. Philosophical Investigations Wiley-Blackwell (1916)
7. Steels, L.: A self-organizing spatial vocabulary. Artificial Life Journal 2, 319 (1995)
8. Baronchelli, A., Felici, M., Loreto, V., Caglioti, E., Steels, L.: Sharp transition towards shared vocabularies in multi-agent systems. Journal of Statistical Mechanics: Theory and Experiment, P06014 (2006)
9. Dall'Asta, L., Baronchelli, A., Barrat, A., Loreto, V.: Nonequilibrium dynamics of language games on complex networks. Physical Review E 74, 036105 (2006)
10. Liu, R.R., Jia, C.X., Yang, H.X., Wang, B.H.: Naming game on small-world networks with geographical effects. Physica A: Statistical Mechanics and its Applications 388(17), 3615–3620 (2009)
11. Baronchelli, A., Loreto, V., Tria, F.: Language Dynamics. Advances in Complex Networks 15, 1203002 (2012)
12. Javarone, M.A., Armano, G.: Emergence of acronyms in a community of language users. The European Physical Journal B 86, 474 (2013)
13. de Saussure, F.: Course in General Linguistics, pp. 1906–1911. Lectures at the University of Geneve (1916)
14. Shannon, C.E.: A mathematical theory of communication. Bell System Tech. Journal 27 (1948)
15. Gray, R.M.: Entropy and Information Theory. Springer, New York (2011)
16. Akcakaya, M., Tarokh, V.: Shannon-Theoretic Limits on Noisy Compressive Sampling. IEEE Transactions on Information Theory 56(1), 492–504 (2010)
17. Zhang, Y., Zhang, Q., Wu, S.: Entropy-based robust spectrum sensing in cognitive radio. IET Communications 4(4), 428–436 (2010)
18. Liu, W., Wenjun, Z., Dong, L., Yao, Q.: Efficient Compression of Encrypted Grayscale Images. IEEE Transactions on Image Processing 19(4), 1097–1102 (2011)
19. Anand, K., Bianconi, G.: Entropy measures for networks: Toward an information theory of complex topologies. Physical Review E 80(4), 045102 (2009)
20. Tenreiro Machado, J.A.: Shannon Entropy Analysis of the Genome Code. Mathematical Problems in Engineering 2012 (2012)
21. Strait, B.J., Dewey, T.G.: The Shannon information entropy of protein sequences. Biophysical Journal 71(1), 148–155 (1996)
22. Shannon, C.E.: Prediction and entropy of printed English. Bell System Tech. Journal 27 (1951)
23. Papadimitrioua, C., Karamanosa, K., Diakonosa, F.K., Constantoudisb, V., Papageorgiou, H.: Entropy analysis of natural language written texts. Physica A: Statistical Mechanics and its Applications 389(16), 3260–3266 (2010)

24. Annick, L., Jean-Luc, B., Pezard, L.: Entropy estimation of very short symbolic sequences. Physical Review E 79(4), 046208 (2009)
25. Barabasi, A.L., Albert, R.: Emergence of scaling in random networks. Science 286, 509–512 (1999)
26. Rogers, T., Gross, T.: Consensus time and conformity in the adaptive voter model. Physical Review E 88(3), 030102 (2013)

Negative Implications of a Power-Law Distribution: A Study on Networks of Scientific Reviewers

Song Qin, Marius C. Silaghi, Ronaldo Menezes, and William Cheung

Abstract. Traditional peer-reviewing is a process whereby submissions by various scientists are selected based on certain criteria passed on to reviewers by organizers of conferences or editors of journals. This process has been used to maintain the quality of the works being presented and also to help grouping reports relevant to a given community (or topic). However, certain scientific opinions and theories compete and have partisans. Common examples of such competitions appear when deciding the most important metric in classification algorithms, what to use as a basis for recommendation algorithms, the best predicting models for a known phenomena, to name a few. The common assumption is that the community will be equally informed about the arguments of all involved studies, in order to come out with objective conclusions. This assumption is reasonable when partisans of each competing opinion can eventually review and recommend for publication the studies that agree with their perspective. In its turn, this can be expected to eventually happen whenever expert reviewers are randomly assigned to corresponding papers. However in recent years we have seen that power-law distributions instead of randomness are present in many social relationships. In this study we investigate what happens in the world of peer-reviewing, more specifically in a network of reviewing relations for an open review journal. We found that a power-law distribution is indeed present, as a small group of reviewers evaluates a significant fraction of all

Song Qin · Marius C. Silaghi
HDSS Lab, Department of Computer Sciences,
Florida Institute of Technology, Melbourne, Florida, USA
e-mail: qsong2008@my.fit.edu, msilaghi@cs.fit.edu

William Cheung
Hong Kong Baptist University, Hong Kong
e-mail: william@comp.hkbu.edu.hk

Ronaldo Menezes
BioComplex Laboratory, Department of Computer Sciences,
Florida Institute of Technology, Melbourne, Florida, USA
e-mail: rmenezes@cs.fit.edu

P. Contucci et al. (eds.), *Complex Networks V*,
Studies in Computational Intelligence 549,
DOI: 10.1007/978-3-319-05401-8_29,

submissions. The problem however is that this is undesirable since these "hubs" have an unmatched influence on what gets published. This experiment presents a first case where arguably the power-law structure of the social network can be considered as an overall negative factor. It also supports an argument for employing the social graph of reviewers as an additional metric of the quality of a journal/conference.

Keywords: Social Networks, Peer-Reviewing Process, Power-Law Distributions.

1 Introduction

Peer reviewing is an essential mechanism of the modern research process. Traditionally, journals have used the peer-review process as a filter to decide which submissions should be selected for publication, potentially based on criteria such as relevance to the main topic of the journal and technical quality. In the past it was difficult to study the properties of reviewing processes due to requirements of anonymity of reviewers. It is generally believed that this process is fair and it is assumed that it avoids bias given the fact that most journals and conferences employ at least 3 individuals to evaluate each publication.

Recently however, several journals and workshops have adopted new models of peer reviewing, where the names of the reviewers are made public [8, 17, 13, 14]. Among them the journal of Biology Direct, which has used open peer-review for several years, yields a large amount of data for verifying assumptions. This allows us to analyze the process and to try to understand peer-reviewing a little better.

While the general public is familiar with "reviewing" for items sold on eBay and Amazon, that type of reviews does not have as purpose the forbidding of the sale of poor quality items. Instead, the poor quality items are still left for sale but they are attached with the relevant information for warning potential buyers.

Traditional peer-reviewing for scientific articles is a very different concept. Reviews of scientific articles are not commonly published with these articles and are not commonly used for warning potential readers of the failures and qualities of the article. Rather, these reviews are meant for helping a chair of conference or journal editors to decide which submissions should be filtered out. The general public (scientists who read these papers) are not aware of any concerns or discussion that took place during the review process.

1.1 Problem

While reviewing aims to be impartial, it is difficult for introspection alone to remove certain biases stemming from the school of thought where the reviewer was herself educated as a scientist. There exist competing schools of thought in economics, science, engineering, and many other fields [15, 10, 7]. A factor that is arguably desirable for the objectivity of a journal, is how well it gives equal chances to researchers from different currents to state their arguments. Based on the assumption of potential bias, it is desirable to have a diverse range of reviewers being able to review and participate in filtering, balancing the chances of the competing schools of thought. While not all schools of thought will review each given submission, each of them should at least get a fair chance of reviewing relevant submissions.

There are four major types of peer-review mechanisms used in reviewing of scientific articles, as well as in (medical) experiments on human subjects. Common *simple blind review* mechanisms are those where reviewers know the name of the authors but authors do not know the names of their reviewers. The motivation is to avoid that reviewers would fear retribution for their reviews. *Double-blind reviews* are those made in such a way that reviewers do not know the name of the paper authors and paper authors do not know the name of their reviewers; they stem from concern of bias for reviewers (e.g. when reviewing an influential person). *Open peer reviews* are mechanisms where reviewers know the names of the authors and authors know the names of their reviewers, while reviews and reviewer names are published with the articles. The motivation is to encourage reviewers to write responsible reviews. *Blinded review* is another mechanism where reviewers do not have access to the names of authors but the names of reviewers are published together with their reviews. The idea is to encourage reviewers to be responsible, by accountability, while helping them to avoid bias (as in the case of the double blind review).

In the aforementioned processes there exist relations between authors and reviewers (even if these relations are sometimes hidden from both parts). According to theories in social network analysis and complex networks, the structure of these relations should tell us something about the process itself. This study looks at the characteristics of the social network of reviewers and its relation to the fairness of the review process.

The online availability of the reviewing information from the Biology Direct Journal gives us a window into the scientific review process. It allows for verifying whether submissions have fair chances to be reviewed by a varied number of researchers, potentially covering multiple schools of thought. What we noticed is that a couple of reviewers eventually reviewed a large fraction of the accepted articles. This took place despite the editor's effort of finding different people to review each work. The issue however is that each submission could have different reviewers that appear to be chosen at random but at the global level the picture is quite different and what emerges when we combine the seemly individual random choices for each paper is a network in which few reviewers are hubs and review quite a lot of papers, a typical social network with hubs and long-tail distributions [1]. Note that this would be less prominent in a conference if we analyze only one edition. Normally conferences enforce a maximum number of reviews per person. For a conference one would have to look at the global picture across many editions (years) of the event.

From the perspective of review processes, we claim that the power-law distribution is detrimental to the process itself because it points to an unbalanced influence for any potential bias that these hub reviewers may have. While a power-law distribution for networks was so far considered as a positive characteristic signaling the stability of the social network, we have identified here an argument for considering the power-law distribution in peer-reviewers networks to be a negative trait. Having journals and conferences publish the networks of reviewers (even if just the structure without actual names) can help developing a new metric for their objectivity.

We discuss the related work in Section 2. Section 3 describes the dataset and how it was modeled as a network. We then look at these networks formed from author-reviewer relations and discuss our findings in Section 4. Finally we conclude with suggestions of how these findings can be used to improve peer-review processes.

2 Related Work

2.1 *Social Network Analysis*

A social network is a structure that represents social interactions and personal relationships. Examples of social networks include: friendship [6], collaboration [16, 5] and email networks [3]. In general a social network can be abstracted as a structure in which the entities are people and the links between these people are extracted from some social relationship. In this paper we address the social network of reviewers. This network is obtained by projecting the bipartite network of reviewers and papers shown in Figure 1 unto the set of reviewers. Two reviewers are connected if they have reviewed the same paper. The strength of their link is given by the number of submission that they have reviewed together.

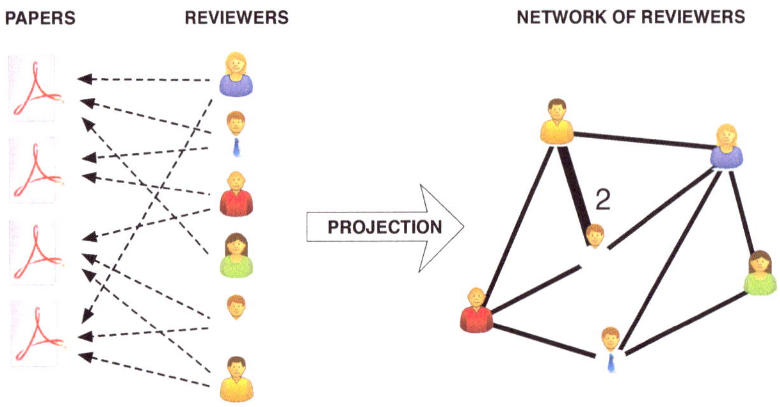

Fig. 1 From a bipartite network in which reviewers are linked to the papers they review, we can project a network of reviewers where reviewers are linked directly if they reviewed the same paper

When looking at networks one can use several metrics to understand the represented phenomena. The metrics we use are the degree distribution, betweenness centrality, closeness centrality and clustering coefficient, explained below:

Degree Distribution: This distribution expresses the probability, $p(k)$, that a node in the network will have k connections. It has been observed that in many real networks [12] their degree distribution roughly follows a power law as given by Equation 1,

$$p(k) = ck^{-\lambda}, \tag{1}$$

where, c and λ are constants. For most of the real networks $2 \leq \lambda \leq 3$. Nodes that have more ties to other nodes may be in advantageous positions.

Betweenness Centrality: The betweenness is a measure of the centrality of a vertex in a network. Betweenness is calculated as the fraction of the shortest paths between vertex pairs, that pass through the vertex of interest. For a network $N = (V, E)$ with n vertices, the betweenness $C_B(v)$ for a vertex v is:

$$C_B(v) = \sum_{s \neq t \neq v \in V} \frac{\alpha_{st}(v)}{\alpha_{st}}, \tag{2}$$

where α_{st} is the number of shortest paths from s to t, and $\alpha_{st}(v)$ is the number of shortest paths from s to t that pass through vertex v. Vertices exhibiting a high level of betweenness are in a position to control information flow in the network.

Closeness Centrality: Centrality is defined as the inverse of the distance from a vertex to all other vertices in the network

$$C_c(i) = \frac{1}{\sum_k d(i,k)}, \tag{3}$$

where $d(i, k)$ is the shortest path between vertices i and k. This metric gives low values for the central nodes and high values for the less central ones. Nodes with high closeness are generally in a position to influence other nodes because they can reach them very quickly.

Clustering Coefficient: This coefficient is a measure of the ratio in which nodes in a graph tend to cluster together. The clustering coefficient, C_i, of node i is given by Equation 4, where, m_i is the number of links between the k_i neighbors of i; the clustering coefficient of the entire network is just the average of all C_i over the number of nodes in the network n. Clustering is relevant to social networks. It can be used to identify small-world networks [19] which are expected to have high clustering and short average path lengths.

$$C_i = \frac{2m_i}{k_i(k_i - 1)}. \tag{4}$$

In this paper we study the aforementioned metrics in relation to a reviewers' network from a peer-review process. However our main discussion in this paper focuses on the drawbacks of the degree distribution to the peer-review process itself.

2.2 Open Peer-Review Process

One of the most common reviewing processes is the *double-blind review*. Under this scheme one publishes only accepted articles and the names of the organizing committee members (names which are supposed to witness to the quality of the reviewing process). Reviewers are not expected to know the names of the authors at the time of the review, and authors will never find out the names of the reviewers of their submission. Reviews are not published and are not digitally signed, and authors have no way to prove that they have received any given review, or even that they have submitted any given article. The camera-ready article eventually published in the proceedings can be completely different from the corresponding article actually evaluated by reviewers (and even the title can be completely changed). As example of venue using double blind review we mention the *International Joint Conference on Artificial Intelligence (IJCAI)*.

Multiple models of open peer-review processes have been proposed and experimented with. The openness varies in terms of *what is revealed*, in terms of *the degree of the revelations*, and in terms of *the articles to which the revelation applies* (e.g., only to accepted articles or to all submitted articles). A classification from the perspective of the object of the revelation contains the following dimensions:

Open/Closed Author Names: This dimension tells whether reviewers are informed about the names of the authors at review time. This information is supposed to help reviewers correctly assess the originality of the submission. An example of open author names reviewing is employed by the *Conference on Principles and Practice of Constraint Programming (CP)*.

Open/Closed Article: Venues may either publish articles, or keep them as part of a closed meeting. Certain workshops do not produce public proceedings and the articles accepted and presented in their forum are not considered published.

Open/Closed Submission: Some venues do publish the submission actually evaluated by reviewers, while others only use the submission as an acceptance criteria for publishing a different article. The actually published (aka *camera-ready*) article is typically assumed to be an improvement of the submitted article based on feedback from reviewers. However, only few journals have mechanisms to ensure that the actually published article has any relation whatsoever with the actually reviewed submission. In practice submitters can change even the title and the list of author names of a submission. An example of venue that publishes both the original submission and the camera ready version is the *Workshop on Decentralized Coordination*.
A limited degree of openness of submissions is offered by some cryptology conferences that give authors digital signature certificates for their submissions, helping them to prove that they have submitted the corresponding articles.

Open/Closed Reviews Summary: At various conferences, a senior committee member is charged with writing a short summary of the reviews, to be privately communicated to the author and to the editor making the acceptance decision.

Under schemes with open reviews summary, an anonymous committee member is charged with writing a public summary of the anonymous reviews for an article (Example: *IEEE Conference on Peer to Peer Computing (P2P)*).

Open/Closed Reviews: With open reviews, the actual reviews received by the article are made publicly available (not necessarily publishing the names of the individuals). This procedure is more common in online journals such as *Philica.com*.

Open/Closed Reviewing Activity: This revelation dimension quantifies whether one publishes the number of articles reviewed by each reviewer (helping to quantify their impact). Most conferences do publish an average of the number of articles reviewed by its reviewers, but note that an average is not significant if the distribution of reviews per reviewer follows a power-law.

Open/Closed Article Reviewers: The names of the reviewers of each article may or may not be published in association with that article. Certain journals publish articles labeled with names of researchers recommending them.

Open/Closed Review Authorship: The dimension of review authorship is used to specify whether the name of each reviewer is published in association with the corresponding review. With open review authorship the reviewers assume the responsibility of their reviews, and they also get credit for improvements suggested in these reviews. Sample venues presenting this feature are the *2013 Workshop on Decentralized Coordination* and the online journal *Biology Direct*.

From the perspective of the degree of revelation, each of the aforementioned items of information can be revealed to any subset of the following groups:

- conference chairs
- reviewers
- authors
- conference audience
- general public

For example, the *2007 Workshop on Material Thinking Design* revealed reviewer names to authors but did not publish them.

Here we focus on the effects of using an *open reviewing activity*. These effects appear also in the case of a stronger revelation that implies the opening of the reviewing activity, such as using *open article reviewers* or *open review authorship*.

There are multiple (more or less logic) qualitative arguments for and against each of these degrees of openness, and there are at least two series of conferences dedicated to the practice of Peer Reviewing. Nevertheless it is a remarkable scientific challenge to design quantitative metrics that can be used for founding a systematic study of this area [13]. We believe that the use of the social graph can move us one step closer to a mechanism in which the general audience can have a better idea of the quality of the peer-reviewing process.

3 Network of Reviewers

Biology Direct[1] is an open peer-reviewed journal where publications and their reviews (also the reviewers' name) are publicly accessible through unique URLs. We downloaded information for about 314 papers (titles) and the corresponding reviewers up to March 2013. The dataset contained 843 reviewers. Once the projection was done as explained earlier, the network contained 843 nodes and 2,512 relations.

We first performed an analysis of the degree distribution of the network and found that two nodes dominated the reviewing process with 192 and 125 reviews each. This is respectively 3 and 2 times as much as the third reviewer who reviewed 65 submissions. The obtained degree distribution is shown in Figure 2.

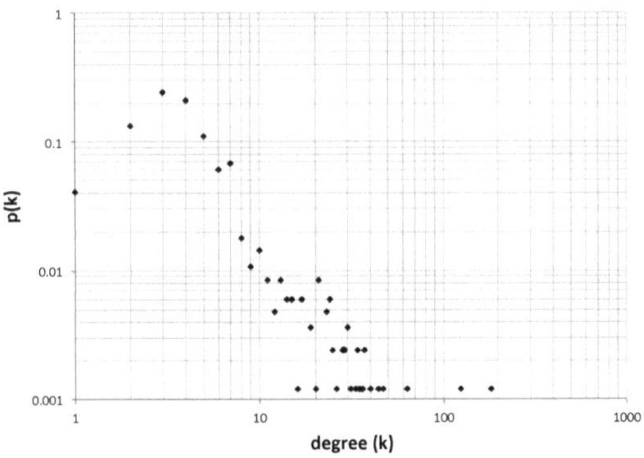

Fig. 2 The degree distribution of reviewers shows a scale-free network in which a couple of nodes are hubs

Next we performed a community analysis to understand whether the hubs are part of the same group of people. Fortunately, in this case they belong to different communities which we believe indicate that they belong to perhaps two different groups of individuals interested in reviewing papers. The community detection algorithm that we used here was proposed by Blondel et al. [4], and it identifies 14 communities. In Figure 3 we can observe the network of reviewers where the size of the node represents the degree and the colors highlight the different communities. Diversity in the community can be used to positively assess the quality of the review process.

As mentioned before, the network of reviewers is built by using nodes to represent the reviewers and arcs to illustrate whether they are connected by reviewing together the same paper. The tool we used for visualization is Gephi [2]. Note that the size of a node is proportional to its degree. Nodes are colored based on the 14 different communities they are in. The employed community detection algorithm is proposed in [4, 9].

[1] `http://www.biology-direct.com/`

Fig. 3 Network of reviewers highlighting the communities they belong to. 14 communities can be observed in this network. The existence of many communities is a positive indicator of the review quality. Communities can be seen as groups of reviewers interested in similar subjects.

In this network we have a high correlation between the degree of a node and its betweenness and closeness so we decided not to show the visualizations for these centralities due to lack of space in the article. The high correlation is due to the fact that the people doing a lot of reviews end up linking different groups and being close to most of the other reviewers. What these high centralities mean is that highly-connected reviewers have an unhealthy chance to exert a strong influence on other reviewers and consequently on the review process. Not only that their reviewing leads to a collection of accepted articles that fits the scientific view of these hubs, but the other reviewers are likely to be indirectly influenced by the reviews of the hubs. In a reviewing process it is common to have reviewers modify their review after they see the reviews of others or participate in discussions. A hub acts as an authority in the process because she has participated in several other discussions.

Last we looked at the clustering coefficient of the nodes in the network and at the average clustering coefficient. Figure 4 shows a network in which the color of the node represents its clustering coefficient; the stronger the color the more clustered the node is.

Note that Figure 4 indicates the existence of a few highly connected groups. These groups are not good for the peer-review process and can indicate the existence of some "mob" phenomena in which a group of individuals may work together to

Fig. 4 Network of reviewers by clustering coefficient. Darker colors represent higher clustering coefficient.

achieve a certain goal—in the case here the goal could be to influence the acceptance/rejection of a paper. Fortunately however this is not generalized throughout the network.

4 Discussion

While each given submission to a venue may be reviewed by a varied set of researchers, the aggregated distribution of reviewers may be less uniform. The aggregated distribution of reviewers can end up with a few reviewers having disproportionate influence. The danger of this phenomenon is increased if one cannot always guarantee the relation between expertise and influence.

The existence of communities is more complicated to understand. Although the existence of communities can be positive for showing a variety of individuals with different interests (as we said before), one has to also be careful because the communities may also mean that we have individuals who review papers together but rarely mix with other groups. When one uses the social graph to assess the quality of a conference, the community structure should be carefully analyzed to help one understand its benefits and drawbacks. Moreover on the drawback side, communities show that the peer-review process is not being vetted by reviewers with diverse backgrounds and interests. We believe a good reviewing network will not have a good resolution on the division of communities, having at most a small number of communities defined (the exact number really depends on the size of the network and on the field of research).

One can also look deeper into the structure of the neighbors of each researcher. Ideally each submission would be reviewed by reviewers that are as disconnected as possible, reducing social contagion (while still being somewhat connected due to their expertise), such as to obtain diverse points of view [18].

A parallel can be drawn between peer-reviewing and genetics. Multiple sets of genes come together to generate a more robust set of genes where the union overshadows individual damaging mutations from each independent set. Similarly, the opinion of multiples reviewers come together to select and influence an article.

With current genome donors one has raised the issue of dangers coming from the disproportionate usage of certain sources [11]. In particular, offspring of the same donors will inherit mutations that can dangerously surface in subsequent generations.

Similarly, there are dangers from disproportionate usage of certain reviewers. Reviewers (as any human) can have preconceived ideas and can subjectively favor certain metrics or scientific views. The relevance of peer-reviewing as a pillar of the modern science is also due to the theory that it can mitigate such subjectivity by joining diverse opinions. Disproportionate involvement of certain reviewers can endanger this property, as their subjectivity will disproportionately impact on the venue and thereby on subsequent generations of researchers.

5 Conclusions

We raise the issue that the power-law distribution generally seen as a positive factor for the stability of social networks can also have negative undesirable connotations. We raise this issue in the context of peer-review social networks, where the fact that certain reviewers are found to be involved in a disproportionate number of articles, conflicts with the objectivity expected from scientific reviewing. Human individuals are intrinsically subject to preconceived ideas, errors and subjective reasoning. Diversity of reviewers is therefore the basis that makes from peer-reviewing the pillar of modern science, where the mixture of multiple views can lead to sounder aggregated result (just as the combination of genes can help defend against damaging mutations in each individual contribution). Just as the defects of a disproportionately frequent donor of genes risk to appear more frequently in subsequent generations, the subjectivity of one disproportionately involved reviewer can impact negatively on generations of researchers.

The availability of data for our study was made possible by the recent trend of openness in journal and conference peer-reviewing. In particular we made an extensive usage of the information on reviews and reviewers made available by the Biology Direct online journal. In the proceedings of this journal we find that a couple of reviewers were involved in reviewing a significant fraction of the articles being published in the venue. We conclude that such openness (at least *openness of reviewer activity*) can be recommended to journals and conferences that want to convince the public about the soundness of their reviewing procedures. Further openness concerning the structure of reviewer neighborhoods (as offered by *openness of article reviewers*) can also be recommended as a way to detect risks of social contagion and detection of undesirable segregation into communities.

While significantly more research is required for establishing a sound scientific foundation to the peer-reviewing procedures found at the foundations of modern science, this study brings a small but clear and objective contribution.

References

1. Barabási, A.-L., Albert, R.: Emergence of scaling in random networks. Science 286(5439), 509–512 (1999)
2. Bastian, M., Heymann, S., Jacomy, M.: Gephi: An open source software for exploring and manipulating networks. In: Intl. AAAI Conference on Weblogs and Social Media (2009)
3. Bird, C., Gourley, A., Devanbu, P., Gertz, M., Swaminathan, A.: Mining email social networks. In: Proceedings of the 2006 International Workshop on Mining Software Repositories, pp. 137–143. ACM (2006)
4. Blondel, V.D., Guillaume, J.-L., Lambiotte, R., Lefebvre, E.: Fast unfolding of communities in large networks. Journal of Statistical Mechanics: Theory and Experiment 2008(10), P10008 (2008)
5. Divakarmurthy, P., Biswas, P., Menezes, R.: A temporal analysis of geographical distances in computer science collaborations. In: 2011 IEEE 3rd International Conference on Social Computing (SocialCom) Privacy, Security, Risk and Trust (PASSAT), pp. 657–660. IEEE (2011)
6. Eagle, N., Pentland, A.S., Lazer, D.: Inferring friendship network structure by using mobile phone data. Proceedings of the National Academy of Sciences 106(36), 15274–15278 (2009)
7. Hongwen, D.: A new school of thought in sequence stratigraphic studies in us: High-resolution sequence stratigraphy. Oil & Gas Geology 2 (1995)
8. Koonin, E., Landweber, L., Lipman, D.: Biology Direct (2013), `http://biologydirect.com/`
9. Lambiotte, R., Delvenne, J.-C., Barahona, M.: Laplacian dynamics and multiscale modular structure in networks. arXiv preprint arXiv:0812.1770 (2008)
10. Moingeon, B., Ramanantsoa, B.: Understanding corporate identity: the french school of thought. European Journal of Marketing 31(5/6), 383–395 (1997)
11. Mroz, J.: One sperm donor, 150 offspring (2011), `nytimes.com/2011/09/06/health/06donor.html`
12. Newman, M.E.: The structure and function of complex networks. SIAM Review 45(2), 167–256 (2003)
13. Peterson, M.J., Silaghi, M.C., Yokoo, M.: Game theoretical modeling and studies of peer-reviewing methods. In: Intl. Symposium on Peer Reviewing, vol. I, pp. 267–272 (2009)
14. Silaghi, M., Qin, S., Cheung, W.: Open peer-review experiment in the decentralized coordination workshop. IEEE Intelligent Informatics Bulletin 13(1) (2013)
15. Spiegel, H.W.: The growth of economic thought. Duke Univ. Press (1971)
16. Barabasi, A., Vicsek, T.: Evolution of the social network of scientific collaborations. Physica A 311, 590–614 (2002)

17. Tonkinwise, C.: Material thinking design reaearch workshop: An experiment in open peer review process at connected: International conference on design education (2007), materialthinking.org
18. Ugander, J., Backstrom, L., Marlow, C., Kleinberg, J.: Structural diversity in social contagion. PNAS 109(16) (2012)
19. Watts, D., Strogatz, S.: Collective dynamics of small-world networks. Nature 393(6684), 440–442 (1998)

Using Complex Network Representation to Identify Important Structural Components of Chinese Characters

Grace Crosley and Mehmet Hadi Gunes

Abstract. It has recently been shown that the Chinese character system can be treated as a complex network, in which the nodes represent written Chinese character components, and a connection between nodes indicates a structural relationship between the two corresponding components. This study explores the complex network formed by written Chinese components. We examine the considerations involved in generating such a network (boundary choice, weighting of edges and nodes, connectivity model, etc.). Treating these considerations as variables, we create several complex network representations of the Chinese writing system, and compare their resulting topologies. By analyzing these networks, we try to identify important written components and component clusters, and thereby gain insight into the best strategies for structure-based written Chinese vocabulary acquisition. Sample networks can be found at `http://cse.unr.edu/~mgunes/Chinese/`.

1 Introduction

Chinese is an important language with approximately 1.2 billion native speakers worldwide. It is also considered to be one of the most difficult human languages to master, in large part because of its complex writing system. Traditionally, mastery of the writing system involves rote memorization of thousands of characters. While memorization will always be required, Chinese could be made more accessible to non-native speakers by developing an organized approach to studying written vocabulary. The central task of this paper is to form complex networks

---

Grace Crosley · Mehmet Hadi Gunes
University of Nevada, Reno
e-mail: `GraceCrosleyCS@hotmail.com`,
`mgunes@cse.unr.edu`

P. Contucci et al. (eds.), *Complex Networks V*,
Studies in Computational Intelligence 549,
DOI: 10.1007/978-3-319-05401-8_30, © Springer International Publishing Switzerland 2014

from the components of written Chinese characters, and apply complex network theory to analyze these networks, with the particular goal of identifying a useful set and ordering of these components for Chinese as a Foreign Language (CFL) learners to study.

Each Chinese character represents one syllable of the Chinese language. The way a Chinese character is written does not necessarily provide any information about its pronunciation, and many characters are written differently but have identical pronunciations. Almost all Chinese words consist of one or two characters; we shall refer to a two-character word as a *compound word*.

In general, every written Chinese character can be ultimately decomposed into one or more of the roughly 200 written elements called *radicals*, many of which are characters in their own right. One or more radicals may combine to what we call a *component*. In turn, one or more components may combine to form a written character, which may combine with another component to form a more complex character. A component may therefore be a single radical, a full-fledged character or something in-between.

About 90% of modern Chinese vocabulary consists of phono-semantic characters, which are made up of a semantic component (typically a radical) that indicates meaning and a phonetic component that indicates pronunciation [1]. Studies have shown that students who use knowledge about phonetic and semantic components to help identify characters are better at reading written Chinese [2]. However, the typical CFL education emphasizes rote memorization of characters, with no formal attention given to the role of radicals or phonetic components [3]. Furthermore, vocabulary lists are typically based on the contents of textbook dialogues, rather than on character frequency or structural similarity of characters. While existing research has suggested that CFL students should learn about written Chinese character components, it has not suggested any set or sequence of components for study.

This paper aims to identify a useful order in which to learn written Chinese character components. Various complex networks will be created in which the nodes are written Chinese components (radicals, characters and, in some cases, intermediate components). A component will be connected to each of its constituent components. The metrics of each complex network will be evaluated, and its hubs and clusters will be examined, to identify radicals and/or intermediate components that are particularly useful for CFL students to learn.

The formation of a complex network from Chinese character components is a relatively new approach. While a couple of recent studies have created and analyzed similar complex networks [4-6], these studies focused on the network-level properties, and did not call attention to any particular nodes or clusters. Complex network analysis has not yet been used to identify important character components or inform Chinese language learning strategies. Our novel approach could form the basis for a much-needed structure-based system for the acquisition of written Chinese vocabulary.

2 Methodology

A publicly available data table from the Wikimedia Commons was selected as a source of raw data about the composition of Chinese characters [7]. The table provides information about the graphical composition of roughly 20,000 traditional Chinese characters. Its advantages are that it is publicly available, deals with traditional characters and has more than sufficient character coverage (given that fewer than 10,000 characters are required for literacy). The disadvantage is that its contributors are anonymous, and therefore its contents may be incomplete or unreliable. As the Wikimedia Commons decomposition list is based on indivisible subcomponents rather than radicals, it was manually modified for this project so that non-radical subcomponents were *broken down* into standard Kangxi radicals. The sources used for this decomposition were `www.zhongwen.com` and `www.wiktionary.com`. A few missing decompositions were also added, using the same sources.

Three separate character frequency lists were used in conjunction with the character composition table. Since each frequency list is based on a unique body of data, the source material may differ greatly as to volume, source, time period and subject matter, and so the character frequencies may differ substantially. Frequency rankings from a selected list were used to limit the boundaries of the network and weight its nodes. The effects of each list on the resulting complex network metrics were compared. Most readily available character frequency lists are based on simplified characters, but three useful traditional character frequency lists were found.

The first list, compiled by Chih-Hao Tsai, is based on a corpus of over 170 million characters gathered from Usenet newsgroups in 1993 and 1994 [8]. It assigns a frequency to each of the 13,060 traditional Chinese characters that are represented in the BIG-5 encoding scheme. The second list, compiled and published by Ho Hsiu-hwang and Kwan Tze Wan, consists of nearly 4 million characters, gathered from *literary texts* from three different decades in Taiwan, Hong Kong and mainland China [9]. For the purposes of this complex network analysis, the mainland China data has been excluded, since it uses simplified characters. The third list, maintained by Patrick Zein, is based on data from various other statistical lists and dictionaries [10]. Zein's list is limited to the 3,000 most frequent characters, and provides the traditional character equivalent to each simplified character in the list. While this sort of conversion from simplified to traditional characters may slightly alter the accuracy of the source data (keeping in mind the fact that a simplified character may correspond to more than one traditional character), it still seemed worthwhile to make use of a frequency list that is based on such a large and varied corpus.

First, any entries with incomplete decomposition data were removed. Then two types of networks were created, one of which emphasized the role of radicals, while the other took complex subcomponents into consideration. In both cases, the set of radicals consists of the standard set of Kangxi radicals. In the ***radical-based network***, vertices belonged to two classes: *radicals* and *characters*. In many cases, a radical is itself a character; in such cases, the radical was represented by two vertices, one classed as a radical and one classed as a character. Each character was connected by arcs to the radical(s) that composed it. In the ***subcomponent-based complex network***, vertices belonged to three classes: *radicals*, *complex subcomponents* and *characters*. If a character was also a radical, it was represented by two vertices – a radical and a character. Likewise, if a character was also a complex subcomponent, it was represented both as a character and as a complex subcomponent. Each character was connected by an arc to its corresponding radical or complex subcomponent. Each complex subcomponent was connected by arcs to its constituent radical(s) and/or complex subcomponent(s).

Each character frequency list was applied to these two basic networks in order to further refine and differentiate them. All but the most frequent n characters from the list were culled from the network. Then each arc connected to each character-vertex in the network was assigned a weight corresponding to that character's normalized frequency of usage. A separate network was created based on the most frequent 1,000; 2,500; and – if available – 5,000 and 10,000 characters from the frequency list.

Individually important radicals and subcomponents (hubs) were identified by examining the network's degree distribution (both weighted and unweighted). The weighted degree distribution gave more importance to radicals and subcomponents that appear in high-frequency characters, while the unweighted degree distribution gave importance to radicals and subcomponents that appear in a greater number of characters. A comparison of the degree distributions of corresponding radical-based and subcomponent-based networks was performed, in the hope that this would provide insight into the relative importance of radicals and subcomponents as units of character study.

In order to identify important clusters, the networks were converted into unimodal format, with the character vertices (and their associated weights) removed. We attempted to identify clusters by finding sets of k-cores with increasingly large values of k. We hoped to find a value of k for each network that would yield many clusters of approximately equal size, each of which could correspond to a textbook lesson. Once a useful-looking set of clusters is identified for a given network, these clusters would be ranked in order of importance by reintroducing the character weights, summing up the weights of the characters connected to each cluster, and assigning this sum as the overall weight of the cluster. Within each cluster, individual radicals and subcomponents would be ranked according to their original weighted degrees.

3 Results

3.1 Giant Components

Each network was dominated by a giant component; only a handful of vertices in each network were not connected to the giant component. This suggests that network-level measurements are representative of the network as a whole.

3.2 *Path Distance*

Path distance was measured only for mixed networks, since the length of each path in the radicals-only networks is 1. Average path distance was fairly consistent among networks. It did decrease slightly as the network size decreased. This is likely because less-common characters are more likely to be complex; longer paths are therefore introduced when less-common characters are added to the network. There was very little variation in path distance between networks based on different frequency lists. The average path distance of two might typically represent a character decomposing into components, which themselves decompose into radicals. However, it must be remembered that paths between all reachable pairs – not just a character and its ultimate radical members - were included in the measurement of the average path length. Thus, the typical number of decompositions done when breaking a character down into radicals is likely higher than two.

Table 1 Average Path Distance

# of Characters	Tsai	Kwan	Zein	Average	Std Dev
1000	1.843	1.816	1.814	**1.824**	0.017
2500	1.955	1.946	1.944	**1.948**	0.006
5000	2.023	2.025	n/a	**2.024**	0.002
10000	2.070	n/a	n/a	**2.070**	n/a

3.3 *Clustering Coefficients*

All other factors being equal, the clustering coefficient of a network that included only radicals was roughly twice as great as the clustering coefficient of a network that included components as well as radicals. In both types of network, the clustering coefficient decreased as the network size decreased. As with average path distance, networks based on different frequency lists showed very little variation in network clustering coefficient.

Table 2 Clustering Coefficients (mixed)

# of Characters	Tsai	Kwan	Zein	Average	Std Dev
1000	0.194	0.178	0.190	**0.187**	0.008
2500	0.210	0.200	0.218	**0.209**	0.009
5000	0.284	0.255	n/a	**0.270**	0.020
10000	0.380	n/a	n/a	**0.380**	n/a

Table 3 Clustering Coefficients (only radicals)

# of Characters	Tsai	Kwan	Zein	Average	Std Dev
1000	0.468	0.479	0.466	**0.471**	0.007
2500	0.558	0.545	0.541	**0.548**	0.009
5000	0.619	0.613	n/a	**0.616**	0.004
10000	0.683	n/a	n/a	**0.683**	n/a

3.4 Clustering

Each sibling network contained only one large cluster; no separate clusters could be identified. As low-degree vertices were pruned during k-core clustering, the large cluster simply got gradually smaller, rather than decomposing into various smaller clusters. This is in accordance with other researchers' observations that Chinese character component networks display disassortative mixing.

3.5 Indegree

Mean indegree was looked at separately for three types of vertex: radicals in a radicals-only network, radicals in a mixed network, and components in a mixed network. By far, the highest mean indegree was found for radicals in a radicals-only network. This is logical, since every arc in the radicals-only networks pointed to a mere 214 (or so) radical vertices. The lowest mean indegree by far was found for components in a mixed network. Mean indegree decreased with the size of the network, but showed little variation between networks that were based on different frequency lists. Within the mixed networks, radicals typically had much higher mean indegree than components. In addition, they had greater variation in indegree; the heavy tail of the indegree distribution in each network was exclusively populated by radicals.

Table 4 Mean Degree (only radicals)

# of Characters	Tsai	Kwan	Zein	Average
1000	13.35	12.75	13.23	**13.11**
2500	32.59	32.92	32.82	**32.78**
5000	67.54	67.87	n/a	**67.70**
10000	138.37	n/a	n/a	**138.37**

Table 5 Mean Degree (mixed)

# of Characters	Tsai	Kwan	Zein	Average
1000	9.07	8.95	9.27	**9.10**
2500	17.93	18.16	18.14	**18.07**
5000	31.35	31.54	n/a	**31.45**
10000	58.05	n/a	n/a	**58.05**

Table 6 Mean Degree (components)

# of Characters	Tsai	Kwan	Zein	Average
1000	1.34	1.32	1.32	**1.33**
2500	1.52	1.52	1.52	**1.52**
5000	1.66	1.66	n/a	**1.66**
10000	1.73	n/a	n/a	**1.73**

3.6 Top Ten Nodes

For the three types of vertices (radicals in a radicals-only network, radicals in a mixed network, and components in a mixed network), the ten highest-ranked vertices were compared across weighted and unweighted networks based on each frequency list.

For radicals in the radicals-only networks, the top ten nodes remained remarkably consistent. Only two of the nodes in this group (3.3% of all nodes) were unique to one frequency list (that is, they appeared in the top-ten list(s) based on one frequency list, but not in the top-ten lists based on other frequency lists).

For radicals in the mixed networks, there were five unique nodes in the sets of top-ten nodes (about 8.3% of the nodes in this group). For components in the mixed networks, there were eight unique nodes in the sets of top-ten nodes, about 13.3%. In other words, the top ten lists are quite consistent regardless of the weighting of the network or the frequency list used to generate it.

Table 7 Top Ten Radicals (only radicals)

Tsai	Tsai - Weighted	Kwan	Kwan - Weighted	Zein	Zein - Weighted
口	口	口	口	口	口
一	一	一	一	一	一
人	人	人	人	人	人
木	丿	丿	丿	丿	丿
丿	木	木	木	木	土
土	土	水	水	土	木
日	日	土	土	日	日
十	十	日	日	十	水
言	言	十	十	水	十
丶	丶	丶	丶	丶	丶

* *Top ten radicals (in terms of indegree) from radicals-only networks. Unique radicals are in red.*

Table 8 Top Ten Radicals (mixed)

Tsai	Tsai - Weighted	Kwan	Kwan - Weighted	Zein	Zein - Weighted
口	口	人	口	口	口
人	人	口	人	人	人
一	一	水	一	一	一
木	木	一	水	水	水
言	言	木	木	木	木
水	水	言	言	言	言
心	心	辵	土	手	手
手	手	土	辵	糸	土
日	日	心	糸	心	糸
土	土	手	手	土	刀

* *Top ten radicals (in terms of indegree) from mixed networks. Unique radicals are in red.*

Table 9 Top Ten Components (mixed)

Tsai	Tsai - Weighted	Kwan	Kwan - Weighted	Zein	Zein - Weighted
古	亼	古	亼	古	古
亼	古	亡	也	各	亼
可	正	丷	古	丷	丷
正	可	亼	寺	亼	各
各	各	可	丷	可	可
六	六	也	可	正	六
僉	僉	各	亡	共	寺
也	也	六	各	六	正
者	丷	共	六	寺	共
主	者	寺	共	丁	也

* *Top ten components (in terms of indegree) from mixed networks. Unique components are in red.*

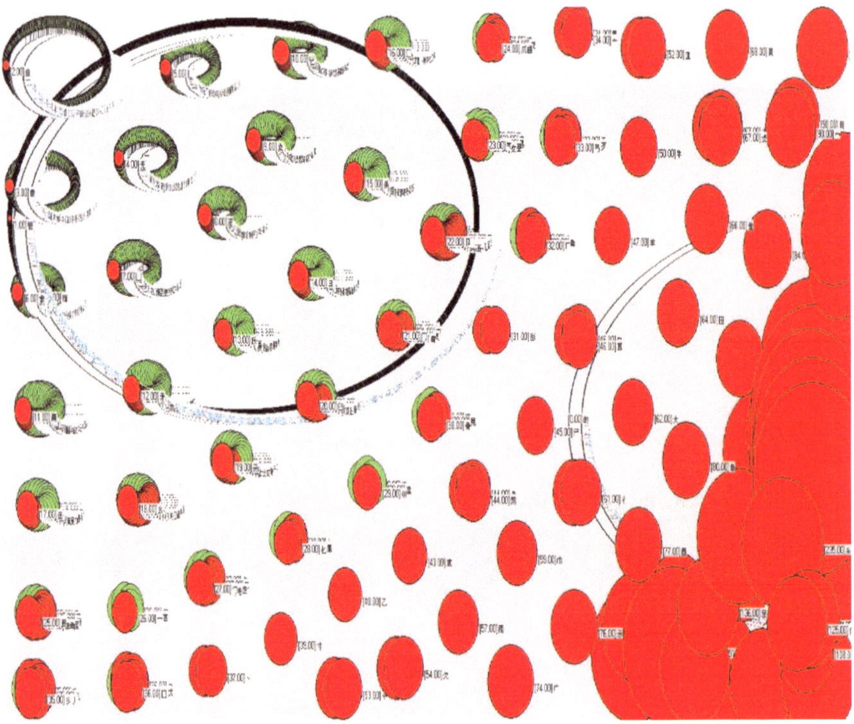

Fig. 1 10,000-character mixed network visualization. (Mixed network using top 10,000 characters from Tsai frequency list. Nodes are grouped by indegree. Nodes with lowest indegree are at top left; nodes with highest indegree are at bottom right. Radical nodes are colored red and component nodes are colored green.

3.7 Consistency in Node Rankings

From each network, radicals (and, separately, components, if present in the network) were ranked based on indegree. The consistency across these rankings was evaluated by taking the average ranking (with standard deviation) of each node, then finding the average standard deviation for all nodes. Rather surprisingly, the average standard deviation was quite high. For component rankings, the average standard deviation was 934.05 – roughly one-tenth the size of the component list. For radical rankings, the average standard deviation was lower – 15.08 – but the size of the radical list is also much lower, with only 214 radicals. It should be noted that rankings based on both weighted and unweighted networks were evaluated together. Possibly the average standard deviation would be substantially lower if weighted and unweighted network rankings were evaluated separately.

4 Conclusions

The reasonably large clustering coefficients of these networks mean that the networks are densely connected, so learning a small number of well-chosen structural components is likely to make a relatively high number of characters available. On the other hand, the lack of discrete clusters in the character decomposition network means that lessons based on clusters of components will not be feasible. Components and radicals must be looked at individually, based on indegree, instead of in clusters.

One approach to learning characters could be to learn radicals and components one at a time, based on their indegree. Students would learn all characters that were composed solely of radicals and components they had already studied. However, this approach may result in learning characters in a piecemeal fashion, where the characters learned in a given lesson sometimes have little in common. In addition, some radicals have such high indegrees that they are not particularly useful as units of study.

Since radicals have such high variation in indegree, another approach to learning characters could involve using components as units of study. However, there are very many components with a low indegree. In addition, as other researchers have shown, phonetic components typically combine with radicals rather than with other components. Therefore, eliminating radicals as a basis for character study would be unhelpful.

We suggest a combination of the preceding two approaches. Students would first learn all 200+ radicals and their meanings without learning any complex characters. Once the radicals had been memorized, components with reasonably high indegree would be used as the basis for vocabulary lessons. Using Tsai's 10,000-character network as a basis, there are many components with an indegree (say, between ten and forty) that would be appropriate for one or two vocabulary lessons.

Under this approach, CFL learners would gain important familiarity with radicals, which are used to organize Chinese dictionaries and are therefore a crucial tool for Chinese language students. In addition, each vocabulary lesson would be tightly organized around a particular written component, allowing students to easily compare and contrast the characters containing that component.

One challenge when designing a course of study will be deciding in what order to present components. The high average standard deviation in the component rankings suggests that the relative importance of each component depends significantly on the corpus of text that is in use.

References

[1] Feldman, L.B., Siok, W.W.T.: Semantic Radicals in Phonetic Compounds: Implications for Visual Character Recognition in Chinese. In: Wang, J., Chen, H., Radach, R., Inhoff, A. (eds.) Reading Chinese Script: A Cognitive Analysis, pp. 19–33. Psychology Press, Wang (1999)

[2] Su, X.: Radical Awareness among Chinese-as-a-Foreign-Language Learners. Ph.D. thesis, Sch. of Tch. Ed., Florida State Univ., Tallahassee, FL (2010)

[3] Morgan, Y.K.: Attitudes toward Hanzi Production Ability among Chinese Teachers and Learners. Ph.D. thesis, Grad. Sch., Purdue Univ., West Lafayette, IN (2012)

[4] Li, J., Zhou, J.: Chinese Character Structure Analysis Based on Complex Networks. Physica A: Statistical Mechanics and its Applications 380, 629–638 (2007)

[5] Wang, J., Rong, L., Jin, T.: An Empirical Study of Chinese Word-Word Language Directed Network. In: IEEE International Conference on Service Operations and Logistics, and Informatics, October 12-15, vol. 1, pp. 498–501 (2008)

[6] Yu, Y., Wang, Z., Gao, W., Gu, G.: Chinese Language Processing with Complex Network Theory. In: 2008 International Conference on Computer Science and Software Engineering, December 12-14, vol. 1, pp. 710–713 (2008)

[7] Wikimedia Commons. Commons: Chinese characters decomposition, `http://commons.wikimedia.org/wiki/Commons:Chinese_characters_decomposition`

[8] Tsai, C.: Frequency and Stroke Counts of Chinese Characters (January 1, 1996), `http://technology.chtsai.org/charfreq/`

[9] Kwan, T.W.: Hong Kong, Mainland China & Taiwan: Chinese Character Frequency – A Trans-Regional, Diachronic Survey (July 7, 2001), `http://humanum.arts.cuhk.edu.hk/Lexis/chifreq/`

[10] Zein, P.H.: The Most Common Chinese Characters (December 2009), `http://www.zein.se/patrick/3000char.html`

Author Index

MIX
Papier aus verantwortungsvollen Quellen
Paper from responsible sources
FSC® C105338

If you have any concerns about our products,
you can contact us on
ProductSafety@springernature.com

In case Publisher is established outside the EU,
the EU authorized representative is:
Springer Nature Customer Service Center GmbH
Europaplatz 3, 69115 Heidelberg, Germany

Printed by Libri Plureos GmbH
in Hamburg, Germany